Smart Innovation, Systems and Technologies

Volume 258

The Smart Innovation, Systems and Technologies book series encompasses the topics of knowledge, intelligence, innovation and sustainability. The aim of the series is to make available a platform for the publication of books on all aspects of single and multi-disciplinary research on these themes in order to make the latest results available in a readily-accessible form. Volumes on interdisciplinary research combining two or more of these areas is particularly sought.

The series covers systems and paradigms that employ knowledge and intelligence in a broad sense. Its scope is systems having embedded knowledge and intelligence, which may be applied to the solution of world problems in industry, the environment and the community. It also focusses on the knowledge-transfer methodologies and innovation strategies employed to make this happen effectively. The combination of intelligent systems tools and a broad range of applications introduces a need for a synergy of disciplines from science, technology, business and the humanities. The series will include conference proceedings, edited collections, monographs, handbooks, reference books, and other relevant types of book in areas of science and technology where smart systems and technologies can offer innovative solutions.

High quality content is an essential feature for all book proposals accepted for the series. It is expected that editors of all accepted volumes will ensure that contributions are subjected to an appropriate level of reviewing process and adhere to KES quality principles.

Indexed by SCOPUS, EI Compendex, INSPEC, WTI Frankfurt eG, zbMATH, Japanese Science and Technology Agency (JST), SCImago, DBLP.

All books published in the series are submitted for consideration in Web of Science.

More information about this series at https://link.springer.com/bookseries/8767

Lakhmi C. Jain · Roumen Kountchev · Bin Hu ·
Roumiana Kountcheva
Editors

Wireless Technology, Intelligent Network Technologies, Smart Services and Applications

Proceedings of 4th International Conference on Wireless Communications and Applications (ICWCA 2020)

Springer

Editors
Lakhmi C. Jain
University of Technology Sydney
Broadway, NSW, Australia

Bin Hu
Changsha Normal University
Changsha, China

Roumen Kountchev
Faculty of Telecommunications
Technical University of Sofia
Sofia, Bulgaria

Roumiana Kountcheva
TK Engineering
Sofia, Bulgaria

ISSN 2190-3018 ISSN 2190-3026 (electronic)
Smart Innovation, Systems and Technologies
ISBN 978-981-16-5167-0 ISBN 978-981-16-5168-7 (eBook)
https://doi.org/10.1007/978-981-16-5168-7

This Springer imprint is published by the registered company Springer Nature Singapore Pte Ltd.
The registered company address is: 152 Beach Road, #21-01/04 Gateway East, Singapore 189721, Singapore

Preface

This book is the second volume of the papers presented at the 4th International Conference on Wireless Communications and Applications (ICWCA2020) which was carried out on December 18–20, 2020, at Hainan University, Sanya, China. The papers cover the contemporary area of wireless technology, intelligent network technologies, smart services and applications and present the latest achievements of the authors.

The presented works correspond to the conference objectives and are based on the investigation and application of wireless communications.

Most of the papers are aimed at remote sensing, analysis and management of complicated systems in education, energy saving and many others.

Special interest attract works are aimed at the contemporary applications, such as robots control; intelligent security system of online cars; energy-saving technology of wireless sensor networks; virtual simulation software based on multimodal sensory theory; in-depth investigation service system for safe road transportation; application of somatosensory interactive technology in teaching; active safety monitoring evaluation system of key operating vehicles; and paradigm for urban safety and risk management with big data. Some investigations are aimed at interesting applications, such as implementation of the identification system for primary protected birds in China; Khmer-named entity recognition based on LSTM-CRF model; bidding strategy in the Dou Dizhu game; method of alleviating depression in pregnant women based on social mobile phone application, and many others.

The aim of the book is to present the latest achievements of the authors to a wide range of readers, such as IT specialists, engineers, physicians, Ph.D. students and other specialists.

Editors

Broadway, Australia — Lakhmi C. Jain
Sofia, Bulgaria — Roumen Kountchev
Changsha, China — Bin Hu
Sofia, Bulgaria — Roumiana Kountcheva
June 2021

Acknowledgments The book editors express their special thanks to chapter reviewers for their efforts and goodwill to help for the successful preparation of the book. Special thanks to Prof. Dr. Lakhmi C. Jain (Honorary Chair), Prof. Dr. Srikanta Patnaik (Program Chair), Prof. Chong Shen and Prof. Kun Zhang (General Chairs), Prof. Dr. Madjid Tavana (International Advisory Chair) and Silai Zhou (Organizing Chair) of ICWCA2020.

The editors express their warmest thanks to the excellent Springer team which made this book possible.

Contents

About the Editors

Prof. Dr. Lakhmi C. Jain KES International, Selby, UK Professor Lakhmi C. Jain, Ph.D., Dr. H.C., M.E., B.E. (Hons), Fellow (Engineers Australia), is with the Liverpool Hope University and the University of Arad. He was formerly with the University of Technology Sydney, the University of Canberra and Bournemouth University.

Professor Jain founded the KES International for providing a professional community the opportunities for publications, knowledge exchange, cooperation and teaming. Involving around 5,000 researchers drawn from universities and companies world-wide, KES facilitates international cooperation and generate synergy in teaching and research. KES regularly provides networking opportunities for professional community through one of the largest conferences of its kind in the area of KES. http://www.kesinternational.org/organisation.php.

His interests focus on the artificial intelligence paradigms and their applications in complex systems, security, e-education, e-healthcare, unmanned air vehicles and intelligent agents.

Prof. Dr. Roumen Kountchev D. Sc. is with the Faculty of Telecommunications, Dept. of Radio Communications and Video Technologies - Technical University of Sofia, Bulgaria. He has 434 papers published in magazines and conference proceedings, 18 books, and 22 patents. A member of Euro Mediterranean Academy of Arts and Sciences; President of Bulgarian Association for Pattern Recognition; Editorial board member of IJBST Journal Group; Editorial board member of: Intern. J. of Reasoning-based Intelligent Systems; Intern. J. Broad Research in Artificial Intelligence and Neuroscience; Editor of books for Springer SIST series.

Bin Hu is an engineer and a counselor at Changsha Normal University, ACM Member, IEEE Member, GCDF, he is the guest editor of EI-indexed journals or book series: International Journal of Reasoning-based Intelligent Systems (IJRIS) , International Journal of Information Systems and Supply Chain Management (IJISSCM), International Journal of Information Systems in the Service Sector (IJISSS) and Smart Innovation, Systems and Technologies (SIST), the technical editor of International Journal of Computer and Communication Technology (IJCCT). He published many articles in high standard journals like Knowledge-based System (KS), Journal of Intelligent and Fuzzy Control (JIFS), Enterprise Information System (EIS), Financial Innovation (FI) and etc.

Dr. Roumiana Kountcheva is Vice President of TK Engineering. She got her Ph.D. at the Technical University of Sofia and became Senior Researcher at TIE, in 1993. She has 183 publications and 5 patents, and presented 16 plenary speeches at international conferences and workshops. She is a member of IRIEM, IDSAI, IJBST Journal Group, and Bulgarian Association for Pattern Recognition. She is a reviewer of WSEAS conferences and journals, and of books for Springer SIST series.

Chapter 1
Design and Implementation of In-Depth Investigation Service System for Road Transportation Safety Accidents Based on Micro-service Architecture

Hong Jia, Wei Zhou, and Haiying Xia

Abstract In order to collect the data of road transportation safety accidents and conduct in-depth investigation and analysis of safety accidents, this paper designs and implements a deep investigation service system of road transportation safety accidents based on micro-service architecture, starting from the functional requirements of road transport industry practitioners and the characteristics of safety accidents. Based on micro-service framework, Java language is used to implement the functions of the system, including personal computer terminal and application terminal. Users use the mobile application as the information acquisition terminal, and guided by the positioning and navigation technology, to provide real-time positioning road transport vehicle accident data collection service for road transport managers, and upload the collected data directly to the cloud. Users can query the relevant accident data in the system personal computer terminal and carry out classified statistical analysis.

1.1 Introduction

At present, our country's accident prevention countermeasures rely on the accident responsibility confirmation and accident investigation report, which mostly focus on the traffic illegal behavior and responsibility identification of people. The lack of investigation and analysis on the causes of human behavior, roads, vehicles, and mainland transportation enterprises will easily lead to the weak system of accident prevention and inaccurate target of measures. Tracing back to the road transportation process, the ability of safety risk prevention and control in key areas and key links of road transportation also needs to be further strengthened, and the risk of production safety accidents in key areas cannot be ignored. It is urgent to establish a unified standard platform to collect road transportation safety accident data, and

H. Jia (✉) · W. Zhou · H. Xia
Research Institute of Highway Ministry of Transport, Ministry of Transport, Beijing, China
e-mail: h.jia@rioh.cn

Key Laboratory of Operation Safety Technology on Transport Vehicles, Ministry of Transport, Beijing, China

L. C. Jain et al. (eds.), *Wireless Technology, Intelligent Network Technologies, Smart Services and Applications*, Smart Innovation, Systems and Technologies 258,
https://doi.org/10.1007/978-981-16-5168-7_1

conduct in-depth investigation and analysis on safety accidents, so as to clarify the real causes behind the accidents to develop relevant standards, policies, laws, and regulations, reduce the incidence of road transport safety accidents, and improve our road transport safety level.

In this paper, based on the functional requirements of road transport industry practitioners and the characteristics of safety accidents, we design and implement a deep investigation service system of road transportation safety accidents based on micro-service architecture. The system includes personal computer (hereafter called PC) terminal and application (hereafter called APP) terminal. Users collect road transportation safety accident data through the system to lay a foundation for in-depth investigation and analysis. Micro-service architecture is a new technology to deploy applications and services in the cloud, which is widely used in various industries [1–5]. Using micro-services to build a system not only uses the horizontal expansion architecture, but also uses the vertical expansion architecture. At the same time, it realizes the combination of application programming interfaces (hereafter called API), making it reusable in the whole business. Micro-service architecture makes continuous deployment easier and easier to update components. A specific micro-service can be re deployed without affecting other service functions.

1.2 System Function Design

1.2.1 User Type

The users of this system are mainly road transport industry management department, system operator, and system administrator. Among them, the grass-roots road transport industry management department can log in the system to enter and query the relevant road transport safety accident data after obtaining the account number, and can query the relevant standards and laws and regulations. The system operator can log in the system and input the road transportation safety accident information in batches or singly. The system administrator can assign authority to system functions and assign account number to system users. After the system is put into operation and obtains certain data basis, the system administrator will be able to conduct big data analysis based on the road transportation safety accident data, so as to provide the basis for scientific decision-making of road transportation management department.

1.2.2 Function Design

The system includes PC terminal and APP terminal. The APP terminal is used as the collection information transmission terminal, and guided by the positioning and navigation technology, to provide real-time positioning road transport vehicle accident data collection service for road transport managers and upload the collected data directly to the cloud. The information filled in by the user through the mobile APP can be queried synchronously on the PC side of the system. Users can query the relevant accident data in the system PC terminal system and make classified statistics. The PC end of the system mainly includes three functions: system management, data acquisition, and large screen display. Among them, the system management module includes user management, department management, role management, parameter management, system log, and other functions, which are mainly used for authority allocation and the addition, deletion, modification, and query of user account; the data acquisition module includes the basic information of accident, vehicle information of accident scene, accident driver information, road information, traffic environment information, and vehicle The large screen display is mainly used for docking with the road transportation emergency support system, including three display pages: the large screen of dispatching platform, the large screen of work order monitoring, and the large screen of resource distribution. The functional architecture of the system PC end is shown in Fig. 1.1.

The main functions of the data acquisition center, including the data acquisition module and the mobile service module, are provided. Data acquisition is mainly divided into four key sectors and eight main categories, covering key elements of transportation such as "people, vehicles, roads, and environment." The four key sections are accident basic information, personnel, and enterprise safety management information, vehicle and collision information, and road and surrounding environment information. The eight main categories are accident basic information, personnel information, enterprise safety management information, vehicle basic information, vehicle use information, collision information, road information, and traffic environment information. The details are given in Table 1.1.

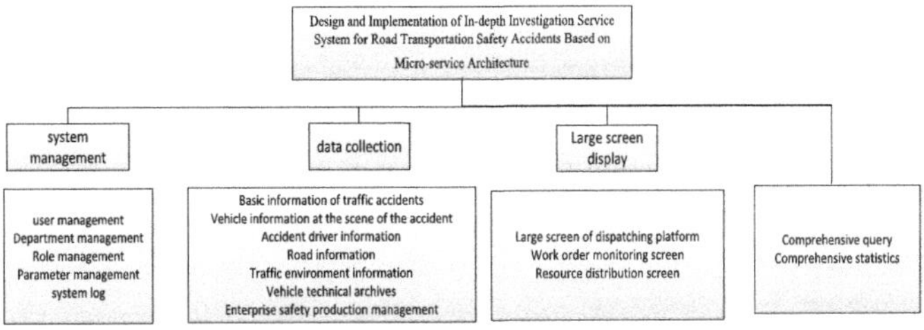

Fig. 1.1 Functional architecture of PC terminal

Table 1.1 Data collection information table

Serial number	Key sector	Category
1	Accident basic information	Accident basic information
2	Personnel and enterprise safety management information	Personnel information, enterprise safety management information
3	Vehicle and collision information	Vehicle basic information, vehicle use information, collision information
4	Road and surrounding environment information	Road information and traffic environment information

1.3 System Architecture Design

The system adopts micro-service technology architecture, which is divided into display layer, business layer, gateway layer, service layer, public technology layer, and infrastructure layer. The details are shown in Fig. 1.2.

The display layer mainly realizes the interface display of user terminal, including system operation and maintenance monitoring interface, PC terminal, APP terminal, and management application interface. The business layer standardizes the data collection items and collection process by formulating standard specifications and operation procedures, so as to reduce the communication cost and improve the cooperation efficiency. Gateway layer is between business layer and service layer. Through unified access, traffic control, protocol adaptation, and other operations, it is convenient to call internal services. The functions of current limiting, security, and fusing are included in this level. The service layer is divided into three layers: the basic service layer, which is mainly responsible for data storage and processing; the transaction framework layer, which is mainly responsible for the registration and scheduling

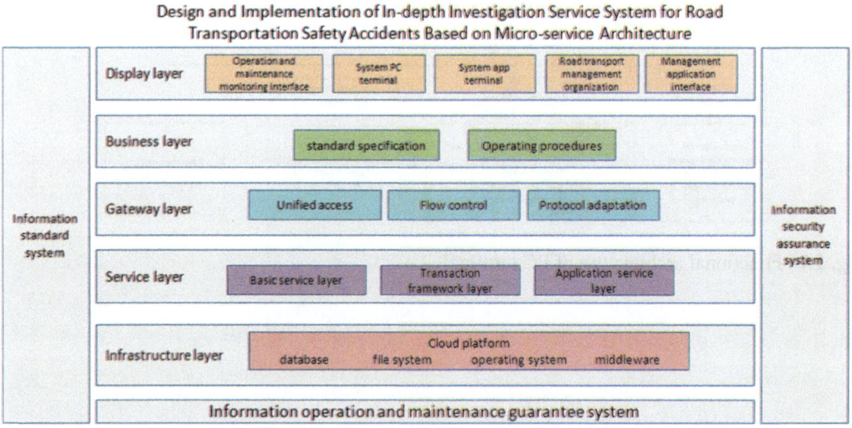

Fig. 1.2 System architecture design

management and distributed transaction processing of micro-services; the application service layer, which mainly implements the API of each micro-service, is used for other micro-services to call directly and service calls of the business layer. SAAS service is the business service that is provided to the public. The docking of related systems is realized at this level. The infrastructure layer mainly provides the infrastructure environment, including network, storage, and basic operation environment, such as database, operating system, file system, and middleware.

1.4 System Function Realization and Advanced Technology

The system is based on alicloud micro-service architecture and PostgreSQL database development and design, and the main development language is Java language. The system software is deployed on three Alibaba cloud servers. The cloud servers are configured with cpu16 cores and 128G memory. At the same time, Alibaba cloud security products such as Web application firewall, situation awareness, and anknight are configured to ensure the data security of the system.

The application development platform adopts Ali micro-service architecture system, which has advanced technology, stability, and flexibility.

Micro-service architecture is very flexible, and different micro-services can be developed with different technologies. Micro-service architecture is reliable, if a function crashes, the whole application will not crash. We can fix this problem in the corresponding micro-services and deploy it immediately. Development based on micro-service architecture can improve development efficiency. Because the code volume of micro-services is much smaller, it is less difficult for new team members to understand and modify the code. The code quality is well maintained, and the IDE reaction speed is faster. The startup time of micro-services is much shorter. All of these factors have greatly increased developer productivity. Using micro-service architecture, it is easy to build complex applications. If we analyze the characteristics of the application correctly, we can decompose it into independent components, which can be deployed independently. Scalability is a major advantage of micro-service architecture, and each micro-service can be scaled individually. Micro-service architecture makes continuous deployment easier and easier to update components. A specific micro-service can be redeployed without affecting other service functions.

1.5 Concluding Remarks

Based on the analysis of the user's functional requirements, this paper designs an in-depth investigation service system of road transportation safety accidents based on micro-service architecture and realizes all the functions of the system by using Java language. The system includes PC terminal and APP terminal. Users use the APP as the information acquisition terminal, and guided by the positioning and navigation technology, to provide real-time positioning road transport vehicle accident data collection service for road transport managers, and upload the collected data directly to the cloud. Users can query the relevant accident data in the system PC terminal system and carry out classified statistical analysis.

Acknowledgements The project is supported by the scientific and technological innovation fund program of the Research Institute of Highway of the Ministry of Transport (2019-C507), the Central Public Research Institutes Special Basic Research Foundation (2020-9067), and the scientific and technological innovation fund program of the Research Institute of Highway of the Ministry of Transport (2018-C0012).

References

1. Xingpeng, L., Chao, S., Baisheng, G., Wei, Z.: Application of intelligent monthly insurance mode based on micro-service architecture. China Equip. Eng. **452**(16), 10–11 (2020)
2. Xiao, Y., Difei, L.: Research on curriculum development of "distributed micro-services" based on spring cloud architecture. Computer Age **338**(08), 91–93 + 97 (2020)
3. Weiming, L., Wei, Y.: A new generation of library service platform based on micro service. Libr. J. **39**(352(08)), 66–74 (2020)
4. Ling, R., Xiuzhe, C, Chao, H., Xiaotao, Z.: Subway internet ticketing system based on micro-service. Autom. Technol. Appl. **39**(301(07)), 36–40 + 45 (2020)
5. Zongyu, Z.: Research on micro-service architecture of new generation operator capability open platform. China New Commun. **22**(14), 73 (2020)

Chapter 2
Survey on Video Information Superimposition and Interaction Method Oriented to Disaster Knowledge of Power Grid

Zhimin He, Lin Peng, Min Xu, Gang Wang, Hai Yu, Xingchuan Bao, Zhansheng Hou, He Wang, and Liang Zhu

Abstract After investigation and analysis, this work adopts a variety of feature fusion methods, based on the business characteristics of power grid equipment, to carry out video information superimposition and interactive technology research. Aiming at the noise characteristics of various power equipment such as power field transformers, it studies the noise reduction of power field audio collection. Key technology; using feature classification-based moving target detection and classification-based moving target tracking technology and continuous learning technology for tracking targets to achieve continuous tracking of power equipment components; based on image vector icon labeling and image overlay technology to achieve map video overlay of textual guidance information.

2.1 Introduction

2.1.1 Research Background

In recent years, various natural disasters have occurred frequently and it has a growing trend. Disasters such as typhoons, heavy rainfall, geological earthquakes, and freezing rain and snow have brought huge losses to the power grid, seriously affecting the safe and stable operation of the power grid and the normal production and operation of the company. Since the 19th National Congress of the Communist Party of China, the Party Central Committee has proposed strengthening, optimizing, and coordinating the construction of national emergency response capabilities, to build a responsible, authoritative and efficient national emergency response capability system that has unified command and consistent power, and to improve production safety, to maintain maintaining public safety, disaster prevention, mitigation and

Z. He (✉) · L. Peng · M. Xu · G. Wang · H. Yu · X. Bao · Z. Hou · H. Wang · L. Zhu
Global Energy Interconnection Research Institute, Nanri Road 8, Nanjing 210003, China

Z. He
State Grid Key Laboratory of Information and Network Security, Beiqijia, Beijing 100000, China

L. C. Jain et al. (eds.), *Wireless Technology, Intelligent Network Technologies, Smart Services and Applications*, Smart Innovation, Systems and Technologies 258,
https://doi.org/10.1007/978-981-16-5168-7_2

relief ability to ensure the safety of people's lives and property and social stability [1]. The documents jointly issued by the Office of the State Council Security Committee, the Office of the National Disaster Reduction Commission, and the Ministry of Emergency Management clearly require the strengthening of basic emergency information management, integrating resources from all parties, promoting information sharing and sharing, strengthening the supervision of disaster accident risks and hidden dangers, and improving safe production and comprehensive prevention. The level of disaster mitigation and disaster relief will promote the formation of an emergency management system with Chinese characteristics featuring unified command, both special and regular, responsive, upper and lower linkage, and peace and war, effectively guaranteeing the safety of people's lives and property and social stability [2].

2.1.2 Purpose and Significance

The State Grid Corporation of China began to develop the emergency command information system in 2008, which has played an active role in the company's emergency response in recent years. The system focuses on the management of emergency information such as emergency plans and emergency teams and has insufficient support capabilities for emergency command and auxiliary decision making such as on-site information collection, information interaction, information fusion, and situation analysis. It has insufficient flexibility in emergency information interaction. After natural disasters such as typhoons, heavy rainfall, geological earthquakes, rain, snow, and freezing occur, information communication between on-site rescuers and on-site rescue headquarters and emergency command centers is mainly through telephone or short messages, which is flexible insufficient degree, and it is difficult to upload and release on-site pictures, audio, and video disaster information.

Therefore, it is necessary to design a method to carry out research on video target labeling and superimposing interaction technology for power grid disaster awareness, to improve the accuracy, timeliness, and systematicness of information interaction between power grid emergency personnel and remote personnel, thereby improving the efficiency of power grid emergency operations.

2.1.3 Overview of Research Levels at Home and Abroad

Information technology is the basis for the development of emergency response systems. At present, many developed countries have established relatively complete emergency response platform systems, and various emergency equipment technologies have become quite mature. In the construction of emergency platform systems in developed countries such as the United States, Japan, and other developed countries, they all attach great importance to the application of digital information.

In 2001, National Institution of Standards and Technology (NIST) began to organize the (TREC Video Retrieval Evaluation) evaluation, and since 2002, it has listed video as a separate task, which marks the beginning of a systematic approach to this field. The evaluation concepts selected by TRECVID are becoming more complex, and at the same time covering more and more. Rare concepts, that is, concepts with a relatively low frequency of occurrence, usually rarely increase in positive examples in the training set. Nevertheless, video annotation is still an emerging field, and there are still many theoretical and technical problems to be solved urgently. Most of the attention of domestic and foreign researchers on video annotation is on the automatic semantic annotation of videos, focusing on the algorithmic research of annotation methods, and on this basis, they have developed video annotation tools for practical applications, such as ELAN (EUDICO Linguistic Annotator), OntoELAN, etc. Because of the different platforms of these annotation tools, there are big differences in functional design.

Related domestic research institutions and universities have also conducted in-depth research on video labeling and classification, and some prototype systems and algorithms have emerged. In 1999, Zhuang Yueting of Zhejiang University and others established a WebMARS video retrieval system. In terms of algorithms, research groups represented by the University of Science and Technology of China, the Institute of Computing Technology of the Chinese Academy of Sciences, Tsinghua University, Beijing Jiaotong University, Peking University, Jilin University, Zhejiang University, Beijing Institute of Technology, and City University of Hong Kong are working on machine learning and context-based video. A lot of work has been done for labeling. Wang et al. have made some improvements in video annotation based on semi-supervised learning. In 2013, Wang Han et al. proposed a joint group weight learning framework to weigh different but related image groups on the Internet and use this knowledge to build video annotations model. At present, the domestic scientific research team has carried out extensive research on video annotation and video classification in specific application fields, such as sports video, news video, and movies, and has achieved good results [3].

2.2 Key Technology Research

Difficult problems encountered by emergency disaster victims require the guidance of the real-time remote command center, and the use of information overlay technology can efficiently guide the emergency rescue operations of on-site operators.

2.2.1 Audio and Video Enhancement Technology in the Complex Environment of the Job Site

In view of the large noise characteristics of various power equipment such as power field transformers, the key technology of power field audio collection and noise reduction is studied, and the audio active noise reduction algorithm based on the noise environment of power grid operation is studied. Study the influence of adjustment and change of parameters such as step genetic factor, instantaneous error energy weight, algorithm convergence step size, and algorithm convergence step size in the adaptive algorithm VSSLMS on reducing noise on the job site [4].

When the mobile video terminal, including wearable devices and handheld devices, performs audio and video interaction, target detection, tracking, and other tasks, due to the jitter of the camera video, there are two kinds of motion vectors in the moving background, the image coordinate system between adjacent frames. The inconsistency makes the output image not very stable, resulting in errors when further observation of the video image is made, and effective and correct information cannot be extracted. Therefore, it is necessary to use motion estimation algorithms and motion filtering and image compensation technology algorithms to enhance mobile video images in order to output stable videos in real-time when the mobile video is jittered in a complex background.

2.2.2 On-Site Target Tracking and Labeling in Mobile Environment

The tracking of the moving target in the mobile environment is through continuous learning of the locked target to obtain the latest appearance characteristics of the target, so as to improve the tracking in time to achieve the best state tracking. That is: Initially, according to the provided one frame of still target image, a comprehensive scan is performed on each frame of the image, to find all the positions of the appearance similar to the target object, and generate positive and negative samples from the results of the detection. With the continuous movement of the target, the system can continuously detect, obtain changes in the target's angle, distance, depth of field, etc., and recognize it in real time. After a period of learning and training, it can accurately capture and track the target and mark it. Specific processing techniques include:

Moving target detection based on feature classification which includes two processing processes, namely the learning process and the decision process. The basic idea of the learning process is to select or construct an image feature that is beneficial to the description of the target type of interest. Through the feature extraction algorithm, a set of labeled image samples are mapped to the feature space to form a feature sample set, then, the sample set is used as input, perform supervised training on the corresponding pattern recognition classifier, and finally get a trained detection classifier. The basic idea of the decision-making process is to first list all the regions in the current image that may contain the target type of interest, and then use the trained classifier to quantify the possibility of targets in these regions, and finally use the decision strategy to evaluate the output of the classifier. Finally realize the detection of the target.

Classification-based moving target tracking, also known as detection-based moving target tracking, is a popular target tracking technology in recent years. The core of this type of algorithm is also to find a boundary between the scene background and the moving target image to separate the foreground target from the local background. Specifically, an online learning feature classifier is used to quantify and classify the feature vector extracted from the current area to be inspected, and the area to be inspected with the optimal classification output is used as the target area. This type of algorithm can continuously update the classifier with the feature vector extracted from the current optimal classification area. This self-feedback mechanism greatly improves the algorithm's ability to adapt to changes in target shape.

2.2.3 Audio and Video Assisted Graphic Guidance Information Overlay Technology

The technical solution of the system supports online annotation and editing functions, which can add text, brushes, two-dimensional vector diagrams, etc. The back-end technical solution experts can add text or annotate vector diagrams at any position of the video screen in real time according to their needs. On-site maintenance personnel can see these annotations simultaneously, which is convenient for remote command. The layer editing process will not affect the live video, synchronous audio playback, and picture quality [5].

There are three commonly used methods for displaying text: including dot matrix display, font image display, and real-time generated text image loading display. They are suitable for different languages and different applications. Each of the three methods has advantages and disadvantages, and is aimed at wearable devices. Due to the particularity of the technical scheme and the complexity of Chinese coding, in the concrete realization of the multi-channel distribution controller superimposing text, we generally adopt the third method, which is to generate files and pictures in real time to load and display. In the technical solution of this subject, the specific idea of multi-channel distribution controller (SFU) superimposing text is as follows: SFU first decodes the terminal video data received by the network into YUV format (or RGB format) and then uses the text display method mentioned earlier. Perform text overlay on YUV (or RGB) image data, encode the superimposed YUV (or RGB) data, and finally send the encoded video data with text display to the network, so that the terminal can display text after receiving its image too.

The technical solution of this subject aims to provide a good graphics rendering platform. The graphics platform needs to support the specific functional requirements of the power industry, such as topological connection, topological color, and power industry business graphic element editors, on the premise of supporting the basic functions of ordinary graphics editors. Using the graphics editor, you can edit different main wiring diagrams, protection wiring diagrams, and other graphics for the specific conditions of different substations. At the same time, the graphic platform should have a good human–computer interaction interface.

The separation of rendering and selection is adopted when drawing and controlling primitives. Its main idea is to draw the rendering part and the selection part separately when drawing a primitive. The rendering part refers to the specific basic primitives that we see with the naked eye after drawing, including its borders, filling pictures, colors, and so on. The selection part is used to control the image element, such as stretching the control point to change the size of the image element, dragging the image element to change its position, and so on. This technical solution uses a multi-threaded manner to control this, and a single rendering thread is responsible for drawing the rendering part, and another selection thread is responsible for drawing the selection part. Then, the user can see the work done by the rendering thread from the naked eye after the user finishes drawing a primitive, that is, draw the primitive that the user wants on the canvas; and the selection thread is also drawn in different ways in the same position. After special processing, the user cannot see the primitive drawn by the selected thread. The work done by the previous selection thread will be used when selecting and dragging the drawn primitives with the mouse. When the mouse moves to a primitive, the part drawn by the selection thread senses the position of the mouse, so that operations such as selection and dragging can be performed. And this kind of induction is judged by the color of the pixel where the mouse is located, that is, the color drawn by the selection thread.

2.3 Conclusion

This article is mainly devoted to the research of core key technologies such as mobile device audio and video enhancement technology, target tracking, and remote collaborative graphic information overlay. Firstly, research about the audio and video enhancement technology in the complex environment for power grid emergency operations can achieve autonomous audio noise reduction, stabilization, and enhancement of video image quality; secondly, the use of feature classification-based moving target detection and classification-based moving target tracking technology and tracking targets continuous learning technology to achieve continuous tracking of power equipment components; finally, based on image vector icon annotation and image overlay technology, to achieve video overlay of graphic guidance information.

Acknowledgements This paper was supported by the science and technology project in State Grid Corporation, which name is 'Research on the Key Technologies of Power Grid Disaster Intelligent Perception and Emergency Command (5700-202019185A-0-0-00)'.

I would like to express my heartfelt gratitude to my colleagues and friends, and at the same time, I want to thank all the scholars who are quoted in this paper.

References

1. Zhou, L.B., Yang, Z.L., Yang, K.L., Su, J.Y.: Research on remote assistance platform for substation operations inspection based on augmented reality technology. Yunnan Water Power **2**, 195–197 (2020)
2. Guo, Z.W., Zhou, Z.J., Yang, H.F., Li, Y., Zhou, Y.: Research and application of substation equipment remote assist based on cloud service platform. China Electr. Power (Technol. Edn.) **6**, 38–41 (2019)
3. Gao, Y.: Design and Implementation of a Web-based Remote Monitoring System for Power Equipment. University of Electronic Science and Technology of China (2013)
4. Jiang, Z.D., Huo, L.P., Jia, S.W., Yu, L.: Design and implementation of superposition for UAV video based on SoC. Computer Measur. Control, 196–199 (2019)
5. Liu, C.L., Wang, L.Y.: Analysis of video information superposition and transmission processing in mobile video monitoring system. Wirel. Internet Technol, 10–11 (2019)

Chapter 3
The Design and Implementation of the Bidding Strategy in the Dou Dizhu Game

Wenjie Huang, Shuqin Li, and Saisai Li

Abstract The Dou Dizhu game is a popular game in China, which has been accepted as a competition in China Computer Game Championship in recent years. The key step of the game is to bid the card, which directly determines the result of the game. In this paper, the related techniques of deep learning have been deeply studied, and the Dou DiZhu game bidding strategy based on deep learning is put forward. Basing on the real game data of a famous online game company in China, this paper gives an innovative method of data calibration and a card representation model and selects the CNN model to train the bidding model. Finally, the bidding system will give a score according to the given initial hand card, and the program is used as AI program to participate in the Chinese University Computer AI Game Championship in 2020. The results show that the bidding strategy proposed in this paper is feasible.

3.1 Introduction

Artificial intelligence is a comprehensive discipline. One of the important fields in artificial intelligence is AI program gaming, and it is an important platform to test the development level of artificial intelligence. According to whether the game information can be completely known, it can be divided into complete information game and incomplete information game [1]. In the complete information game, Google's "Alphago Zero" reaches [2] to a new height in the field of computer games. It provides a general solution for complete information games such as go, chess, and so on. Computer chess AI players gradually reach the level of professional players, and even more than human beings, which draw a great attention in deep learning. Fighting against the Dizhu is an interesting, extensive, and professional poker game. As a typical problem of incomplete information game, more and more experts in the

W. Huang · S. Li (✉) · S. Li
School of Computer, Beijing Information and Science and Technology University, Beijing 100101, China
e-mail: lishuqin_de@126.com

Sensing and Computational Intelligence Joint Lab, Beijing Information and Science and Technology University, Beijing 100101, China

L. C. Jain et al. (eds.), *Wireless Technology, Intelligent Network Technologies, Smart Services and Applications*, Smart Innovation, Systems and Technologies 258,
https://doi.org/10.1007/978-981-16-5168-7_3

field of artificial intelligence pay attention to it. Bidding card is an important part before the beginning of the game round. Whether its strategy is good or bad can lead to win or lose in the game. Li and his team have used Adaboost algorithm to make classifier to formulate bidding strategy and obtained the highest 75% accuracy [3] of imitating human bidding. The deep neural network (Convolutional Neural Networks, CNN) [4] is an important supervised learning method, which has proved to be a powerful model and has reached great achievement in the progress of face recognition [5], image processing, speech recognition, and game AI. Depth model CNN gradually developed from traditional image processing to an efficient method of extracting high dimension features from raw data. Therefore, this paper uses the deep neural network model to study the bidding problem in Dou DiZhu game, which has obtained theoretical significance and practical value.

3.2 Dou DiZhu Game

Dou DiZhu is a poker game played by three players [6]. A set of 54 cards, each player received 17 cards, leaving three cards. At the beginning of the game, players first bid, the player who bid the highest can get the cards and first hand qualifications to play, the other two sides are farmers. Farmers cooperate and fight against the Dizhu. The Dizhu has the first turn. If one of the farmers or the Dizhu hand out the hand card first, he will wins (as long as one farmer hand out the hand card first, both farmers win). The rule of playing cards is that cards with great power can control small power. For the Dou Dizhu game, the size of the card type gets different scores. The larger the score, the greater the card force. The rocket has the largest score, followed by the bomb. Apart from rockets and bombs, the other types of cards must meet the same card type. The total number of sheets and scores must be higher than the previous card surface score. The bombs are also according to the value of the card size. When the card type is same, the comparison of card size is defined as Red Joker > Black Joker > 2 > A > K > Q > J > 10 > 9 > 8 > 7 > 6 > 5 > 4 > 3 (no distinguishing color). Bombs, single card, double cards, and triple cards are compared by the value of the size, according to the maximum card score to compare the size. The aircraft with wings and four with two are calculated by the size of the three and four cards‘ parts, and the wings card does not affect the size of the card. Before the card distribution, there is the bidding part, where the selection has 3 points, 2 points, 1 point, and 0 point (not calling to become a Dizhu). When the player gets a certain initial hand card, the current hand card will be valued according to the number of main cards, the number of loose cards, the combination of the card type and other factors, and the score will be estimated. There are several factors to bid the cards, the number of main cards includes “A,” “2,” “X,” “D,” “bombs”; the number of loose cards is a single card that must be played separately; the combination degree of the card type, which can be understood as the flexibility of the card, and the card type can be combined to form a rich form of cards.

3.3 Dou Dizhu Bidding Strategy

Basing on the CNN, the basic design idea of the convolution bidding system designed in this paper is: First, obtain a large number of real person card playing data in the actual game platform and eliminate the noise. Second, the hand card type tagging can be used to train data. Then, a convolutional neural network structure is designed, the hand cards of the three players are taken as the input data of the model, the bidding task is transformed into a multi-classification problem, and the probability of which kind of scores the hand cards belong to is obtained. The highest probability corresponds to the result of the bid. Finally, the system is integrated, the trained model is imported, and the test results are analyzed. The design is shown in Fig. 3.1.

3.3.1 Data Collection

The data used in this paper is the real data of two dozen and one game provided by a famous online game company in China. A total of 5,401,939 raw data record the id, hand card record, card information, call score, and card playing record of the three players during the game. The separator between the data uses 0×01 and 0×02 in the ASCII code to record all the information of the game part. The raw data is shown in Fig. 3.2

The raw data is stored in the text file, and one set of data contains the complete process (CSBP) of the platform per game, that is, the account number (C), the initial hand card and the base card (S), the bidding process (B), and the playing process (P).

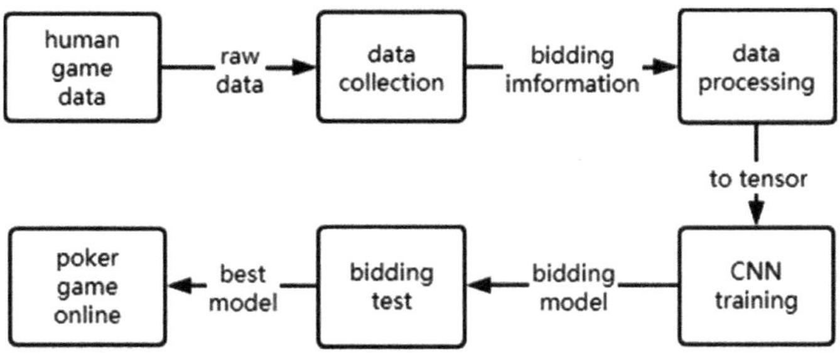

Fig. 3.1 Flowchart of game bid strategy

```
g-t111161106571STXg-t447752814991STXlirenlong654SOH
4466689TTTJQAAA22STX3335556789QKKKKXDSTX3457778899JJJQQ22STX4TASOH
0,bid,1,3STX1,twice,1,2STX2,twice,1,0STX0,twice,1,2SOH
0,1,4448STX2,1,3777STX2,1,88STX0,1,22STX1,1,KKKKSTX1,1,56789STX1,1,
```

Fig. 3.2 Raw data storage format

All data are separated by two separators, where SOH is the character corresponding to number 01 in the ASCII; SXT is the character corresponding to 02. SOH used to split the above CSBP processes, SXT used to segment different content within different processes, as shown in Fig. 3.2, for ease of observation, display the data into four rows in different processes, so the data is stored in 1 row.

The first line uses SXT to split different account information, dividing the initial hand card with the base card by using the SOH and the second line; the first three positions in the second line are the initial hand cards of three players, "4TA" represents the cards preparing to distribute to the Dizhu; the third line is bidding, "bid" represents "call the Dizhu," "twice" means double the score, the last player who apply for double score will become Dizhu. Through Fig. 3.2, the No. 0 player becomes Dizhu; the fourth line shows the process of playing cards, "0, 1, 4448" represents for "the No. 0 player play 4448," With the order No. "0, 1, 2, 0, 1, 2 ..." begins with the Dizhu. According to the demand, the initial hand card information and score should be extracted from the three positions of "0, 1, 2." However, if only the original hand card information is used as the feature and the original call information is used as the label, there will be a large error, because the evaluation of the opponent card comes from the bidding information given by the player. These bidding information cannot fully represent the initial hand card level gap. Section 3.2 describes how to translate the initial call information into an evaluation label that reasonably reflects the initial hand card.

3.3.2 Hand Card Marker

In this paper, convolutional neural network is used in the bidding system, and it is difficult to mark and prepare the training data. According to the rules of fighting against the Dizhu, you can choose not to call the Dizhu (0 points), 1 point, 2 points, and 3 points. Although the original data contains the player's bidding process and the winning or losing status of the game ending, the player level in the game record is different, so we think that the current hand card cannot be directly marked by the current player position from the raw data. Here, we provide a method to mark the score of the hand card, and it is shown in Table 3.1.

We set some state when marking the hand according to the raw data: (1) When the hand card is the position of the Dizhu and Dizhu player wins the game, it can be regarded as "3 points," which can be regarded as the potential of the current hand card, the ability to fight against the farmers, and the probability of "3 points" can be increased when meeting the similar hand card; (2) When the hand card is the position of the farmer and the farmer itself wins the card, it can be regarded as "2 points," which can be regarded as the potential of the current hand card, the ability to fight against the Dizhu, and the probability of "2 points" can be increased when meeting the similar hand card; (3) The rest of the hand cards, regardless of their position, shall be marked as "1 point" as long as the result is loss, and the division of "not called" operation is abandoned here, because "1 point" and "not called" operation

Table 3.1 Method of marking the hand card

Hand card position	Outcome	Bid score
Dizhu	The Dizhu on	3
Farmer 1, 2	The Dizhu won	1
Dizhu	Farmers win	1
Farmers 1	Farmer 1 wins	2
Farmers 1	Farmer 2 wins	1
Farmers 2	Farmer 1 wins	1
Farmers 2	Farmer 2 wins	2

are very similar, and they have the same weak card force. If other players at the same time chose "do not call" operation, then it means their hand card is weak.

In order to balance the different label data, it is necessary to screen and subtract the generated data randomly to balance all kinds of training data. Considering the difference of player level in raw data, some players may not dare to call "3 points," although they have strong hand cards, and some even if they have weak hand cards, they still risk calling "3 points." According to this, when a farmer win the game, the other farmer's card and Dizhu's card and their label are removed to keep a balance.

3.3.3 CNN Bidding Model

When designing the format to input data into convolutional neural network, the color is not considered. The card face is from "A" to "K," and the size trump card is regarded as two cards. For this reason, the dimension of each card is set to 1 * 15, as is shown in Fig. 3.3. Use the corresponding number directly to represent the number of cards shown in Fig. 3.4.

The card type "10" is represented by "T," the card type "small joker" is represented by "X," and the card type "big joker" is represented by "D" in the figure.

A	2	3	4	5	6	7	8	9	T	J	Q	K	X	D

Fig. 3.3 Meaning of each dimension

4	4	4	4	4	4	4	4	4	4	4	4	4	1	1
0	2	3	1	0	1	1	0	0	2	2	3	1	0	1
4	2	1	3	4	3	3	4	4	2	2	1	3	1	0

Fig. 3.4 Number of cards for each dimension

Fig. 3.5 CNN model representation

The total number of "A" to "K" in Fig. 3.4 is 4, the number of the 2 kinds of jokers is one, the first line represents the total number of each hand card, the second line represents the number of each type in a given hand card, and the third line represents the number of cards that do not exist in a given initial hand card.

Convolutional neural network used in bidding system is input as hand card information, and the dimension of situation specification is [15 × 3], shown in Fig. 3.4. The structure contains five layers of convolution layer and two layers of full connection layer. All convolutional cores are [3 × 3] sized. The number of the cores in each layers is 32, 64, 128, 192, 256. The first fully connected layer has 256 nodes, the output layer contains three nodes, represents the probability of "3," "2," "1." According to the requirements of the competition, 3 points of model output will be regarded as bidding 3 points, 2 points of model output will be regarded as bidding 1 point, and 1 point of model output will be regarded as bidding 0 point (not calling to be the Dizhu). The convolutional neural network structure is shown in Fig. 3.5.

3.4 Experimental Results and Analysis

The server of the training environment is configured as Ubuntu 16.04.2LTS operating system; NVIDIA GeForce GTX TITAN X graphics card 12 GB memory; Tensorflow version 1.0.0. The data is a real-time game record for the live combat platform. A total of over 5 million data, including 3 million Dizhuwon, 2 million farmers won.

3.4.1 Training Effectiveness of the Model

The optimal structure of the model is obtained after many tuning training. The hyperparameter table is shown in Table 3.2.

Table 3.2 Hyperparameter configuration

Parameter name	Parameter	Value
Number of training rounds	num_epochs	500
Learning rate	learning_rate	0.001
Single round	batch_size	32
Loss function	Criterion	Cross entropy loss
Optimizer	Optimizer	Adam

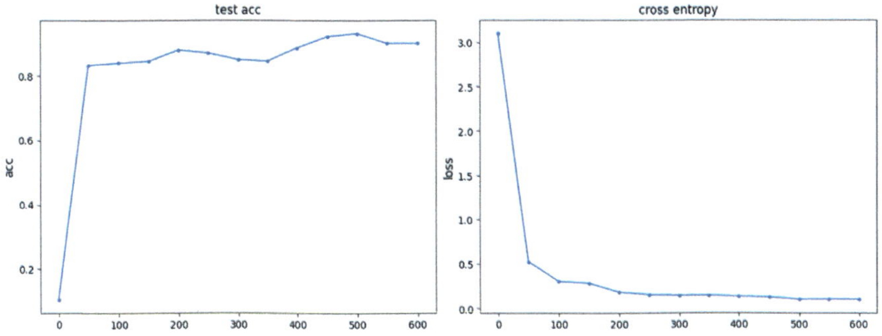

Fig. 3.6 Call accuracy and loss

For the bidding test, the training accuracy chart and the training loss chart of the model experiment are shown in Fig. 3.6, and the accuracy rate is improved to 93.2%, which has a high precision of bidding cards.

3.4.2 Model Bidding Experiment

The bidding system was used to participate in the Chinese University Computer Games Championship. The starting 17 cards by the hand card will be transformed into 0–53 coding sequence. And according to the requirements of the competition in the "DEAL A+ hand card code" formate, the hand card in A (or B, C, one of the three game locations) is 44557TTJJJQQKKKAD, and the final result is called the Dizhu called 3 points, as shown in Fig. 3.7. "DEAL B3," in the coded form of a hand card that needs to be called 35,8,41,34,30,18,6,11,39,44,43,38,53,42,32,4,31, The output is "BID B3" indicates that when the starting as 44557TTJJJQQKKKAD in the B position, CNN call system gives a call score of 3. Obviously, this score will be directly selected as the Dizhu in the game.

The bidding system is used in the exe which contain the whole part of Do Dizhu AI. The platform open the exe file and makes the exe interact with the platform during the competition. Figure 3.8 shows the game of the platform. During the match, the platform will open the three players' fighting master AI and completes the match.

```
DDZAI (1)
C:\Users\Administrator\anaconda3\envs\aiplay\python.ex
DEAL B35,8,41,34,30,18,6,11,39,44,43,38,53,42,32,4,31
OK DEAL
BID WHAT
BID B3
```

Fig. 3.7 Bid 3 for test card

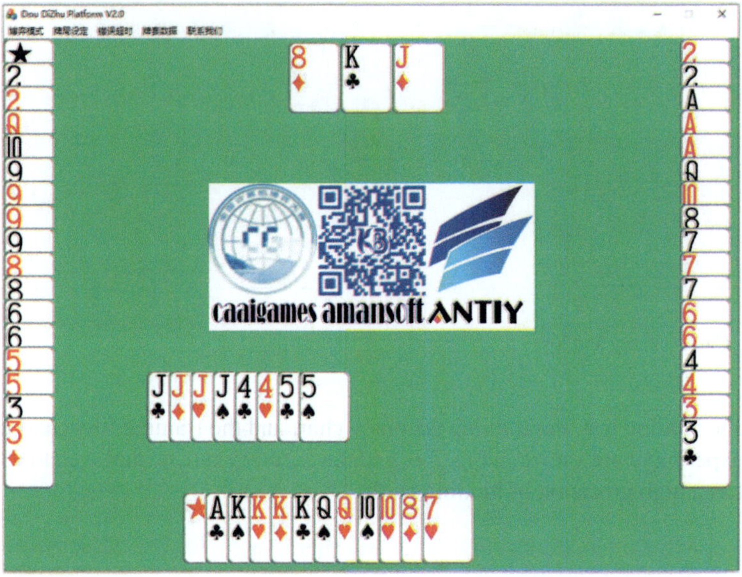

Fig. 3.8 Game chart

At the left, right and bottom of the platform represent the hand card and the card information of the AI program of the three players, and the "8, K, J" above represents the card that has been distributed to Dizhu. The platform's interactive information is displayed in the communication box shown in Fig. 3.9. Our design of the bidding system can give a more accurate initial hand card evaluation in the competition, which contribute to the award a lot.

We extract a total of 100,000 real information from the original data as a comparison, using the bidding system to evaluate the same 100,000 sets of initial hand cards, the accuracy of which is more than 80%. Considering that human players will affect their bidding scores in real poker environments because of their opponent's score, this accuracy indicates that the system has basically gained the ability of bidding like human being.

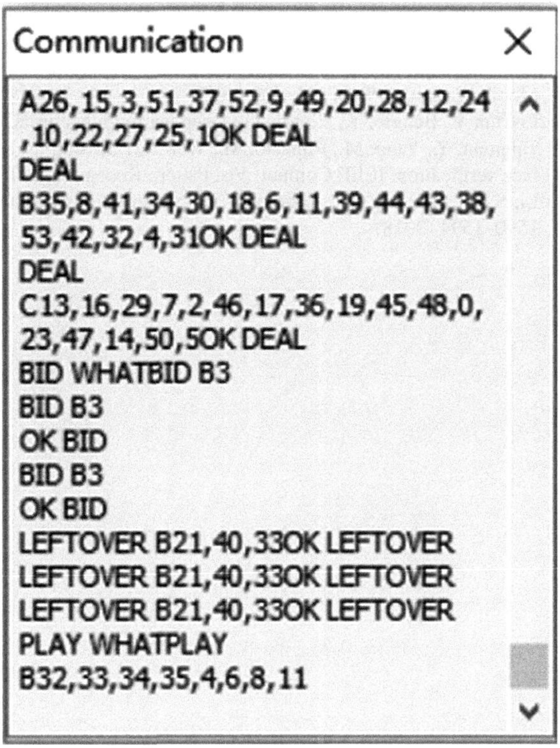

Fig. 3.9 Communication between AI program and platform

3.5 Summary

This paper combines the CNN model and the bidding system to make the CNN neural network obtain the characteristics of the initial hand card through the self-designed hand card input format. Finally, the CNN neural network is used to predict the result of bidding the initial hand card. This is used in the competition effectively.

Acknowledgements This work is supported by Key potential projects of Promoting Research Level program at Beijing Information Science and Technology University. (NO. 5212010937), by Normal projects of General Science and Technology research program (NO. KM201911232002), by Construction Project of computer technology specialty (NO. 5112011019), and by Normal projects of promoting graduated education program at Beijing Information Science and Technology University (NO. 5112011041)

References

1. Kreps, D.M., Wilson, R.: Reputation and imperfect information. J. Econ. Theor. **27**(2), 253–279 (1982)

2. Silver, D., Schrittwieser, J., Simonyan, K., Antonoglou, I., Huang, A., Guez, A., Hassabis, D.: Mastering the game of go without human knowledge. Nature **550**(7676), 354–359 (2017)
3. Cao, H., Li, S.: Research on the DouDiZhu's playing strategy based on XGBoost. Int. J. Wirel. Mobile Comput. **19**(4), 391–400 (2020)
4. LeCun, Y., Bengio, Y., Hinton, G.: Deep learning. Nature **521**(7553), 436–444 (2015)
5. Taigman, Y., Yang, M., Ranzato, M., Wolf, L.: Closing the gap to human-level performance in face verification. IEEE Comput. Vis. Pattern Recogn. **5**, 6, 1701–1708 (2014)
6. Li, S., Li, S., Ding, M.: Landlord's equal cards force generation algorithm. J. Eng. **2018**(16), 1590–1594 (2018)

Chapter 4
Research on the Evaluation of "Special Delivery Classroom" to Promote the Children Development in Weak Rural Schools

Xiaoxia Li and Wanxia Zhu

Abstract Around the "high-quality and balanced development of compulsory education," the experiments and practical explorations have been carried out all over the country, including special delivery class, famous teacher class, and famous school classroom. However, there have been a lack of professional researches on the impact of "three classrooms" on rural children. Based on this, this study carried out an evaluation study on the theme of "special delivery classroom and rural children development." This study constructs an index system for rural children development, then tests 1068 students in rural schools in Chongyang County, Hubei Province. Finally, based on the evaluation results, some suggestions are put forward.

4.1 Introduction

Under the background of the new era, the "high quality and balance" oriented by "fairness and quality" is an important task and goal of the development of compulsory education, which is highly valued by the party and the government [1]. In order to implement the national call for "high-quality and balanced development of education," local governments at all levels have carried out a lot of beneficial explorations in helping rural weak schools "open up all, enough and good classes." Some counties in China have actively carried out the experimental and practical exploration of "educational informatization helps rural children development," and has achieved certain results. For example, at the end of 2015, the "special delivery classroom" mode has been implemented in Chongyang County, Hubei Province, which was widely concerned by the society and news media. The "special delivery classroom" refers to a mode that in order to solve the problems of "incomplete, insufficient, and poor" quality courses such as music and art in weak rural schools, the network is used to realize the interconnection between teachers and students in many weak rural schools, and the lecturer teachers teach students from multiple teaching points

X. Li (✉) · W. Zhu
Northwest Minzu University, Lanzhou, China
e-mail: 397056748@qq.com

L. C. Jain et al. (eds.), *Wireless Technology, Intelligent Network Technologies, Smart Services and Applications*, Smart Innovation, Systems and Technologies 258,
https://doi.org/10.1007/978-981-16-5168-7_4

(generally 2 or 3). At present, there is a lack of evidence-based scientific evaluation research on the "special delivery classroom" model to promote the rural children development.

4.2 The Component Dimensions and Evaluation Framework of Rural Children Development

4.2.1 The Component Dimensions of Rural Children Development

This study mainly uses the methods of literature, in-depth interview and Delphi expert consultation to construct the component dimensions of rural children development. The final evaluation index of rural children development consists of three first-level dimensions, and each level dimension consists of several secondary dimensions. The three first-level dimensions are psychological development, behavior development, and art subject basic literacy. The secondary dimension of psychological development dimension is self-esteem, core self-evaluation, life satisfaction, and emotional resilience; the secondary dimension of behavior development dimension is interpersonal communication ability, daily behavior, and interest specialty; the secondary dimension of art subject basic literacy dimension is music and art subject basic literacy.

4.2.2 Evaluation Framework

This study mainly uses the research method of "education quasi experiment" to evaluate the effect of "special delivery classroom" mode on promoting the rural children development. By comparing the results of each evaluation dimension between the experimental group and the control group, this paper analyzes the influence of the "special delivery classroom" mode on children development.

4.3 Evaluation Tools, Survey Sample and Reliability and Validity of the Scale

4.3.1 Source and Revision of Evaluation Tools

The self-esteem scale is a Chinese version of the self-esteem scale (SES) (Rosenberg 1965) compiled by Xiangdong Wang [2]. The core self-evaluation scale comes from the core self-evaluation scale revised by Dr. Wenzhu Chen on the basis of the Chinese version of CSES revised by JianZheng Du and tested by the students in rural teaching spots [3]. The life satisfaction scale is derived from the life satisfaction scale of young students compiled by Dr. Wenzhu Chen on the basis of the life satisfaction scale of young students compiled by Xinggui Zhang and then revised and tested the life satisfaction scale of rural teaching students. The emotional resilience scale was developed by Dr. Wenzhu Chen on the basis of the adolescent emotional resilience questionnaire compiled by Zhang and Lu [4], and revised and tested the emotional resilience scale of rural students. The interpersonal communication ability scale is derived from doctoral thesis of Wenzhu Chen. On the basis of the sub-scale of social intelligence scale for primary school students compiled by Zaihua Liu, Dr. Wenzhu Chen revised and tested the interpersonal ability scale of rural students. Based on Achenbach, Rutter, Conners children's behavior scale, and "primary school students daily behavior standard," the daily behavior scale was revised and tested by Dr. Wenzhu Chen. The interest specialty scale was developed by Dr. Wenzhu Chen and tested by the rural students. The art subject basic literacy scale self-designed.

4.3.2 Survey Sample

In the survey, this study evaluated the respondents through the combination of tablet computer test and paper-based questionnaire. Finally, 1068 valid student samples were collected.

4.3.3 Reliability and Validity of the Scale

In this study, confirmatory factor analysis and internal consistency reliability analysis were carried out on the data of formal survey. The results of model fitting index and internal consistency coefficient of confirmatory factor analysis of self-esteem scale were as follows: The ratio of chi-square to degree of freedom (CMIN/DF), Root Mean Square Residual (RMR), Goodness of Fit Index (GFI), Adjusted Goodness of Fit Index (AGFI), Normed Fit Index (NFI), Comparative Fit Index (CFI), Root Mean Square Error of Approximation (RMSEA), and Cronbach's Alpha [5] were, respectively, 0.830, 0.01, 0.999, 0.995, 1.000, 0, 0.651. The internal consistency coefficient of "core self-evaluation" scale was 0.667. The results of model fitting index and internal consistency coefficient of confirmatory factor analysis of life satisfaction scale were as follows: CMIN/DF, RMR, GFI, AGFI, NFI, CFI, RMSEA, and Cronbach's Alpha were, respectively, 2.386, 0.025, 0.987, 0.977, 0.970, 0.982, 0.036, and 0.771. The results of model fitting index and internal consistency coefficient of confirmatory factor analysis of emotional resilience scale were as follows: CMIN/DF, RMR, GFI, AGFI, NFI, CFI, RMSEA, and Cronbach's Alpha were, respectively, 1.512, 0.026, 0.995, 0.989, 0.981, 0.993, 0.022, and 0.607. The results of model fitting index and internal consistency coefficient of confirmatory factor analysis of behavior development scale were as follows: CMIN/DF, RMR, GFI, AGFI, NFI, CFI, RMSEA, and Cronbach's Alpha were, respectively, 2.314, 0.022, 0.990, 0.979, 0.974, 0.985, 0.035, and 0.788. The results of model fitting index and internal consistency coefficient of confirmatory factor analysis of music subject basic literacy scale were as follows: CMIN/DF, RMR, GFI, AGFI, NFI, CFI, RMSEA, and Cronbach's Alpha were, respectively, 2.859, 0.038, 0.996, 0.984, 0.984, 0.990, 0.041, and 0.654. The results of model fitting index and internal consistency coefficient of confirmatory factor analysis of art subject basic literacy scale were as follows: CMIN/DF, RMR, GFI, AGFI, NFI, CFI, RMSEA, and Cronbach's Alpha were, respectively, 0.932, 0.016, 0.999, 0.996, 0.998, 1.000, 0, and 0.624. It can be seen that the above scales have good model fitting degree and internal consistency reliability.

4.4 Evaluation Results and Analysis

4.4.1 The Difference Test Results of Variable Results in the Dimension of "Psychological Development"

In this study, the "psychological development" dimension includes four sub-dimensions: self-esteem, core self-evaluation, life satisfaction, and emotional resilience. In the four dimensions, the level of the intervention group of "special delivery classroom" mode is significantly higher than that of the traditional mode intervention group (see Table 4.1).

4.4.2 The Difference Test Results of Variable Results in the Dimension of "Behavior Development"

In this study, the "behavior development" dimension includes three sub-dimensions: interpersonal communication ability, daily behavior, and interest specialty. The level of "interpersonal communication ability" and "interest specialty" in the intervention group of "special delivery classroom" mode was significantly higher than that of traditional group. However, there was no difference between the two groups in the dimension of "daily behavior" (see Table 4.2).

Table 4.1 Difference and significance test results of "psychological development" in children development under "special delivery classroom" project intervention and traditional mode intervention

Variable	Group	Sample number	Mean	Standard deviation	*P* value
meanSE	Special delivery classroom	576	2.964	0.641	0.000
	Traditional classroom	492	2.818	0.581	
meanCSE	Special delivery classroom	576	2.941	0.853	0.000
	Traditional classroom	492	2.756	0.836	
meanLIS	Special delivery classroom	576	3.098	0.558	0.000
	Traditional classroom	492	2.942	0.552	
meanER	Special delivery classroom	576	3.060	0.571	0.001
	Traditional classroom	492	2.944	0.566	

Note (1) meanSE—the average score of self-esteem scale; meanCSE—the average score of core self-evaluation scale; meanLIS—the average score of life satisfaction scale; meanER—the average score of emotional resilience scale. (2) The scale in the table is Likert four-point scale

Table 4.2 Difference and significance test results of "behavior development" in children development under "special delivery classroom" project intervention and traditional mode intervention

Variable	Group	Sample number	Mean	Standard deviation	*P* value
meanIC	Special delivery classroom	576	3.097	0.721	0.000
	Traditional classroom	492	2.940	0.708	
meanDB	Special delivery classroom	576	3.145	0.782	0.114
	Traditional classroom	492	3.070	0.756	
meanIS	Special delivery classroom	576	3.245	0.687	0.001
	Traditional classroom	492	3.103	0.711	

Note (1) meanIC—the average score of interpersonal communication ability scale; meanDB—the average score of core daily behavior scale; meanIS—the average score of interest specialty scale. (2) The scale in the table is Likert four-point scale

Table 4.3 Difference and significance test results of "art subject basic literacy" in children development under "special delivery classroom" project intervention and traditional mode intervention

Variable	Group	Sample number	Mean	Standard deviation	*P* value
meanBML	Special delivery classroom	576	3.835	0.836	0.001
	Traditional classroom	492	3.671	0.799	
meanBAL	Special delivery classroom	576	4.241	0.744	0.000
	Traditional classroom	492	3.976	0.859	

Note (1) meanBML—the average score of music subject basic literacy scalc; meanBAL—the average score of music subject basic literacy scale. (2) The scale in the table is Likert four-point scale

4.4.3 The Difference Test Results of Variable Results in the Dimension of "Art Subject Basic Literacy"

In the dimension of "art subject basic literacy," the "music and art subjects basic literacy" in the intervention group of "special delivery classroom" mode were significantly higher than those of traditional group (see Table 4.3).

4.5 Conclusions and Suggestions

From the evaluation results, the evaluation results of all dimensions (including the first-level dimension and the second-level dimension) of the "special delivery classroom" mode intervention group were higher than those of the traditional mode intervention group. Except for the second dimension of "daily behavior," the results of other dimensions are significant.

Based on the results and conclusions of this study, the following suggestions are put forward for the practical work of "informatization helps rural children development." Educational administrators at all levels, principals and teachers of rural schools, and relevant researchers engaged in rural education research should realize that "information technology has great potential in promoting the rural children development," and attach importance to the role and significance of information technology, especially the innovative teaching mode based on information technology, in promoting the development of rural children. It is suggested that the local education bureaus should carry out the theoretical, and practical exploration of "informatization helps rural children development" according to local conditions, based on the "input–output" evaluation results based on the "special delivery classroom" mode.

Acknowledgements This study was supported by the Fundamental Research Funds for the Central Universities 2021 (31920210119); the foundations of the education and teaching reform project of Northwest Minzu University in 2019 (2019XJJG-51); and the Fundamental Research Funds for the Central Universities 2016 (31920160079).

References

1. Fan, X., Guo, Q.: Improvement of education quality in rural small-scale schools: from the perspective of cluster development. Educ. Res. **40**(03), 90–98 (2019)
2. Wang, X., Wang, X., Ma, H.: Handbook of mental health assessment scale. Revised edition. China J. Mental Health, 52–53 (1999)
3. Wang, J., Chen, W.: Observation and Evaluation of Information Assisted Development of Rural Teaching Point Students—Taking Xianan Experimental Area as an Example. China Electric Education (2018)
4. Zhang, M., Lu, J.: Research report on adolescent emotional resilience questionnaire. Psychol. Sci. (2010)
5. Cohen, J., Cohen, J.W., Cohen, J.E., Cohen, J.D.: Statistical power analysis for the behavioral science. Technometrics **31**(4), 499–500 (1988)

Chapter 5
Research on the Map Construction of SLAM Mobile Robot Based on the Two-Dimensional Code Correction

Ye Chao, Zhang Yi, Yu Bin, and Xing Bin

Abstract Since the emergence of the concept of Simultaneous Localization and Mapping (SLAM) in the 1980s, due to its powerful functionality, it has been widely used in mobile robots, unmanned driving, and other fields. Various related theories are constantly being improved, especially laser SLAM. However, in complex applications in real life (such as long and narrow corridors and unstructured work scenarios), there are still problems such as inaccurate positioning and map ghosting. In response to this problem, this paper uses a two-dimensional code artificial road sign as an absolute positioning label, participates in the map optimization laser SLAM process, strengthens the constraint between the mobile robot's pose and the map, and constructs related sensor models. Finally, the article verifies the feasibility and practicability of the method through experiments.

5.1 Introduction

Mobile robot, also known as Automated Guided Vehicle (AGV), is a high-performance mobile transportation equipment, mainly responsible for the handling and movement of goods, and is used in industry and logistics to build efficient and fast logistics systems [1, 2]. However, the poor positioning accuracy during the operation of AGV has always restricted its application in many industrial environments. At present, there are many navigation and guidance methods commonly used by AGVs, such as magnetic guidance, ribbon guidance, laser navigation, GPS navigation, inertial navigation, and QR code label navigation. Among them, the magnetic guidance method [3] is a relatively mature guidance method at present. The advantages are low technical cost, but simple functions, low positioning accuracy, and fixed path, which is difficult to update later; literature [4] uses laser guidance AGV has high

Y. Chao (✉) · Y. Bin · X. Bin
College of Advanced Manufacturing Engineering, Chongqing University of Posts and Telecommunications, Chongqing, China

Z. Yi · Y. Bin · X. Bin
Chongqing Industrial Big Data Innovation Center Co. Ltd., Chongqing, China

L. C. Jain et al. (eds.), *Wireless Technology, Intelligent Network Technologies, Smart Services and Applications*, Smart Innovation, Systems and Technologies 258,
https://doi.org/10.1007/978-981-16-5168-7_5

positioning accuracy, but in some narrow corridors and unstructured work scenes, there are still problems such as inaccurate positioning and map ghosting. Literature [5] pointed out that visual guidance AGV takes into account the characteristics of low cost and high precision. But its technical cost is high, and signal processing is difficult.

In summary, the single-sensor mobile robot navigation and positioning are difficult to achieve better results, so this article finally uses a multi-sensor data fusion SLAM method. Because the two-dimensional code recognition is convenient and easy and the visual positioning can provide reliable positioning accuracy, this paper uses the two-dimensional code as an artificial road sign, as a positioning label for absolute positioning, combining various sensor models, and optimizing the correlation of laser SLAM through theoretical experiments. Algorithm to further improve the positioning and navigation accuracy of laser SLAM, so that it can better meet the actual industrial needs of mobile robots.

5.2 Lidar and Its Error Model

At present, the commonly used sensors of SLAM algorithm are lidar and camera, which correspond to laser SLAM algorithm and visual SLAM algorithm, respectively. Although lidar is much more expensive than cameras, it is currently more widely used in industry because of its high measurement accuracy and stable measurement performance. Laser SLAM has developed relatively rich theoretical research results so far, but there are still problems in actual applications such as insufficient positioning accuracy and inaccurate map construction, which have brought certain obstacles to robots in performing related tasks.

The lidar model used by the laboratory mobile robot is RPLIDAR_S1. It scans the surrounding environment with a fixed angular resolution during work. After scanning 360°, a set of point information (including distance and included angle) of the surrounding obstacles relative to the current position can be obtained. These points can be used in the lidar coordinate system which is expressed in polar coordinates. A group of data points collected by lidar at time t can be expressed in the form of a set $\{p_t\}$, as shown in Eq. (5.1).

$$\{p_t\} = \left\{\left(\rho_t^i, \alpha_t^i\right) | i = 1, 2, 3 \ldots, n\right\} \tag{5.1}$$

In the formula, ρ_t^i is the distance of the point obtained by the ith measurement of the laser beam when the lidar is scanning at time t; α_t^i is the measurement angle corresponding to point ρ_t^i; n is the scanning at time t the total number of measurements at the time.

Figure 5.1 is the observation model of lidar, where $w_x O w_y$ is the world coordinate system, $L_x O L_y$ is the lidar coordinate system, assuming that the vector of the lidar scanning pose at time t in the world coordinate system is expressed as x_L, as shown in Eq. (5.2).

Fig. 5.1 Laser observation model

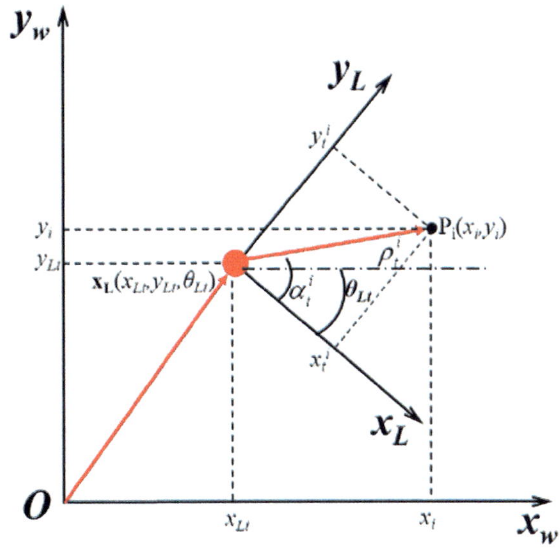

$$x_L^T = (x_{Lt} \quad y_{Lt} \quad \theta_{Lt}) \tag{5.2}$$

Suppose that the laser observes the environmental feature point P_i for the ith time, and the pose information of this point measured by the radar is (ρ_t^i, a_t^i), which is expressed in the form of Cartesian coordinates, as in Eq. (5.3).

$$\beta_t^i = \begin{bmatrix} x_t^i \\ y_t^i \end{bmatrix} = \rho_t^i \begin{bmatrix} \cos\alpha_t^i \\ \sin\alpha_t^i \end{bmatrix} \tag{5.3}$$

Assuming that the actual coordinates of the feature point P_i in the world coordinate system are (x_i, y_i), and the observation equation of the lidar can be expressed in the form of formula (5.4).

$$\begin{bmatrix} \rho_t^i \\ \alpha_t^i \end{bmatrix} = \begin{bmatrix} \sqrt{(x_{Lt} - x_i)^2 + (y_{Lt} - y_i)^2} \\ \arctan \frac{y_{Lt} - y_i}{x_{Lt} - x_i} \end{bmatrix} + v_{t,i} \tag{5.4}$$

Equation (5.4) is the observation equation of the radar in the world coordinate system. Solving this equation can get the pose of the mobile robot in the world coordinate system.

However, the data observation model of the above formula is under ideal conditions without considering the noise error of the lidar. In actual use, the lidar will be affected by random noise and deviation. If the influence of noise is taken into account, the distance of the ith measurement point collected by the lidar at time t can be expressed as the following form.

$$\rho_{\rm t}^{i} = D_{\rm t}^{i} + \varepsilon_{\rho} \tag{5.5}$$

Equation (5.5) $\rho_{\rm t}^{i}$ is the measured value of laser distance; $D_{\rm t}^{i}$ is the true value of distance measurement; ε_{ρ} is the noise term, and its value obeys Gaussian distribution.

Similarly, the angle information measured by the laser is also affected by noise. Adding the noise term, the radar measures the angle of the ith measurement point collected at time t, which can be expressed as.

$$\alpha_{\rm t}^{i} = \varphi_{\rm t}^{i} + \varepsilon_{a} \tag{5.6}$$

In the formula, $\alpha_{\rm t}^{i}$ is the measured value of the laser angle; $\varphi_{\rm t}^{i}$ is the true value of the angle; ε_{a} is the noise item of the radar angle measurement, and its value obeys the Gaussian distribution with zero mean.

Substituting (5.5) and (5.6) into Eq. (5.3) can obtain the Cartesian coordinate form of the measuring point in the case of noise considerations [6], as shown in Eq. (5.7).

$$\beta_{\rm t}^{i} = \begin{bmatrix} x_{\rm t}^{i} \\ y_{\rm t}^{i} \end{bmatrix} = \rho_{\rm t}^{i} \begin{bmatrix} \cos\alpha_{\rm t}^{i} \\ \sin\alpha_{\rm t}^{i} \end{bmatrix} = \left(D_{\rm t}^{i} + \varphi_{\rm t}^{i}\right) \begin{bmatrix} \cos\left(\varphi_{\rm t}^{i} + \varepsilon_{a}\right) \\ \sin\left(\varphi_{\rm t}^{i} + \varepsilon_{a}\right) \end{bmatrix} \tag{5.7}$$

5.3 Integration of QR Code Information

The typical framework of a modern SLAM system is shown in Fig. 5.2. The front end of SLAM is mainly used to estimate the local consistency of robot motion, and the back end is used to estimate the global consistency of robot motion. For laser SLAM, data association is the core work. Data association is the one-to-one correspondence between the laser radar scan data and the objects or features in the real world to establish an inter-related relationship to determine the relative pose of the robot.

After the feature matching is successful, it is necessary to estimate the transformation of the robot's relative pose. The common algorithm steps are as follows, by constructing the least square objective function, using an iterative method to solve, as shown in Eqs. (5.8) and (5.9).

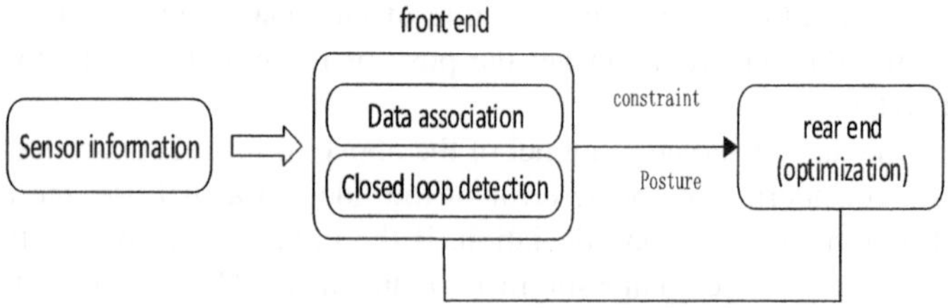

Fig. 5.2 Modern SLAM system framework

$$\hat{P}_t^i = RP_{t+1}^i + t \tag{5.8}$$

$$\underset{R,t}{\arg\min}\ I = \sum_{i=1}^{n} \left| P_t^{J(i)} - \hat{P}_t^i \right|^2 \tag{5.9}$$

In the formula, P_{t+1}^i is the feature extracted from the scanned data at time $t + 1$; $\hat{P}_t^i$ feature P_{t+1}^i is projected to the coordinates in the coordinate system at time t; $P_t^{J(i)}$. The feature that matches P_{t+1}^i extracted from the scan data at time t.

However, in complex applications in real life (such as long and narrow corridors and unstructured work scenarios), there are still problems such as inaccurate positioning and map ghosting. Therefore, this article uses the two-dimensional code as the artificial road sign, as the positioning label for absolute positioning, combined with the laser radar observation model, and optimizes the relevant algorithms of laser SLAM through theoretical experiments and other methods to further improve the positioning and navigation accuracy of laser SLAM.

The camera adopts a QR code dedicated camera (AGV50F-15, Scanning frequency is 50 Hz), which is installed at the geometric center of the robot, 15 cm from the ground, and contains QR code processing library, which can obtain the distance and pose information of the QR code label relative to the AGV. The specific observation model is as follows. The world coordinate system (reference coordinate system) is established with the center of the QR code as the origin, and the camera coordinate system is established with the camera imaging plane as the origin. One point in the world coordinate system (x_w, y_w, z_w) can be in the camera coordinate system. Expressed as:

$$\begin{bmatrix} X \\ Y \\ Z \end{bmatrix} = [R|t] * \begin{bmatrix} X_w \\ Y_w \\ Z_w \\ 1 \end{bmatrix} \tag{5.10}$$

Among them, the R and T matrices are the external parameter matrices of the camera, where R is the rotation matrix and T is the translation matrix. The conversion relationship from the world coordinate system to the pixel coordinate system is shown in formula (5.11):

$$Z\begin{bmatrix} \mu \\ v \\ 1 \end{bmatrix} = \begin{bmatrix} 1/\mathrm{dx} & 0 & \mu_0 \\ 0 & 1/\mathrm{dy} & v_0 \\ 0 & 0 & 1 \end{bmatrix} * \begin{bmatrix} f & 0 & 0 \\ 0 & f & 0 \\ 0 & 0 & 1 \end{bmatrix} * [R|t] * \begin{bmatrix} X_w \\ Y_w \\ Z_w \\ 1 \end{bmatrix} \tag{5.11}$$

5.4 Correction Graph Optimization Algorithm Principle

The optimization-based SLAM theory algorithm has attracted more and more attention from the academic community. The theory expresses the robot's pose and its observation data in the form of a graph structure and then continuously optimizes the configuration of the graph. The final graph configuration is regarded as the SLAM problem solution. The article briefly introduces the general SLAM method of graph optimization modeling in the second section. This section will continue to analyze how to fuse the observation information of the QR code and participate in the front-end graph construction process.

Closed-loop detection is an important part of the graph optimization SLAM method, because it can greatly improve the inconsistency of the constructed map. However, the laser closed loop is difficult to match the correct closed loop in a monotonous scene, especially in a narrow corridor or a symmetrical environment. Because the two-dimensional code contains uniquely identified information, it can overcome the aforementioned shortcomings and find the correct closed loop, so this article adopts the graph optimization SLAM method that incorporates the two-dimensional code information. The improved graph optimization framework is shown in Fig. 5.3. This article adds the detection of the QR code within the framework of the traditional laser odometer model. When a new QR code is detected, the current node and QR code information are saved. When the same two-dimensional code is repeatedly detected, a closed-loop signal is generated for the closed-loop detection of the front end to improve the robustness of the closed-loop detection.

If the location of the landmark points is determined at the same time in the positioning process, a feature map with landmark points as the characteristic can be established. In the front end of graph optimization, the relationship between nodes and edges is established, and the back end needs to optimize the configuration of the graph. Let $C = (C_1, C_2, \ldots C_3)$ is the node pose we need to optimize:

$$C_i = \left[t_i^T, \theta_i\right]^T = [x_i, y_i, \theta_i]^T \tag{5.12}$$

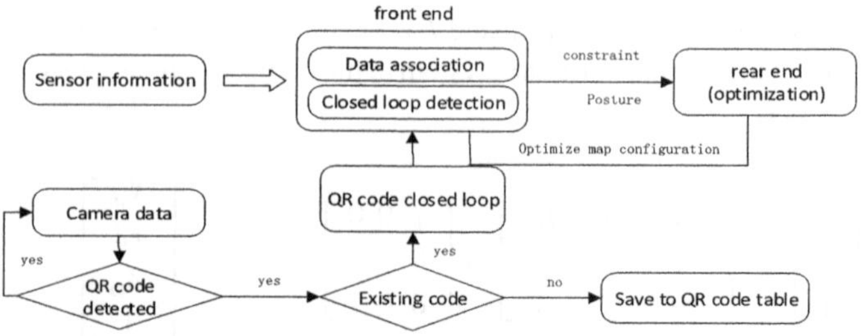

Fig. 5.3 QR code improvement framework

In the formula, t_i, θ_i are the position coordinates and direction angle of the robot on the map, respectively.

For the constraints between the ith and jth nodes, the relative transformation relationship between the two nodes is:

$$\widehat{Z_{ij}}(C_i, C_j) = \begin{bmatrix} R_i^t(t_j - t_i) \\ \theta_j - \theta_i \end{bmatrix} \tag{5.13}$$

R_i is the cosine matrix related to θ_i.The total error function of all edges can be obtained:

$$F(c) = \sum_{ij} e_{ij}^T \vartheta_{ij} e_{ij} = \sum_{ij} F_{ij} \tag{5.14}$$

where ϑ_{ij} represents the covariance matrix between c_i, c_j. Our optimization goal is to find the optimal pose configuration to minimize the overall error function:

$$c^* = \arg\min F(c) = \arg\min \sum_{ij} e_{ij}^T \vartheta_{ij} e_{ij} \tag{5.15}$$

According to the initial node pose estimation, the Gauss–Newton iteration strategy is used to obtain the numerical solution of the above formula. Let the initial estimate be $\hat{c}$, c_1, c_2 are the initial estimates of the ith and jth nodes, respectively. The iterative strategy is to linearize the error function near the initial estimated value, seek the optimal solution of the linear system to approximate the estimation, and gradually approach the optimal solution of the nonlinear system through iteration.

$$e_{ij}(c_i, c_j) = e_{ij}(\widehat{c_i} - \Delta c_i, \widehat{c_j} - \Delta c_j) = e_{ij}(\hat{c} + \Delta c) \approx e_{ij} + J_{ij} \Delta c \tag{5.16}$$

where $J_{ij} - e_{ij}(c)$ is the Jacobian matrix at $\hat{c}$. The new approximate solution after solving is:

$$c^* = \hat{c} + \Delta c \tag{5.17}$$

After finding H and b at the initial point in each iteration, sparse Cholesky can be used to decompose the linear equation, and then $c*$ can be updated with c. The actual solution process can use the CSparse library [7] to provide an understanding method.

5.5 Correction Graph Optimization Algorithm Principle

After constructing a map through the SLAM algorithm, the robot can use the map to perform navigation tasks. The higher the positioning accuracy of the SLAM algorithm, the more accurate the map constructed, and the higher the accuracy of the robot when performing navigation tasks. This paper uses ROS as the software platform to realize the graph optimization SLAM that integrates the observation information of the QR code; finally, the QR code is used as the initial estimation of the robot's pose, and the location of the QR code in the map is set as the navigation target to complete the navigation task. The robot's navigation accuracy is used to indirectly measure the effect of SLAM algorithm positioning and map construction.

According to different robot navigation strategies, the working environment is divided into two types. Among them, for continuous storage shelves, the two-dimensional code navigation method is directly adopted [8]. As shown in Fig. 5.4, the shelves are continuously and evenly distributed, and the QR code is pasted in the middle of the shelves at a certain distance.

The second is the graph optimization SLAM for QR code correction, which is the main work of this article. The specific experimental environment is shown in Fig. 5.5. The QR code labels are arranged in advance at key positions of the path to assist the graph optimization SLAM closed-loop detection. The above two kinds of experimental tests use the two-dimensional code mobile robot independently developed by the laboratory, and the experimental environment is part of the laboratory simulation.

Fig. 5.4 Dense storage environment

Fig. 5.5 Complex corridor

Figures 5.6 and 5.7 are the mapping results of SLAM experiment two. Figure 5.6 shows the result of pure laser mapping. As the composition progresses, due to the sensor error, the map appears inconsistent within the marked frame. And Fig. 5.7 is the result of optimizing the laser SLAM according to the graph fused with the observation information of the QR code used in this article. The green punctuation

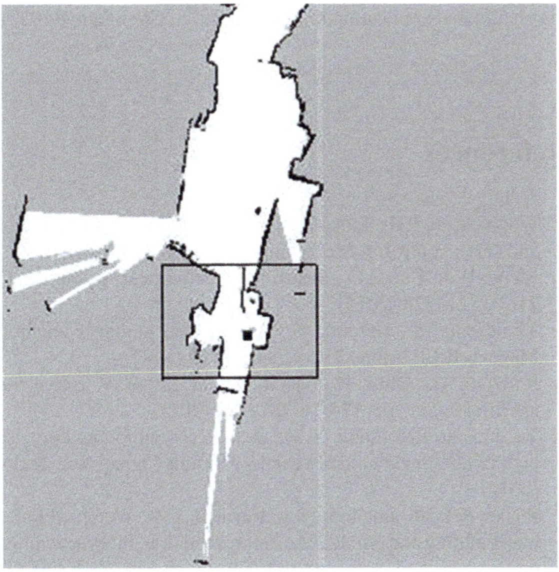

Fig. 5.6 Laser mapping

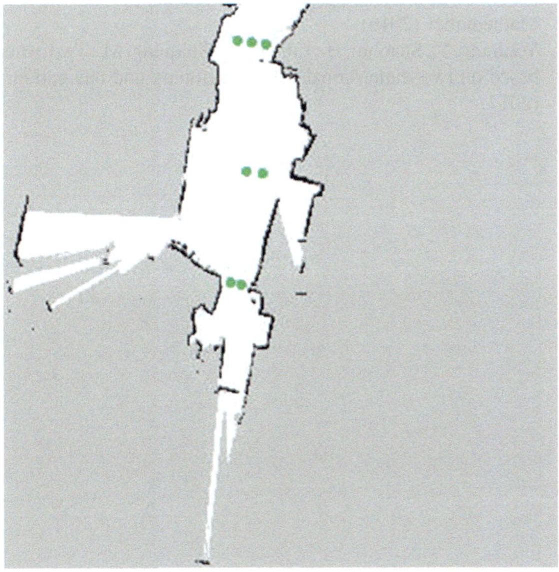

Fig. 5.7 Two-dimensional code fusion mapping

is the QR code pasted on the ground. When the same QR code is detected for the second time, the closed-loop signal is activated to complete the optimization of the laser closed-loop, and the pose is corrected in the process of map optimization. The subsequent trajectory has a higher degree of overlap with the original trajectory, so the map consistency will be better.

Acknowledgements This article is supported by The demonstration project of technological innovation and application in Beibei District, Chongqing (2020-5).

References

1. Wu, X., Lou, P.H., Tang, D.B.: Hybrid motion control of automatic guided vehicle path tracking and servo control. J. Mech. Eng. **47**(3), 43–48 (2011)
2. Meng, W.J., Liu, Z.Q.: Research on path tracking control of visual-guided AGV. Control Eng. **21**(3), 321–325 (2014)
3. Kamewaka, S., Uemura, S.: A magnetic guidance method for automated guided vehicles. Magnet. IEEE Trans. **23**(5), 2416–2418 (1987)
4. Ronzoni, D., Olmi, R., Secchi, C., et al.: AGV global localization using indistinguishable artificial landmarks **124**(1), 287–292 (2011)
5. Yu, J., Lou, P.H., Qian, X.M., et al.: Recognition and precise measurement of visual guidance path of automatic guided vehicle. J. South China Univ. Technol. Nat. Sci. Edn. **40**(3), 143–149 (2012)
6. Pfister, S.T., Roumeliotis, S.I., Burdick, J.W.: Weighted line fitting algorithms for mobile robot map building and efficient data representation. In: Proceedings of IEEE International Conference on Robotics and Automation. IEEE, Piscataway, N.J. (2003)
7. Davis, T.A.: Direct Methods for Sparse Linear Systems (Fundamentals of Algorithms 2) Direct Methods for Sparse Linear Systems, pp. 420–460. Society for Industrial and Applied Mathematics (2016)
8. Youliang, Y., Shaohui, H., Lihong, Z., Zhiqiang, M.: Two-wheel differential steering AGV design based on two-dimensional code positioning and navigation. Logistics Technol. **40**(10), 40–42 (2017)

Chapter 6
Research on Intelligent Security System of Online Car Based on Computer Vision

Hetao Sun and Jianyong Zhong

Abstract The online car is the product of the "Interonline+ " era. As a new city intelligent transportation mode, it breaks the old interest pattern and the barrier of taxi industry based on franchise. In recent years, the number of online car Hailing has increased sharply. However, due to the lack of effective supervision means, security problems have gradually emerged, such as rape, robbery, intentional injury, intimidation, abuse of passengers and other incidents. In order to solve this problem, this project proposes an intelligent security system for online car Hailing based on machine vision, constructs an advanced machine vision solution framework, and introduces graphics and image processing technology and machine learning abnormal behavior intelligent warning model. At the same time, combined with the embedded technology, it can automatically warn the abnormal behavior in the online car Hailing and quickly access the manual intervention and alarm, so as to timely detect the security threats and potential dangers in the process of taking the bus, so as to realize the purpose of protecting the safety of the online car Hailing passengers.

6.1 Introduction

The online car is the product of the "Interonline plus" era. As a new city intelligent transportation mode, it breaks the old interest pattern and the barrier of taxi industry based on franchise. In recent years, the number of online car Hailing has increased sharply. However, due to the lack of effective supervision means, security problems have gradually emerged, such as rape, robbery, intentional injury, intimidation, abuse of passengers, and other incidents [1]. In order to solve this problem, this project proposes an intelligent security system for online car Hailing based on machine vision, constructs an advanced machine vision solution framework, and introduces graphics and image processing technology and machine learning abnormal behavior intelligent warning model. At the same time, combined with the embedded technology, it can automatically warn the abnormal behavior in the online car Hailing

H. Sun (✉) · J. Zhong
Jiaxing Vocational and Technical College, Jiaxing 314036, China
e-mail: zkweng19@163.com

L. C. Jain et al. (eds.), *Wireless Technology, Intelligent Network Technologies, Smart Services and Applications*, Smart Innovation, Systems and Technologies 258,
https://doi.org/10.1007/978-981-16-5168-7_6

and quickly access the manual intervention and alarm [2], so as to timely detect the security threats and potential dangers in the process of taking the bus, so as to realize the purpose of protecting the safety of the online car Hailing passengers [3–5]. The number of online car Hailing has increased dramatically, and the coverage period and coverage area have also greatly increased. Security problems have emerged gradually, such as rape, robbery, intentional injury, intimidation, abuse, beating passengers, drunk driving, drunk driving, and driving after poisoning [6, 7].

From the perspective of online car Hailing platform, CEO Cheng Wei of Didi company realized that Didi needed to make up lessons after he walked alone in Wulin by way of subsidy. In 2018, the company's overall strategy has shifted to "cultivating internal strength," from simple growth indicators to safety, experience and efficiency. However, Meituan and Gaode saw that Didi company used a lot of funds to seek strategic transformation to break through the bottleneck, and other aspects were weak. At this time, Meituan and Gaode entered the online car Hailing market in the form of capital to start a new round of "subsidy war." Our country has entered the stage of consumption upgrading. The simple price war makes enterprises too busy to cope with it. It is difficult to calm down to improve the management level and service quality, which is not conducive to the healthy development of the industry. After the high price subsidy of online car Hailing industry was stopped, there was a lack of other effective means of competition in the industry, and there was a lack of ways to improve user experience. Meituan and Gaode were in a dilemma. Didi company was tired of transformation and upgrading and needed to update its competitive chips. Solving such security problems will undoubtedly be an effective way to enhance competitiveness and lead the industry to upgrade security [8].

6.2 The Proposed Framework

Computer vision system uses machine instead of human eyes to make various measurements and judgments. As a branch of computer science, it integrates optics, machinery, electronic computer, and other technologies. The core of this project is to apply machine vision to monitor abnormal human behavior in online car Hailing. The intelligent security system of online car Hailing based on machine vision uses the advanced machine vision algorithm architecture, combines big data analysis and machine learning abnormal behavior intelligent early warning model to early warn the high abnormal behavior in the car, and quickly carry out manual intervention and alarm, so as to timely discover and avoid the safety threats and safety hazards in the process of riding. The overall framework of the project is shown in Fig. 6.1.

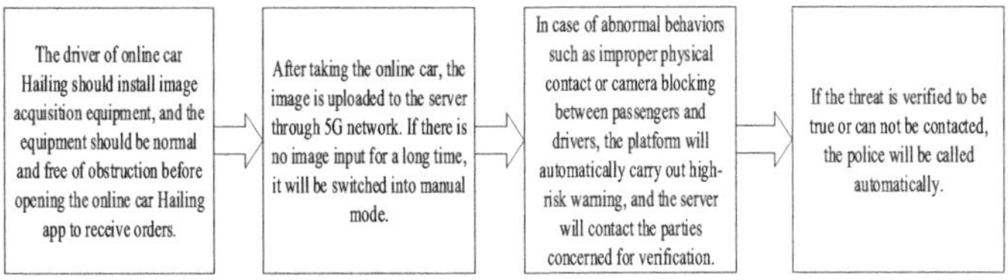

Fig. 6.1 Framework of our proposed method

6.2.1 *Software Module of the System*

From the perspective of fixed scene and single recognition object, we hope to develop a better recognition method for human abnormal behavior in specific scenes. Therefore, this project designs a recognition method of abnormal behavior in vehicle with has high recognition accuracy, low computational complexity, and strong portability, which is based on fusion features.

Firstly, the input video sequence is divided into several spatiotemporal volumes. Secondly, the motion difference image (MDI) is extracted by inter-frame difference. For each MDI, the histogram features of gradient direction are extracted and normalized as the overall feature descriptor. At the same time, the scale invariant feature is extracted from each frame transform (SIFT) local features, where $D(x, y, \sigma)$ represents the motion difference image, k is the difference coefficient, $G(x, y, k\sigma)$ and $L(x, y, k\sigma)$ represent the differential processed image, *and* $I(x,y)$ indicates the original image:

$$\begin{aligned} D(x, y, \sigma) &= [G(x, y, k\sigma) - G(x, y, k\sigma)] * I(x, y) \\ &= L(x, y, k\sigma) - L(x, y, \sigma) \end{aligned} \tag{6.1}$$

Through bag of visual word (BoVW) model to encode SIFT features and then fuse the features. Finally, support vector machine (SVM) is used to judge the abnormal behavior of the fused features, where ω indicates hyper-plane normal vector, C indicates the number of classes, and ξ indicates the regularization coefficient:

$$\begin{aligned} &\min_{w,b} \frac{1}{2}\|\omega\|^2 + C\sum_{i=1}^{N} \xi_i \\ &\text{s.t.} \quad y_i(\omega^T X_i + b) \geq 1 - \xi_i \end{aligned} \tag{6.2}$$

The process framework of this method is shown in Fig. 6.2.

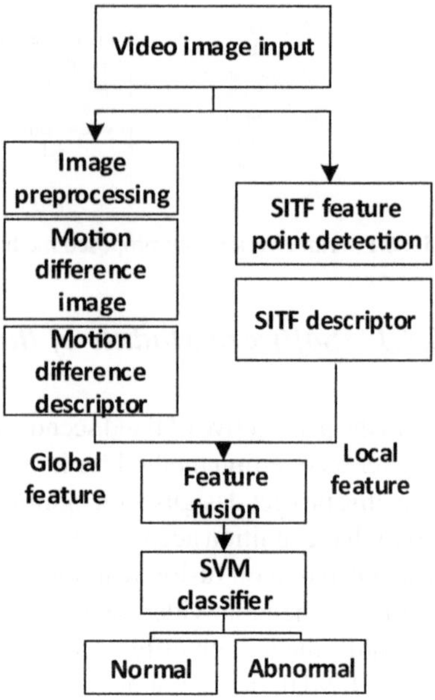

Fig. 6.2 Process of human abnormal behavior recognition framework

6.2.2 Hardware Module of the System

In the design of vehicle data acquisition terminal, tiny210 core board is used. The core board is equipped with S5PV210 processor. At the same time, it is equipped with a 512 MB NAND flash chip and four 32-bit DDR2 DRAM. The 5G wireless communication adopts sim7600ce communication module, with the maximum uplink speed of 50 Mbps and the downlink maximum speed of 150 Mbps, and supports TCP/IP and PPP communication protocols. Sim7600ce is directly connected with USB2 port of the processor, and the working voltage is 4.2 V. With the powerful data transmission ability of 4G wireless communication module, the sim7600ce can communicate with the server-side network car Hailing security center in real time. Beidou positioning module adopts um220 module of Hexin satellite communication company. The module supports BD2 and GPS independent positioning and multi-system joint positioning and uses serial communication mode to communicate with s5pv210 processor. The overall block diagram of system hardware is shown in Fig. 6.3.

6.3 Experimental Result

In the experiment, we use four video clips of driving behavior collected by ourselves to test. The test hardware environment is CPU i7 9700, RAM 32G, and GPU GTX1080ti. The experimental result is shown in Fig. 6.4. The green parts indicate the normal frames in one clip. On the contrary, the red parts indicate the abnormal frames

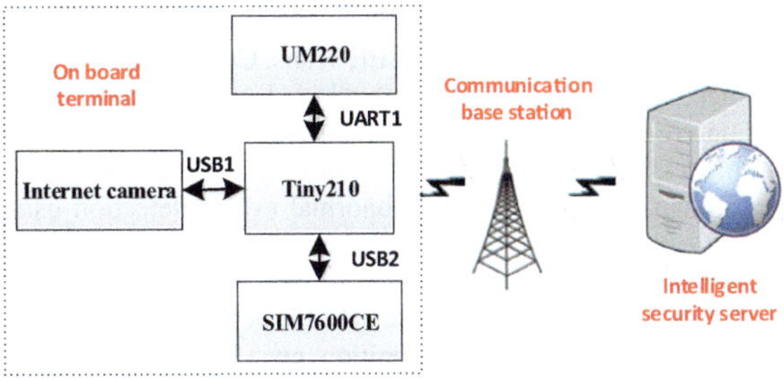

Fig. 6.3 Overall block diagram of system hardware

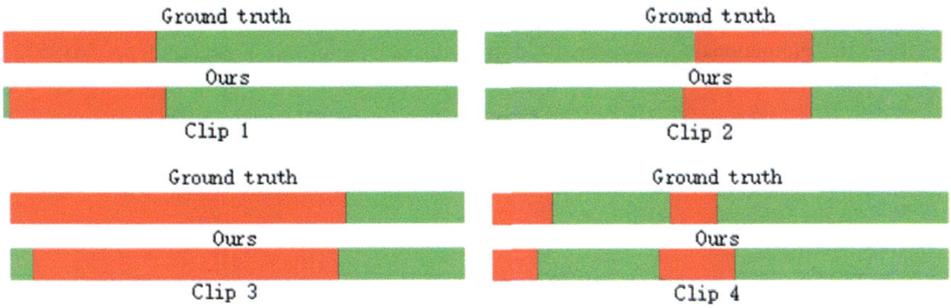

Fig. 6.4 Experimental result on four video clips

in the clips. The second row of each sub-figure represent the proposed method's performance in four clips. It is not difficult to find that the proposed method can detect the occurrence of abnormal behavior more accurately.

6.4 Conclusions

This paper proposes an intelligent security system of online car based on computer vision. Through video image acquisition, combined with machine vision technology, this project can identify and early warn the abnormal behavior of human body in the car, and quickly carry out manual intervention and alarm, so as to avoid the risk at the first time and ensure the life and property safety of online car Hailing passengers.

References

1. Sharif, M.H., Djeraba, C.: An entropy approach for abnormal activities detection in video streams. Pattern Recogn. **45**(7), 2543–2561 (2012)
2. Wang, T., Snoussi, H.: Histograms of optical flow orientation for visual abnormal events detection. In: IEEE Ninth International Conference on Advanced Video & Signal-based Surveillance, pp. 311–316. IEEE (2012)

3. Zhang, Y., et al.: Social attribute-aware force model: exploiting richness of interaction for abnormal crowd detection. IEEE Trans. Circ. Syst. Video Technol. **25**(7), 1231–1245 (2015)
4. Chen, Y., et al.: Abnormal behaviors detection using particle motion model. In: International Conference on Graphic & Image Processing International Society for Optics and Photonics, pp. 1563–1569 (2015)
5. Capman, F., Valmy, B.D.: Abnormal events detection using unsupervised one-class SVM—application to audio surveillance and evaluation. In: 2011 8th IEEE International Conference on Advanced Video and Signal Based Surveillance (AVSS), pp. 2346–2353. IEEE (2011)
6. Sultani, W., Choi, J.Y.: Abnormal traffic detection using intelligent driver model. In: International Conference on Pattern Recognition, pp. 1256–1266. IEEE (2010)
7. Zhang, Z., Mei, X., Xiao, B.: Abnormal event detection via compact low-rank sparse learning. IEEE Intell. Syst. **31**(2), 29–36 (2016)
8. Wang, W., Zhang, P., Wang, R.: Abnormal video sections detection based on inter-frames information. In: International Conference on Multimedia & Ubiquitous Engineering, pp. 563–572. IEEE (2009)

Chapter 7
Research on the Design of Mobile APP Product—Taking 'Wei Weather' APP as an Example

Zhaoyuan Song and Shui Jin

Abstract Through the summary and combing of the guiding ideology and experience in the design concept and design practice process in the six aspects of functional design, visual design, interaction design, interface design, and icon design in the "Wei weather" mobile phone APP, a mobile APP product is obtained. Every link in the design process should be based on demand, everything should be user-centric, and user experience should be emphasized.

7.1 Introduction

With the rapid development of Internet technology and the popularization of smart phones, mobile phones have become an indispensable part of people's daily lives. At the same time, mobile apps have also been widely used, making people's lives, work, and learning more convenient and efficient. In a digital age, people can use mobile phones to socialize, shop, obtain information, etc. Weather apps have greatly changed the way of obtaining weather information in the past and can disseminate weather information more timely and conveniently, and integrate social, sharing, and life information features greatly improve the quality of life and travel of users.

"Wei Weather" is a weather app designed for people who often go out for work, business trips, and travel. From the needs of users, it focuses on disseminating weather information, integrating social networking, sharing, and other functions. The purpose is to improve people's quality of life in the Internet age, new technology, new ideas, and new aesthetics. This article will take the "Wei weather" APP as an example and conduct the following research on the design of mobile APP products.

Z. Song · S. Jin (✉)
Beijing Institute of Fashion Technology, Beijing 10029, China
e-mail: jinshui@126.com

L. C. Jain et al. (eds.), *Wireless Technology, Intelligent Network Technologies, Smart Services and Applications*, Smart Innovation, Systems and Technologies 258,
https://doi.org/10.1007/978-981-16-5168-7_7

7.2 Functional Design

The design of the functional modules should be user-centric, and in the current situation where there are many APP products and the market is becoming saturated, it conforms to the user's habits and satisfies the user's function of obtaining weather information. The main functional modules of "Wei Weather" are:

7.2.1 Log in to the Registration Module

This module meets the functions of new user registration and user login and can realize the storage and update function of user information. According to the registration and login information, it can meet the needs of users of different genders, occupations, and ages for weather information.

7.2.2 "My" Module

It includes functions such as location, footprint, message, etc. Users can locate and understand the surrounding weather information in real time and find surrounding destinations to meet the needs of work and life. Footprints display travel information for users. The message bar can accept platform pushes and nearby people's information. Messages and users can socialize and ask for information.

7.2.3 "Home" Module

The homepage is the core functional module, which must meet the user's needs for weather information. This module focuses on displaying weather information, divided into four categories: temperature, precipitation, wind, and humidity. It is as simple and concentrated as possible to accurately and quickly convey weather information to users, to clearly communicate the temperature of each period of the day, the highest temperature and the lowest temperature today, whether there is precipitation and the expected precipitation, wind speed, air humidity, and avoid excessive classification to cause a messy experience for users, and improve the quality of users' outings.

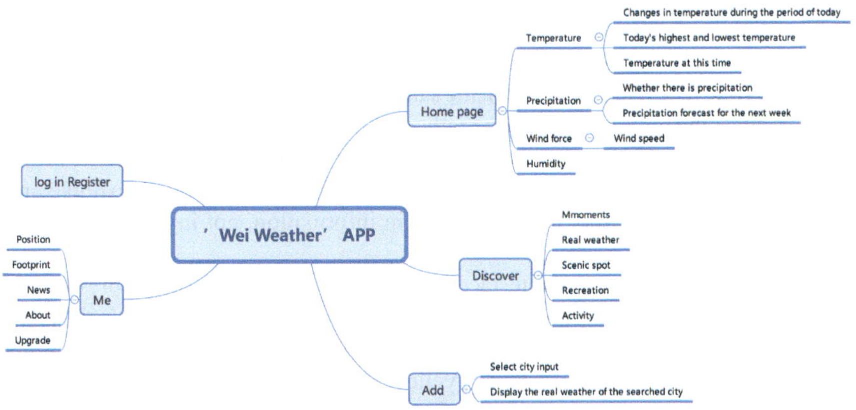

Fig. 7.1 Functional information architecture diagram

7.2.4 "Discover" Module

In addition to the function of disseminating weather information, users also have certain needs for other functions. "micro-weather" focuses on disseminating weather information, integrating social networking, sharing, and other functions. Under the "Discover" module, there are moments, live weather, scenic spots, entertainment, life, and five categories. In the "moments," users can share pictures and videos to provide information to other users near the same place to meet users' needs for real weather. Share pictures in real time to understand the real weather conditions and provide the conditions of nearby scenic spots to meet the needs of users for leisure and travel after work trips. The entertainment and life category bar provides users with nearby food, hotels, convenience stores, coffee shops, and other places. Screening is carried out to meet the user's work and life needs and reduce the time for selection.

7.2.5 "Add" Module

This module is used to search and add destinations before traveling and to add cities in advance to understand weather information to prepare for traveling (Fig. 7.1).

7.3 Visual Design

The visual design of APP includes the interface layout, color, text, visual style of icons in APP products, etc. "micro-weather" is a weather APP that serves people who often go out for work, business trips, and travel. Therefore, in the design process, consider the age, habits, use experience, use scenarios, and emotional needs of the user group. Everything is user-centric [1]. A simple page layout is adopted in the interface layout. The fixed icons are in the same position on all pages. Important icons always display, the bottom navigation method is used to reduce the menu bar as much as possible. The first-level page uses a large picture display layout to have a strong visual impact, and the second-level page uses a grid display layout to be simple and clear, which facilitates users to quickly obtain information, not information mixed and dazzling.

Color is an important part of interface design. Color will leave a first impression on users when they use APP for the first time. Different colors will give people different feelings: red represents celebration, enthusiasm, and vitality, and orange represents youth, happiness, and happiness. Sense, yellow represents sunshine, cartoons, blue is atmosphere, calmness, reason, technology, etc. The color conveys the nature and positioning of the product. As a weather APP product, the sky, rain, and other elements will soon be associated with "weather," as well as technology, network, and other attributes. Therefore, the color design of "micro-weather" is mainly blue, which conforms to the attributes of weather apps also give users a sense of sensibility, safety, and business and increases their attractiveness to target users.

As an important carrier for conveying information, text also plays an important role in visual design. Different fonts, font sizes, and colors will give users a different experience and feeling. "Wei weather" uses elegant black fonts with beautiful structure and fonts. The opening of the middle palace enhances the applicability of the text and is more suitable for the display of the screen; the round and block shape of the font is good, the display is clear, and the visual experience is comfortable.

The bottom of the icon is transparent, and the design form adopts a flat design style, which is easy to understand, simple and direct, and pursues artistic style on the basis of expressing the basic meaning based on the principle of simplicity and ease of use, so that users can read and use comfortably experience.

7.4 Interaction Design

Interaction design has changed the tradition of taking objects as objects in industrial design, graphic design, and space design, directly taking human behavior as design objects [2]. The five elements of interaction design include people, actions, environment, tools, and purpose. In the interaction design, the relationship between the user and the product should be considered, how to trigger feedback, reduce unnecessary steps, and complete the interactive task the fastest, Improve user experience. With the gradual increase of APP products, people have become more and more proficient in the use of APP and have a certain basis for some interactive trigger gestures. There are many types of gestures: general gestures, simple gestures, complex gestures, etc., most of which are triggered gestures serve functions, and some gestures are purely for fun and artistry. Based on the usage scenario and the user population, "Wei Weather" should try to conform to the user's usage habits in the interaction design, use simple interactive gestures, reduce the error rate in the operation, and be user-centric. On the core function page of the "Home" page, a long picture display format is adopted. Users can swipe with one hand to learn more. Single click to switch the page, double click to zoom in on the interface information, long press to perform to share information, and these interactive trigger gestures can be completed in one action with one finger. The learning cost is low, and the operation is simple. It fully takes into account the use scene and purpose of the user when going out, which is the most convenient and quickest in the unstable use environment to get weather information.

7.5 Interface Design

Interface design is an important part of mobile APP products. The interface is a medium for users to communicate with APP products, as well as a way of expressing product cultural connotations and emotions. The interface design directly determines the quality of the user experience. In the interface design, not only the design layout, text, color, etc., of the interface itself must be considered, but also the user, including the user's experience and learning cost. A good interface design is simple to operate, clear and easy to understand, and the information is scientific and reasonable, which makes users feel happy and comfortable. While improving user efficiency, it also brings users emotional and emotional satisfaction. "micro-weather" interface architecture The design includes three always-displayed icons: "Home" at the bottom of the navigation and "My," "Locating," and "Add" at the top of "My" to prevent users from getting lost and quickly return to the home page to find the information they want. The information arrangement adopts the method of big picture display and grid display. The large picture display has a strong visual impact. The weather information is displayed in order of importance on the background picture. The grids are displayed neatly, which facilitates users to quickly obtain information. Based on ergonomics, the law of human visual movement is mainly "F" type, that is, when browsing information, the horizontal movement of the eyes is faster than the vertical movement. The eyes are used to moving from left to right and from bottom to top. The observation rate of the content in the interface or plane is in order, the top left is the best, the second is the top right, the bottom left is again, and the worst is the bottom right [3]. George A. Miller proposed in 1965 that people can only remember 5–9 things or process 5–9 pieces of information at a time (that is, 7 plus or minus 2), but later researchers such as psychologists Alan Baddeley and Nelson Cowan showed: The number of information that can be remembered or processed is "4" [4]. "micro-weather" reduces the number of navigation bars at the bottom in the interface design, avoids unnecessary memory for users, and takes the user's needs for weather information as the starting point to reduce the need for other unnecessary navigation bars and functions.

In the interface design, it is necessary to go through low-fidelity design first and then carry out high-fidelity design on the basis of low-fidelity. The low-fidelity frame defines the overall frame of the interface for later high-fidelity production and improvement, which is convenient and quick. Some scholars put forward the three principles of user interface design: placing the interface under the user's control; reducing the user's memory burden; and maintaining the consistency of the interface [5].

7.6 Icon Design

The icons include startup icons and function icons. The startup icon is the facade of the APP, which is particularly important to the user's first impression. The startup icons of WeChat, Alipay, NetEase Cloud Music, and other APPs have their own characteristics. Users can quickly remember the product and be accurately grasp the function and tone of the product. The startup icon of "micro-weather" adopts the current popular flat design style, with a blue rounded rectangular frame as the background, abstracting the outline of the sun and cloud elements to summarize the simple atmosphere.

In the design of function icons, the design of the icons basically adopts the basic icon elements that users are familiar with, and the icons are mainly linear icons when they are not clicked, and the click state uses a color-filled form (such as the bottom navigation icon) in some commonly used icons. The higher-level function icons are displayed in a face-to-face manner, which is more three-dimensional than linear. For people who are out for work and business trips, the weather will not prevent them from going out. Therefore, in the icon design of the weather category, no more detailed distinctions are made between heavy rain, moderate rain, and light rain. It just reminds and informs users that there is you need to go out and bring an umbrella for protection. The basic functional icons are slightly optimized in detail to match the overall page experience. For example, the lines of the icons are unified with a thickness of 3 px, the overall specific gravity is slightly larger and heavier, and the corners and endpoints are rounded to make the graphics more rounded and better. Integrate into the whole question page design.

7.7 Motion Design

Motion design is the visual top level of interface design. It is an important carrier that guides and responds to user operations and realizes deep interaction with users. Motion effect design can attract users' attention with an interesting visual effect and improve the user experience when using APP. Too little or no motion effect will make users boring when using APP products, but too much and too dazzling. The dynamic effects will cause users to pay too much attention to the dynamic effects and ignore the main information. Therefore, the dynamic effects should be arranged reasonably in the animation design, and the dynamic effects should not be overwhelming. "micro-weather" clearly expresses the level changes of the interface in the animation design, so that users can clearly know their own level and understand the transition changes of the interface. The specific design is switch pages of different levels horizontally, and enter the first-level page its child page fades in from right to left, the child page also fades in from right to left when entering the next level of child page, and it fades in from left to right when returning to the parent page, adopting the design conventions formed in APP products, respecting a habit developed by the user to complete the interaction simply and efficiently. Efficiently feedback the user's operation, so that the user clearly knows their current state, "Wei weather" involves the acquisition of the target location information, after the user enters the destination, a confirmation box to check the information will appear to avoid location errors and confusion If the two positions are the same, there will be a red reminder sign and a vibrating box, giving the user clear feedback. To sublimate the user experience and promote the emotional expression of the APP, "Wei Weather" is a weather APP product that focuses on weather information. The first thing that appears after entering the home page is the weather conditions of the destination or the place where it is located, combined with the weather conditions. An immersive dynamic design, which is specifically expressed as: rainy dynamic effect is the effect of raindrops falling slowly and occasionally wetting the phone screen, and sunny dynamic effect is the sun and a beam of sunlight turns from left to right, conveying the weather interestingly. The information makes users understand and at the same time adds bright spots to the APP [6]. Effectively help guide users and reduce users' learning costs. There are countless APP products in the market, and many users are often confused. When logging in and registering, they will use dynamic progress bars and guiding icons to help users understand the progress of tasks quickly. Complete tasks and reduce user anxiety. As the "soul" of APP products, motion design is the most eye-catching tool, and it needs to be the best in a limited area and time.

7.8 To Sum up

With the rapid development of Internet technology and the continuous popularization of smart phones, mobile APP products are countless in terms of types and quantities, and more and more people are beginning to use mobile APPs to work, live, study, etc. The user groups cover all kinds of Age group. Mobile APP product design includes information architecture design, function design, interface design, icon design, interaction design, motion design, etc. Each link should be user-centric, starting from the needs of users, and ending in the user experience of users. The author believes that the emergence of mobile APP products conforms to the trend of the times, and the purpose is to improve the convenience of people's lives. Therefore, in the design of mobile APP products, we must start from the demand, focus on the user, and pay attention to the user experience. Serve us better. This article hopes to provide some new ideas for the design of other mobile phone APP products by analyzing the design of the "Wei weather" mobile phone APP.

Acknowledgements This work was supported by the Special Fund for the Construction of High-level Teacher's Group of Beijing Institute of Fashion Technology, Project No. biftxz201905.

References

1. Bai, W.T., Liu, Z.J.: Demand analysis and design principles of user interface. J. Dalian Maritime Univ. **04**, 86–88 (2004)
2. Xin, X.Y.: Interaction design: from physical logic to behavioral logic. Decoration **01**, 58–62 (2015)
3. Dong, H.J.: Research on the visual design of mobile app interface. Art Technol. **30**(08), 291 (2017)
4. Weinschenk, S.: 100 Things Every Designer Needs to Know About People. People's Posts and Telecommunications Press (2013)
5. Tu, X.D.: Talking about UI design. Comput. Knowl. Technol. **6**(07), 1706–1707 (2010)
6. Yang, W.G.: On the application and dimensional analysis of motion design in user interface design. Art Design (Theory) **2**(05), 47–49 (2016)

Chapter 8
A Low-complexity Tensor Completion Scheme Combining Matrix Factorization and Smoothness

Leiming Tang, Chuang Yang, Zheng Wang, and Xiaofei Zhang

Abstract In this paper, the low-complexity tensor completion (LTC) scheme is proposed to improve the efficiency of low-rank tensor completion with competitive performance, which consists of the smooth matrix factorization (SMF) model and the corresponding alternating direction method of multiples (ADMM)-based solution. As for the SMF model, on one hand, we adopt the matrix factorization into the model of low-rank tensor completion for complexity reduction. On the other hand, we introduce the smoothness by total variation regularization and framelet regularization to guarantee the completion performance. To solve the SMF model, an ADMM-based solution is further proposed to realize the efficient and effective low-rank tensor completion. Finally, simulation results are presented to confirm the system gain of the proposed LTC scheme in both efficiency and effectiveness.

8.1 Introduction

As a natural form to express more complicated intrinsic structures of higher-mode data, tensor has many applications in various research fields, such as video inpainting, and signal reconstruction. Low-rank tensor completion (LRTC) aims to recover the incomplete tensor data, which exploits the low rank prior of tensor to estimate the missing entries. Specifically, the fundamental problem of LRTC heavily lies on how to minimize the targeting tensor rank, which has not been officially defined yet. To this end, the Tucker-rank is generally accepted to serve as the alternative of tensor rank. Although minimizing Tucker-rank is still NP-hard, it can be relaxed in the convex way by sum of nuclear norm (SNN) since the tightest surrogate of the matrix rank is the nuclear norm. Based on the property of SNN, Liu et al. [1] proposed three methods (SiLRTC, HaLRTC, and FaLRTC) for tensor completion. Unfortunately, minimizing SNN requires to calculate singular value decomposition (SVD) iteratively, which is complexity exhausted. Meanwhile, only considering the low-rank prior may results in inevitable performance degradation with the decrement of the sampling rate.

L. Tang · C. Yang · Z. Wang (✉) · X. Zhang
Nanjing University of Aeronautics and Astronautics, Nanjing, China
e-mail: z.wang@nuaa.edu.cn

L. C. Jain et al. (eds.), *Wireless Technology, Intelligent Network Technologies, Smart Services and Applications*, Smart Innovation, Systems and Technologies 258,
https://doi.org/10.1007/978-981-16-5168-7_8

Besides the low-rank structure, smoothness also plays an important role in LRTC for tensor recovering. In order to improve the performance, the total variation (TV) [2] can be used to impose piecewise smoothness constraints. Yokota et al. [3] proposed LRTV-PDS to minimize the tensor nuclear norm and TV simultaneously to improve the tensor completion quality. In particular, the TV regularization usually leads to staircase effect [4], which may cause the possible loss of information and geometric features in practice. For this reason, the framelet regularization can be further applied to avoid the possible performance degradation.

In this paper, the low-complexity tensor completion (LTC) scheme is proposed to improve the efficiency of LRTC. The proposed LTC mainly consists of two parts—the smooth matrix factorization (SMF) model and the alternating direction method of multiples (ADMM)-based solution. As for the SMF model, firstly, the matrix factorization is introduced into SNN to exploit the low-rank structure of the entire tensor. By factorizing matrices, the computational complexity in calculating SVD can be significantly reduced, which leads to efficiency improvement. Secondly, to guarantee the effectiveness of tensor completion, the TV is applied to utilize the smoothness along the space dimension. Finally, to address the staircase effect of TV, the framelet regularization is further used to preserve the information due to its redundancy. Given the proposed SMF model, we also give an effective ADMM-based algorithm to solve it. As shown in numerical experiments, our proposed LTC scheme can achieve a better trade-off between performance and complexity for LRTC.

8.2 Low-rank Tensor Completion

The purpose of LRTC is to recover the incomplete tensor. Mathematically, the LRTC problem can be written as

$$\min_{\mathcal{X}} \operatorname{rank}(\mathcal{X}), \text{ s.t. } \mathcal{X}_\Omega = \mathcal{T}_\Omega. \tag{8.1}$$

Here $\mathcal{X}$ is the targeting tensor, $\mathcal{T}$ is the observed tensor, and Ω is the index set of available entries.

Based on Tucker-rank, Liu et al. [1] developed a theoretical framework for LRTC and proposed SNN, a definition of the nuclear norm for tensors. The corresponding weight value ω_i of the mode-i unfolding matrix $X_{(i)}$ is introduced, and the LRTC model can be written as

$$\min_{\mathcal{X}} \sum_{i=1}^{N} \omega_i \left\| X_{(i)} \right\|_*, \text{ s.t. } \mathcal{X}_\Omega = \mathcal{T}_\Omega. \tag{8.2}$$

In order to solve this LRTC model, Liu et al. [1] proposed three corresponding algorithms (SiLRTC, HaLRTC, and FaLRTC) for tensor completion.

8.3 The Proposed SMF Model

In this paper, the proposed LTC scheme is presented, which consists of the SMF model and the ADMM-based algorithm. We first introduce the SMF model in this section. Considering a three-way tensor $\mathcal{X} \in \mathbb{R}^{n_1 \times n_2 \times n_3}$, the proposed SMF model is as following:

$$\min_{\mathcal{X}} \sum_{i=1}^{3} \omega_i \|R_i\|_* + \lambda_1 \left\| W X_{(3)}^T \right\|_{1,1} + \lambda_2 \left\| D_S X_{(3)} \right\|_{1,1},$$
$$\text{s.t. } \mathcal{X}_\Omega = \mathcal{T}_\Omega, X_{(i)} = L_i R_i, L_i \in \text{St}(I_i, s_i), i = 1, 2, 3. \tag{8.3}$$

Here λ_1 and λ_2 are regularization parameters, $\mathcal{X}$ is the object tensor, $\mathcal{T}_\Omega$ is the incomplete observed tensor, and Ω is the set of indices of available data in $\mathcal{T}$, $L_i \in \text{St}(I_i, s_i)$, $R_i \in \mathbb{R}^{s_i \times \prod_{j \neq i} I_j}$, $\text{St}(I_i, s_i)$ denotes the Stiefel manifold, and $s_i > r_i$ is a given upper bound of rank. Besides W indicates framelet transformation, D_S denotes the difference matrix, and $l_{1,1}$-norm is the sum of absolute values of the matrix elements.

Typically, the SMF model contains two main terms—SNN with matrix factorization and smoothness constraints, which are detailed in the following subsections.

8.3.1 *SNN with Matrix Factorization*

The first term, SNN with matrix factorization, can improve the efficiency. Given a low-rank tensor $\mathcal{X} \in \mathbb{R}^{I_1 \times I_2 \times \cdots \times I_N}$ with rank $(r_1, r_2, \ldots, r_N)$, the mode-i unfolding $X_{(i)}$ can be factorized into the form $X_{(i)} = L_i R_i,\ i = 1, \ldots, N$. Then we have the following property of tensor nuclear norm via matrix factorization:

$$\left\| X_{(i)} \right\|_* = \|L_i R_i\|_* = \|R_i\|_*, \quad i = 1, \ldots, N. \tag{8.4}$$

Therefore, the tensor nuclear norm minimization problem can be rewritten with smaller scale matrices to reduce computational complexity.

The goal of introducing SNN is to exploit the globally multidimensional structure, which is the basic of LRTC. Based on the introduced SNN, matrix factorization is further applied to save calculation cost. For example, the computational complexity of SVD of X ($X \in \mathbb{R}^{m \times n}$, $\text{rank}(X) = r$) is $O(m^2 n + mn^2)$. By introducing matrix factorization, the cost of computing SVD can be reduced to $O(m^2 s + ms^2)$, $r < s \ll n$ [5].

8.3.2 *Smoothness Constraints*

The smoothness constraints contain the total variation regularization and the framelet regularization, which are used for a better performance of tensor completion.

The total variation regularization $\left\| D_S X_{(3)} \right\|_{1,1}$ is used to exploit piecewise smoothness along the mode-3 unfolding of $\mathcal{X}$, where D_s is the difference matrix. The TV regularization is used to make the third dimension of the recovered tensor smooth to improve the efficiency of tensor completion.

The framelet regularization $\left\| W X_{(3)}^T \right\|_{1,1}$ retains details in spatial domain, where W denotes the framelet transform matrix with orthogonality and $X_{(3)}$ indicates the unfolding matrix of $\mathcal{X}$. The framelet regularization can ensure the spatial mode of the recovered data to be smooth and preserve the details due to its redundancy.

8.4 The ADMM-Based Algorithm

In this section, we design an effective ADMM-based algorithm to resolve the SMF model (8.3). In particular, we separate blocks of variables in (8.3) by introducing matrices M and N. Then the augmented Lagrangian function of (8.3) becomes:

$$\begin{aligned} L(L_i, R_i, \mathcal{X}, M, N) = &\sum_{i=1}^{3} \left(\omega_i \| R_i \|_* + \frac{\beta_1}{2} \| X_{(i)} - L_i R_i \|_F^2 \right) \\ &+ \lambda_1 \| M \|_{1,1} + \langle W X_{(3)}^T - M, \Psi \rangle + \frac{\beta_2}{2} \| W X_{(3)}^T - M \|_F^2 \\ &+ \lambda_2 \ll \| N \|_{1,1} + \langle D_S X_{(3)} - N, \Theta \rangle + \frac{\beta_3}{2} \| D_S X_{(3)} - N \|_F^2, \\ &\text{s.t. } L_i \in \mathrm{St}(I_i, s_i), \quad i = 1, 2, 3, \mathcal{X}_\Omega = \mathcal{T}_\Omega, \end{aligned} \tag{8.5}$$

where Ψ and Θ are the Lagrange multipliers, β_1, β_2, and β_3 are the penalty parameters. Based on ADMM, we can split the problem (8.5) into the following subproblems which are easier to deal with in smaller sizes.

The first one optimizes the variable L_i, which can be written as:

$$L_i^{k+1} = \arg\min_{L_i} \| X_{(i)}^k - L_i R_i^k \|_F^2, \text{ s.t. } L_i \in \mathrm{St}(I_i, s_i). \tag{8.6}$$

By solving the least squares problem with the orthogonality constraint, the optimal L_i with all other variables fixed is given. The optimal solution is given by QR-decomposition:

$$L_i^{k+1} = QR(X_{(i)}^k (R_i^k)^T). \tag{8.7}$$

The second subproblem optimizing the variable R_i can be calculated as follows:

$$R_i^{k+1} = \arg\min_{R_i} \ \omega_i \|R_i\|_* + \frac{\beta_1}{2} \left\| X_{(i)}^k - L_i^{k+1} R_i \right\|_F^2, \tag{8.8}$$

which has an explicit solution

$$R_i^{k+1} = \mathrm{SVT}_{\omega_i/\beta_1}((L_i^{k+1})^T X_{(i)}^k). \tag{8.9}$$

Here $\mathrm{SVT}_\tau(\cdot)$ is a singular value thresholding operator defined by $\mathrm{SVT}_\tau(X) = U\mathrm{diag}[\max(\sigma - \tau, 0)]V^T$.

The third subproblem about M related to framelet regularization can be written as:

$$M^{k+1} = \arg\min_{M} \ \lambda_1 \|M\|_{1,1} + \left\langle W(X_{(3)}^k)^T - M, \Psi \right\rangle + \frac{\beta_2}{2} \left\| W(X_{(3)}^k)^T - M \right\|_F^2. \tag{8.10}$$

This problem has an explicit solution

$$M^{k+1} = S_{\frac{\lambda_1}{\beta_2}} \left(W(X_{(3)}^k)^T + \frac{\Psi}{\beta_2} \right), \tag{8.11}$$

where $S_\mu(\cdot)$ is a soft-thresholding operator

$$S_\mu(x) = \begin{cases} 0, & \text{if } |x| \le \mu, \\ \mathrm{sign}(x)(|x| - \mu), & \text{if } |x| > \mu. \end{cases} \tag{8.12}$$

The fourth subproblem about N concerning total variation regularization can be calculated as follows:

$$N^{k+1} = \arg\min_{N} \ \lambda_2 \|N\|_{1,1} + \left\langle D_S X_{(3)}^k - N, \Theta \right\rangle + \frac{\beta_3}{2} \left\| D_S X_{(3)}^k - N \right\|_F^2, \tag{8.13}$$

We can update N as follows:

$$N^{k+1} = S_{\frac{\lambda_2}{\beta_3}} \left(sX_{(3)}^k + \frac{\Theta}{\beta_3} \right). \tag{8.14}$$

The final subproblem optimizing $\mathcal{X}$ can be formulated as:

$$\mathcal{X}_{\Omega^c}^{k+1} = \arg\min_{\mathcal{X}} \sum_{i=1}^{3} \left(\frac{\beta_1}{2} \left\| X_{(i)} - L_i R_i \right\|_F^2 \right)$$

$$+\frac{\beta_2}{2}\left\|WX_{(3)}^{kT}-M+\frac{\Psi}{\beta_2}\right\|_F^2+\frac{\beta_3}{2}\left\|D_SX_{(3)}^{k}-N+\frac{\Theta}{\beta_3}\right\|_F^2. \quad (8.15)$$

We can update $\mathcal{X}$ as follows:

$$\mathcal{X}_{\Omega^c}^{k+1}=\left(\sum_i \text{fold}_i(\text{inv}(\beta_1 I+\beta_2 I+\beta_3 D_s^T D_s)\right.$$
$$\left.*\left(\beta_1 L_i^{k+1}R_i^{k+1}+\beta_2\left[W^T\left(M-\frac{\Psi}{\beta_2}\right)\right]^T+\beta_3 D_s^T\left(N-\frac{\Theta}{\beta_3}\right)/N\right)\right)_{\Omega^c}, \quad (8.16)$$

According to ADMM, we update the multipliers Ψ and Θ as:

$$\Psi^{k+1}=\Psi^k+\beta_2\left(W(X_{(3)}^{k+1})^T-M^{k+1}\right),$$
$$\Theta^{k+1}=\Theta^k+\beta_3\left(D_SX_{(3)}^{k+1}-N^{k+1}\right). \quad (8.17)$$

Algorithm 1: The Proposed ADMM-based Algorithm for Solving the SMF Model in (3)

Input: The incomplete tensor $\mathcal{T}$, index set Ω, parameters λ_1, λ_2, β_1, β_2 and β_3.

Output: The recovered tensor $\mathcal{X}$.

1: initialize: $\mathcal{X}_\Omega^0=\mathcal{T}_\Omega$,

$L_i^0=\text{eye}(I_i,s_i)$, $R_i^0=\text{rand}(s_i,\prod_{i\neq j}I_j)$, $i=1,2,3$. M^0, N^0, Ψ^0 and Θ^0 initialized to 0, $k=0$.

2: while not converged and $k<k_{max}$ do

3: Update L_i^{k+1}, R_i^{k+1} via (7), (9)

4: Update M^{k+1}, N^{k+1}, $\mathcal{X}^{k+1}$ via (11), (14), (16)

5: Update Ψ^{k+1}, Θ^{k+1} via (17)

6: end while

7: return $\mathcal{X}^{k+1}$

To summarize, the proposed iterative ADMM-based algorithm for solving the SMF model in (8.3) is outlined in Algorithm 1.

8.5 Numerical Experiments

In this section, we conduct several experiments on the video data to evaluate our proposed LTC scheme. We also compare it with the modern methods, including

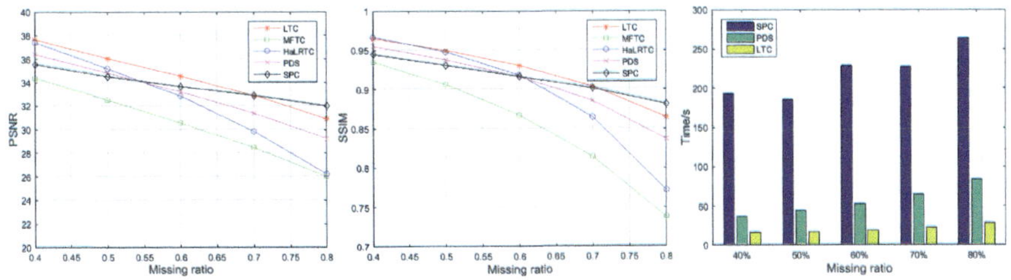

Fig. 8.1 Comparison of PSNR, SSIM, running time by five LRTC methods on the video

MFTC [5], HaLRTC [1], LRTV-PDS [3], and SPC [6]. To measure the quality of the completed tensor data, peak signal-to-noise ratio (PSNR) and structural similarity index (SSIM) are employed as the quality metrics.

In this experiment, the video "suzie" is tested, and only the first 30 frames of the video are used for tensor completion. The size of testing video is $144 \times 176 \times 30$.

Figure 8.1 shows the performance of tensor completion for video "suzie" with different MRs by the proposed LTC and the comparing modern methods in PSNR, SSIM, and running time. The MRs are set as 40%, 50%, 60%, 70%, and 80%, respectively. Apparently, PDS, SPC, and LTC perform better than MFTC and HaLRTC when the MR is high. While ensuring the performance, LTC achieves about 90% and 50% running time reduction compared to SPC and PDS, respectively. It means that introducing matrix factorization can reduce the running time of recovering incomplete videos effectively.

8.6 Conclusion

In this work, we propose a low-complexity tensor completion scheme. Our model takes advantage of SNN to exploit the low-rankness, TV, and framelet to preserve details and utilize the smoothness and matrix factorization to improve the efficiency. In order to solve the SMF model, we develop an efficient ADMM-based algorithm. The experimental results on synthetic data and real-world data demonstrate both the efficiency and effectiveness of the proposed algorithm.

References

1. Liu, J., Musialski, P., Wonka, P., Ye, J.: Tensor completion for estimating missing values in visual data. IEEE Trans. Pattern Anal. Mach. Intell. **35**(1), 208–220 (2012)
2. Rudin, L.I., Osher, S., Fatemi, E.: Nonlinear total variation. Based noise removal algorithms. Phys. D Nonlin. Phenomena **60**(1–4), 259–268 (1992)
3. Yokota, T., Hontani, H.: Simultaneous tensor completion and denoising by noise inequality constrained convex optimization. IEEE Access **7**, 15669–15682 (2019)

4. Dobson, D.C., Santosa, F.: Recovery of blocky images from noisy and blurred data. SIAM J. Appl. Math. **56**(4), 1181–1198 (1996)
5. Liu, Y., Shang, F.: An efficient matrix factorization method for tensor completion. IEEE Signal Process. Lett. **20**(4), 307–310 (2013)
6. Yokota, T., Zhao, Q., Cichocki, A.: Smooth PARAFAC decomposition for tensor completion. IEEE Trans. Signal Process. **64**(20), 5423–5436 (2016)

Chapter 9
DOA Tracking with Multi-Bernoulli Filter for Two-Parallel Linear Array: Reconstruct MUSIC as Pseudo-Likelihood

Xudong Dong, Xiaofei Zhang, Jun Zhao, Meng Sun, and Jianfeng Li

Abstract To solve the problem of the multi-sources direction-of-arrival (DOA) tracking by using two-parallel line arrays, a multi-target Multi-Bernoulli (MeMBer) filtering based on random finite set (RFS) is proposed. Unlike traditional DOA tracking algorithms, the number of sources in this paper is time-varying. The proposed method modifies and reconstructs the MUSIC pseudo-spectrum as its likelihood function, which greatly improves the weight of sampled particles in the high likelihood area. Two experimental simulation scenarios are provided. One is that the number of sources is fixed and known, and the other is that the number of sources is time-varying and unknown. Simulation results verify the effectiveness of the proposed method in estimating the number of time-varying sources and tracking their dynamics.

9.1 Introduction

2D direction-of-arrival (DOA) estimation has been widely used in the field of radar, sonar, and wireless communication [1–3]. Many signal processing methods, such as MUSIC algorithm [4, 5], ESPRIT [6] algorithm, propagator method (PM) algorithm [7], and DOA matrix algorithm [8–10] have been proposed for DOA estimation. The above-mentioned methods are based on the assumption of stationary sources. However, in reality, the source is moving, and its number may be unknown and

X. Dong · X. Zhang (✉) · M. Sun · J. Li
College of Electronic Information Engineering, Nanjing University of Aeronautics and Astronautics, Nanjing 210000, China
e-mail: zhangxiaofei@nuaa.edu.cn

X. Zhang
Key Laboratory of Dynamic Cognitive System of Electromagnetic Spectrum Space, Nanjing University of Aeronautics and Astronautics, Ministry of Industry and Information Technology, Nanjing 211106, China

J. Zhao
College of Electronic and Information Engineering, Tongji University, Shanghai 201800, China

L. C. Jain et al. (eds.), *Wireless Technology, Intelligent Network Technologies, Smart Services and Applications*, Smart Innovation, Systems and Technologies 258,
https://doi.org/10.1007/978-981-16-5168-7_9

time-varying. For moving sources, it is necessary to develop a multi-source tracking methods to estimate their number and capture their DOA.

In [11], the conception of random finite set (RFS) is proposed, and a series of multi-target tracking methods based on RFS are provided in [12–14]. Multi-Target Multi-Bernoulli (MeMBer) filtering has been used for DOA tracking in [15, 16]. In this paper, we provide an improved 2D MeMBer method for two-parallel uniform linear arrays (ULAs). The contributions of this method are as follows: (1) It works in the dimension of single source space, avoiding the problem of data association of filters. (2) Its tracking performance is superior to the traditional DOA tracking method. (3) The issues of time-varying target number can be solved.

9.2 Data Model

As show in Fig. 9.1, $N(k)$ narrowband signals $s_n(k)$, $n = 1, \ldots, N(k)$ with DOA $\boldsymbol{\theta}_n(k) = [\phi_n(k), \varphi_n(k)]^T$, impinge on two-parallel ULAs equipped with 2 M array elements at discrete time k, $N(k)$ denotes the number of sources at time k. $\phi_n(k)$ and $\varphi_n(k)$ represent the elevation and azimuth of DOA. According to [2], the steering matrix of X-axis and Y-axis are

$$\mathbf{A}_x = \left[\mathbf{a}_y(\phi_1, \varphi_1), \ldots, \mathbf{a}_x(\phi_n, \varphi_n), \ldots, \mathbf{a}_x\left(\phi_{N(k)}, \varphi_{N(k)}\right)\right]^T \tag{9.1}$$

$$\mathbf{A}_y = \left[\mathbf{a}_y(\phi_1, \varphi_1), \ldots, \mathbf{a}_y(\phi_n, \varphi_n), \ldots, \mathbf{a}_y\left(\phi_{N(k)}, \varphi_{N(k)}\right)\right]^T \tag{9.2}$$

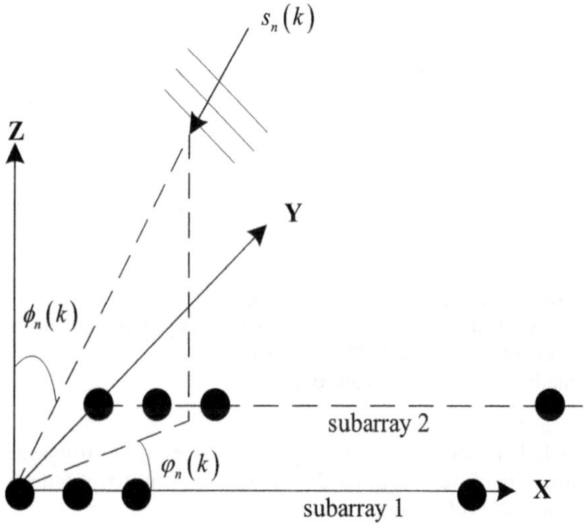

Fig. 9.1 Two-parallel uniform linear arrays model

$\boldsymbol{a}_x(\phi_n, \varphi_n) = \left[1, \ldots, e^{j2\pi d \sin\phi_n \cos\varphi_n /\lambda}, \ldots, e^{j2\pi(M-1)d \sin\phi_n \cos\varphi_n /\lambda}\right]^T$ is the steering vector of X-axis, $\mathbf{a}_y(\phi_n, \varphi_n) = \left[1, e^{j2\pi d \sin\phi_n \cos\varphi_n /\lambda}\right]^T$ is the steering vector of Y-axis, $(\cdot)^T$ is the transpose operation,d denotes the array spacing, and λ is the wavelength of receive signals. Then the outputs of subarray 1 and 2 are

$$\mathbf{z}_1(k) = \mathbf{A}_x\mathbf{s}(k) + \mathbf{n}_1(k) \tag{9.3}$$

$$\mathbf{z}_2(k) = \mathbf{A}_x\mathbf{\Phi}\mathbf{s}(k) + \mathbf{n}_2(k) \tag{9.4}$$

$\mathbf{\Phi} = \operatorname{diag}\left(e^{j2\pi d \sin\phi_1 \sin\varphi_1/\lambda}, \ldots, \right) e^{j2\pi d \sin\phi_{N(k)} \sin\varphi_{N(k)}/\lambda}$, $\mathbf{s}(k) = \left[s_1(k), \ldots, s_{N(k)}(k)\right]^T$ is the signal vector, $\operatorname{diag}(\cdot)$ denotes the diagonalization operator, $\mathbf{n}_1(k)$ and $\mathbf{n}_2(k)$ represent additive white Gaussian noise of subarray 1 and 2 with zero mean and variance σ_v^2, respectively, which are uncorrelated with signals.

By combining (9.3) and (9.4), we can get:

$$\begin{aligned} \mathbf{Z}_k &= \begin{bmatrix} \mathbf{z}_1(k) \\ \mathbf{z}_2(k) \end{bmatrix} = \begin{bmatrix} \mathbf{A}_x \\ \mathbf{A}_x\mathbf{\Phi} \end{bmatrix}\mathbf{s}(k) + \begin{bmatrix} \mathbf{n}_1(k) \\ \mathbf{n}_2(k) \end{bmatrix} \\ &= \left[\mathbf{A}_y \odot \mathbf{A}_x\right]\mathbf{s}(k) + \mathbf{n}(k) = \mathbf{A}\mathbf{s}(k) + \mathbf{n}(k) \end{aligned} \tag{9.5}$$

$\odot$ is Khatri-Rao product, the data covariance matrix [4] can be expressed as

$$\mathbf{R}_k = \mathbb{E}\left\{\mathbf{Z}_k\mathbf{Z}_k^H\right\} = \mathbf{A}\mathbb{E}\left\{\mathbf{s}_k\mathbf{s}_k^H\right\}\mathbf{A}^H + \mathbb{E}\left\{\mathbf{n}_k\mathbf{n}_k^H\right\} = \mathbf{A}\mathbf{S}_k\mathbf{A}^H + \sigma_v^2\mathbf{I}_{2M}. \tag{9.6}$$

where $(\cdot)^H$ is the conjugate transpose, $\mathbf{I}_{2M}$ is a $2M \times 2M$ identity matrix, and $\mathbb{E}$ denotes the expectation operator. By performing eigenvalue decomposition (EVD) on (9.6), we can get the noise subspace $\mathbf{U}_N$, composed of the eigenvectors corresponding to the remaining $2M - N(k)$ eigenvalues.

9.3 MeMBer DOA Tracking Algorithm

Because DOA between adjacent time steps is highly correlated, it is necessary to combine time information to estimate DOA of sources. In this section, we adopt the collaborative turning right (CTR) to simulate the motion characteristics of the source.

9.3.1 Sources Motion Model

Suppose the nth source moves with a speed of $\dot{\theta}_n(k) = \left[\dot{\phi}_n(k), \dot{\varphi}_n(k)\right]^T$ rad/s, $n = 1, \ldots, N(k)$. Combining DOA $\boldsymbol{\theta}_k$ and motion speed $\dot{\theta}_k$, we get the source state $\mathbf{x}_k = \left[\phi_n(k), \dot{\phi}_n(k), \varphi_n(k), \dot{\varphi}_n(k)\right]^T$. The CTR is as follow:

$$\mathbf{x}_k = \mathbf{F}_k \mathbf{x}_{k-1} + \mathbf{G}_k \mathbf{v}_k \tag{9.7}$$

with the coefficient matrices $\mathbf{F}_k$ and $\mathbf{G}_k$:

$$\mathbf{F}_k = \begin{bmatrix} 1 & \sin(\Delta T\omega)/\omega & 0 & (\cos(\Delta T\omega) - 1)/\omega \\ 0 & \cos(\Delta T\omega) & 0 & -\sin(\Delta T\omega) \\ 0 & -(\cos(\Delta T\omega) - 1)/\omega & 1 & \sin(\Delta T\omega)/\omega \\ 0 & \sin(\Delta T\omega) & 0 & \cos(\Delta T\omega) \end{bmatrix} \tag{9.8}$$

$$\mathbf{G}_k = \begin{bmatrix} \Delta T^2/2 \\ \Delta T \end{bmatrix} \otimes \mathbf{I}_2 \tag{9.9}$$

where $\otimes$ is Kronecker product, ΔT represents the time step, ω is turning rate, and $\mathbf{v}_k$ denotes the white Gaussian process with $\mathbf{v}_k \sim \mathcal{N}(\mathbf{0}, \boldsymbol{\Sigma}_k)$ and $\boldsymbol{\Sigma}_k = \text{diag}\left\{\delta_\phi^2, \delta_\varphi^2\right\}$.

9.3.2 Improve Likelihood Function Model

According to [4], the pseudo-spectrum of MUSIC can be written as

$$P_{\text{MUSIC}}(\boldsymbol{\theta}_k) = \frac{1}{\left[\mathbf{a}_y(\boldsymbol{\theta}_k) \otimes \mathbf{a}_x(\boldsymbol{\theta}_k)\right]^H \mathbf{U}_N \mathbf{U}_N^H \left[\mathbf{a}_y(\boldsymbol{\theta}_k) \otimes \mathbf{a}_x(\boldsymbol{\theta}_k)\right]} \tag{9.10}$$

$\mathbf{U}_N$ is the noise subspace. Similar to (9.10), for array measurement $\mathbf{Z}_k$ and state $\mathbf{x}_k$, we define the likelihood function as follows:

$$g(\mathbf{Z}_k|\mathbf{x}_k) = |P_{\text{MUSIC}}(\cdot)|^\xi = \left|\frac{1}{\left[\mathbf{a}_y(\mathbf{cx}_k) \otimes \mathbf{a}_x(\mathbf{cx}_k)\right]^H \mathbf{U}_N \mathbf{U}_N^H \left[\mathbf{a}_y(\mathbf{cx}_k) \otimes \mathbf{a}_x(\mathbf{cx}_k)\right]}\right|^\xi \tag{9.11}$$

where $\mathbf{cx}_k = [1, 0, 1, 0]$ denotes the elevation and azimuth, $|\cdot|$ denotes the module of a complex value. ξ is the exponential weighting factor, which can solve the problem of particle effectiveness reduction caused by obesity of likelihood function in low signal-to-noise ratio (SNR) scene.

9.3.3 MeMBer DOA Tracking Algorithm

In this part, a 2D-DOA tracking algorithm called 2D-MeMBer algorithm is proposed. We provide a Sequential Monte Carlo (SMC) implement of 2D-MeMBer method.

Prediction: Assuming that the update MeMBer parameter sets at time $k-1$ is $\left\{r_{k-1}^{(i)}, p_{k-1}^{(i)}\right\}_{i=1}^{T_{k-1}}$, $i = 1, \ldots, T_{k-1}$, where $p_{k-1}^{(i)} = \left\{\left(\mathbf{x}_{k-1}^{(i,j)}, \omega_{k-1}^{(i,j)}\right)\right\}_{j=1}^{N_{k-1}^{(i)}}$, $r_{k-1}^{(i)}$ is existence probability of the ith multi-target parameter sets, $N_{k-1}^{(i)}$ is the total particles. According to [14], the predicted multi-target parameter sets $\left\{\left(r_{P,k|k-1}^{(i)}, p_{P,k|k-1}^{(i)}\right)\right\}_{i=1}^{T_{k-1}} \cup \left\{\left(r_{\Gamma,k}^{(i)}, p_{\Gamma,k}^{(i)}\right)\right\}_{i=1}^{T_{\Gamma,k}}$ is defined as:

$$r_{P,k|k-1}^{(i)} = r_{k-1}^{(i)} \cdot \sum_{j=1}^{N_{k-1}^{(i)}} \omega_{k-1}^{(i,j)} \cdot P_{sur,k}\left(\mathbf{x}_{k-1}^{(i,j)}\right) \tag{9.12}$$

$$p_{P,k|k-1}^{(i)}(\mathbf{x}) = \sum_{j=1}^{N_{k-1}^{(i)}} \omega_{P,k|k-1}^{(i,j)} \delta_{\mathbf{x}_{P,k|k-1}^{(i,j)}}(\mathbf{x}) \tag{9.13}$$

$$r_{\Gamma,k}^{(i)} \text{ denotes the parameter given by newborn model} \tag{9.14}$$

$$p_{\Gamma,k}^{(i)}(\mathbf{x}) = \sum_{j=1}^{N_{B,k}^{(i)}} \hat{\omega}_{\Gamma,k}^{(i,j)} \delta_{\mathbf{x}_{\Gamma,k}^{(i,j)}}(\mathbf{x}) \tag{9.15}$$

where $P_{sur,k}(\cdot)$ is the survival probability and $N_{B,k}^{(i)}$ is the newborn particles. $\mathbf{x}_{P,k|k-1}^{(i,j)}$ is calculated by (9.7), $j = 1, \ldots, N_{k-1}^{(i)}$, $\omega_{P,k|k-1}^{(i,j)} = \omega_{k-1}^{(i,j)} / \left[\sum_{j=1}^{N_{k-1}} \omega_{k-1}^{(i,j)} \cdot P_{\mathrm{sur},k}\left(\mathbf{x}_{k-1}^{(i,j)}\right)\right]$. $\mathbf{x}_{\Gamma,k}^{(i,j)} \sim \mathcal{N}(\cdot,\cdot)$, $\omega_{\Gamma,k}^{(i,j)} = 1/N_{B,k}^{(i)}$, $j = 1, \ldots, N_{B,k}^{(i)}$.

Update:: Assuming that the predicted MeMBer parameter sets at time k is $\left\{\left(r_{k|k-1}^{(i)}, p_{k|k-1}^{(i)}\right)\right\}_{i=1}^{T_{k|k-1}}$, $i = 1, \ldots, T_{k|k-1}$, and $p_{k|k-1}^{(i)} = \left\{\left(\mathbf{x}_{k|k-1}^{(i,j)}, \omega_{k|k-1}^{(i,j)}\right)\right\}_{j=1}^{N_{k|k-1}^{(i)}}$, $T_{k|k-1} = T_{k-1} + T_{\Gamma,k}$ is the total predicted hypotheses, $N_{k|k-1}^{(i)} = N_{k-1}^{(i)} + N_{B,k}^{(i)}$ is the total prediction particles.

Then the updated MeMBer parameter sets $\left\{\left(r_k^{(i)}, p_k^{(i)}\right)\right\}_{i=1}^{T_{k|k-1}}$ can be represented by:

$$r_k^{(i)} = \frac{r_{k|k-1}^{(i)} \cdot \left\langle g\left(\mathbf{Z}_k|\mathbf{x}_{k|k-1}^{(i,j)}\right), p_{k|k-1}^{(i)}\left(\mathbf{x}_{k|k-1}^{(i,j)}\right)\right\rangle}{1 - r_{k|k-1}^{(i)} + r_{k|k-1}^{(i)} \cdot \left\langle g\left(\mathbf{Z}_k|\mathbf{x}_{k|k-1}^{(i,j)}\right), p_{k|k-1}^{(i)}\left(\mathbf{x}_{k|k-1}^{(i,j)}\right)\right\rangle} \tag{9.16}$$

$$p_k^{(i)}(\mathbf{x}_k) == \sum_{j=1}^{N_k^{(i)}} \omega_k^{(i,j)} \delta_{\mathbf{x}_k^{(i,j)}}(\mathbf{x}_k) \tag{9.17}$$

$g\left(\mathbf{Z}_k|\mathbf{x}_{k|k-1}^{(i,j)}\right)$ is calculated from (9.11), $\mathbf{x}_k^{(i,j)} = \left\{\mathbf{x}_{P,k|k-1}^{(i,j)}\right\}_{j=1}^{N_{k-1}^{(i)}} \cup \left\{\mathbf{x}_{\Gamma,k}^{(i,j)}\right\}_{j=1}^{N_{B,k}^{(i)}}$, $j = 1, \cdots, N_{k|k-1}^{(i)}$.

$$\omega_k^{(i,j)} = \omega_{k|k-1}^{(i,j)} \cdot g\left(\mathbf{Z}_k|\mathbf{x}_{k|k-1}^{(i,j)}\right) / \left(\sum_{i=1}^{T_{k|k-1}} \sum_{j=1}^{N_{k|k-1}^{(i)}} g\left(\mathbf{Z}_k|\mathbf{x}_{k|k-1}^{(i,j)}\right) \cdot \omega_{k|k-1}^{(i,j)}\right).$$

After applying resampling algorithm and pruning operations [14], we get the update Multi-Bernoulli parameter sets

$$\left\{r_k^{(i)}, \left(\mathbf{x}_k^{(i,j)}, \omega_k^{(i,j)}\right)_{j=1}^{N_k^{(i)}}\right\}_{i=1}^{T_k} \tag{9.18}$$

where T_k represents the number of update Multi-Bernoulli parameter sets and $N_k^{(i)}$ denotes the total particles by resampling method.

9.4 Simulations

In this simulations, we use the joint root mean square error (RMSE) as the performance evaluation standard, as follows:

$$\mathrm{RMSE} = \frac{1}{\mathrm{MC}} \sum_{j=1}^{\mathrm{MC}} \left(\sqrt{\frac{1}{N(k)} \sum_{n=1}^{N(k)} \frac{1}{K} \sum_{k=1}^{K} \left(\hat{\phi}_{kn,j} - \phi_{kn}\right)^2 + \left(\hat{\varphi}_{kn,j} - \varphi_{kn}\right)^2} \right) \tag{9.19}$$

where $\hat{\phi}_{kn,j}$ and $\hat{\varphi}_{kn,j}$ are the estimated values of the elevation and azimuth of nth source at jth MC simulations for time k. MC denotes Monte Carlo running number. In order to more effectively analyze the performance of the proposed algorithm, we consider two experimental scenarios, which are as follows:

Experiment 1. Consider a CTR motion scenario with 3 sources, whose number is known and fixed. The array elements $M = 10$, and SNR $= 10\,\mathrm{dB}$, $L = 200$, $\mathrm{MC} = 100$, $\xi = 4$, $\mathbf{\Sigma}_k = \mathrm{diag}(4, 4) * 1e - 4$, the survival probability is assumed to be constant $p_{\mathrm{sur},k}(\mathbf{x}) = 0.99$, $\left\{r_{\Gamma,k}^i, \left(\mathbf{x}_{\Gamma,k}^{(i,j)}, \omega_{\Gamma,k}^{(i,j)}\right)_{j=1}^{N_{B,k}}\right\}_{i=1}^{3}$ is the newborn Multi-Bernoulli parameter sets, $r_{\Gamma,k}^{(i)} = 0.02, 0.02, 0.03$, $\mathbf{x}_{\Gamma,k}^{(i,j)} \sim \mathcal{N}(\mathbf{m}_i, \mathbf{P}_i)$, where $\mathbf{m}_1 =$

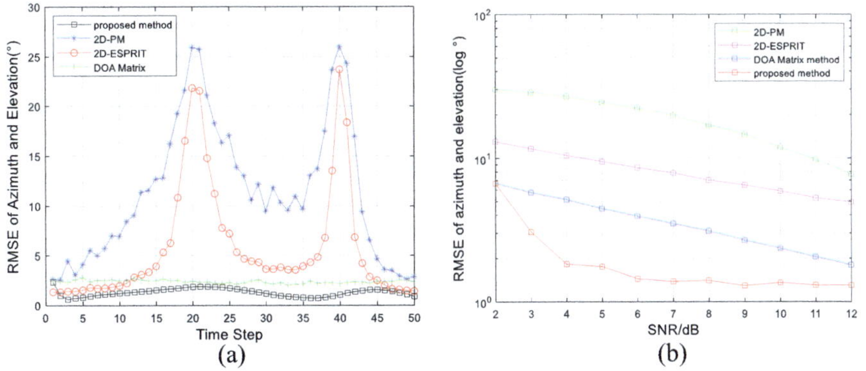

Fig. 9.2 **a** RMSE of real-time tracking. **b** RMSE comparison at different SNR

[20; 0; 30; 0], $\mathbf{m}_2 = [30; 0; -40; 0]$, $\mathbf{m}_3 = [40; 0; -70; 0]$, $\mathbf{P}_i = [4; 0.04; 4; 0.04]$, $i = 1, 2, 3$ and $N_k^{(i)} = N_{B,k}^{(i)} = 600$.

Figure 9.2a depicts the RMSE of moving sources, and it shows that the tracking error of the proposed method is the smallest. Figure 9.2b exhibits the estimation performance comparison between the traditional DOA tracking methods and proposed method, such as 2D-PM algorithm, 2D-ESPRIT algorithm, and DOA-matrix method versus different SNRs. Compared with traditional DOA tracking methods, the proposed method owns better estimation performance.

Experiment 2. The number of sources caused by birth and death varies with time, and the sources' duration is 1–50, 7–30, 10–50, 15–40 s, $\left\{r^{i}_{\Gamma,k}, \left(\mathbf{x}^{(i,j)}_{\Gamma,k}, \omega^{(i,j)}_{\Gamma,k}\right)_{j=1}^{N_{B,k}}\right\}_{i=1}^{4}$, $r^{(i)}_{\Gamma,k} = 0.2, 0.2, 0.3, 0.3$, $\mathbf{x}^{(i,j)}_{\Gamma,k} \sim \mathcal{N}(\mathbf{m}_i, \mathbf{P}_i)$, and $\mathbf{m}_1 = [20; 0; 30; 0]$, $\mathbf{m}_2 = [-30; 0; -40; 0]$, $\mathbf{m}_3 = [40; 0; -35; 0]$, $\mathbf{m}_3 = [60; 0; 80; 0]$, $\mathbf{P}_{1,4} = [4; 0.04; 4; 0.04]$, $\mathbf{P}_{2,3} = [4; 0.25; 4; 0.25]$. The exponential weighting factor $\xi = 4$ in the proposed method and $\xi = 3$ in MeMBer-Capon algorithm, which use Capon pseudo-spectrum to calculate the likelihood function of MeMBer algorithm. Other experimental parameters are similar to experiment 1.

Cardinality estimation performance is shown in Fig. 9.3. As can be seen from Fig. 9.3 that the proposed method can capture the emergence of newborn sources and the disappearance of existing source and can accurately estimate the source number. However, the MeMBer-Capon method underestimates the number of sources, resulting in poor performance in Fig. 9.3b. Figure 9.3b also shows that the estimation performance of proposed approach is better than the MeMBer-Capon method.

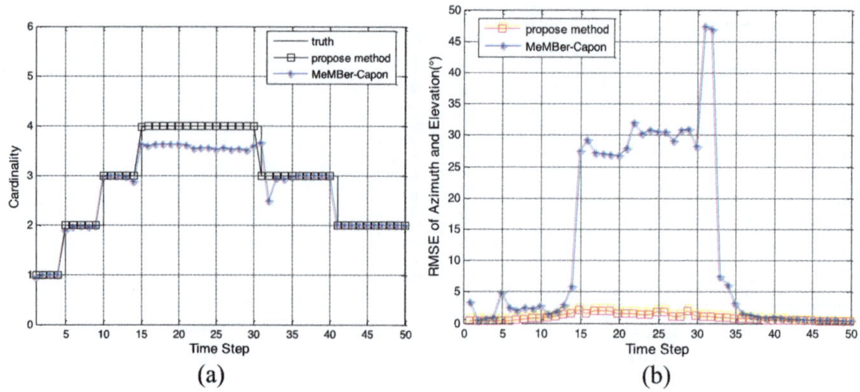

Fig. 9.3 **a** Cardinality estimation. **b** RMSE of time step

9.5 Conclusion

A SMC 2D-MeMBer method for DOA tracking is provided in this article. The proposed method modifies the MUSIC pseudo-spectrum as its likelihood function. By exponential weighting the likelihood function, the sampled particles are improved. Simulations result indicates that the proposed method provides better tracking performance compared with the traditional DOA tracking methods and can accurately estimate the quantity of sources and their states.

References

1. Kang, H.Y., Kim, Y.S., Chang, J.K.: Spatially close signals separation via array aperture expansions and spatial spectrum averaging. ETRI J. **26**(1), 45–47 (2004)
2. Kedia, V.S., Chandna, B.: A new algorithm for 2-D DOA estimation. Signal Process. **60**(3), 325–332 (1997)
3. Rajagopal, R., Rao, P.R.: Generalised algorithm for DOA estimation in a passive sonar. Radar Signal Process. IEE Proc. **140**(1), 12–20 (1993)
4. Schmidt, R.O.: Multiple emitter location and signal parameter estimation. IEEE Trans. Antennas Propag. **34**(3), 276–280 (1986)
5. Kunda, D.: Modified MUSIC algorithm for estimating DOA of signal. Signal Process. **48**(1), 85–90 (1996)
6. Roy, R., Kailath, T.: ESPRIT—estimation of signal parameters via rotational invariance techniques. IEEE Trans. ASSP **37**(7), 984–995 (1986)
7. Tayem, N., Kwon, H.M.: L-shape 2-dimensional arrival angle estimation with propagator method. IEEE Trans. Antennas Propag. **53**(1), 1622–1630 (2005)
8. Yin, Q.Y., Newcomb, R.W., Zou, L.H.: Estimating 2-D angles of arrival via two parallel linear arrays. In: International Conference on Acoustics, Speech, and Signal Processing, Glasgow, UK, pp. 2803–2806 (1989)
9. Dai, X.R., Zhang, X.F., Wang, Y.F.: Extended DOA-matrix method for DOA estimation via two parallel linear arrays. IEEE Commun. Lett. **23**(11), 1981–1984 (2019)

10. Li, J.F., Zhang, X., Chen, H.: Improved two-dimensional DOA estimation algorithm for two-parallel uniform linear arrays using propagator method. Signal Process. **92**(12), 3032–3038 (2012)
11. Mahler, R.P.S.: Statistical Multisource-Multitarget Information Fusion. Artech House, Inc. (2007)
12. Vo, B.T., Clark, D., Vo, B.N., Ristic, B.: Bernoulli forward-backward smoothing for joint target detection and tracking. IEEE Trans. Signal Process. **59**(9), 4473–4477 (2011)
13. Ristic, B., Vo, B.T., Vo, B.N., Farina, A.: A tutorial on Bernoulli filters: theory, implementation and applications. IEEE Trans. Signal Process. **61**(13), 3406–3430 (2013)
14. Vo, B.T., Vo, B.N., Cantoni, A.: The cardinality balanced multi-target multi-bernoulli filter and its implementations. IEEE Trans. Signal Process. **57**(2), 409–423 (2009)
15. Zhong, X.H., Premkumar, A.B.: Particle filtering approaches for multiple acoustic source detection and 2-D direction of arrival estimation using a single acoustic vector sensor. IEEE Trans. Signal Process. **60**(9), 4719–4733 (2012)
16. Zhong, X.H., Premkumar, A.B., Madhukumar, A.S.: Particle filtering and posterior Cramér-Rao bound for 2-D direction of arrival tracking using an acoustic vector sensor. IEEE Sensors J. **12**(2), 366–377 (2012)

Chapter 10
Fuzzy-ADRC Strategy to DFIG Low-Voltage Ride Through

Shengqing Li, Xin Yao, and Jingyu Liu

Abstract To prevent the disconnection of ZigBee wireless communication wind turbines caused by voltage drop and achieve the implementation of low-voltage ride through (LVRT) in doubly fed induction generator (DFIG), this paper proposes fuzzy active disturbance rejection control (Fuzzy-ADRC) strategy under unbalance power grids, which uses generalized differential with fuzzy operations to realize the fast no-overshooting tracing with real-time parameter modifications. The proposed scheme does not depend on system's precise parameters, and it can avoid the overflow of rotor side when power grid malfunction, curb the negative sequence current and transient current oscillation. Finally, the feasibility and effectiveness of the proposed method are verified by computer simulation.

10.1 Introduction

The stator side of DFIG unit becomes more sensitive to voltage fault because it connects grid directly. The drop-off voltage could cause severe electromagnetic transients and lead to the overflow of stator and rotor and the fluctuation of DC bus voltage [1–5], so study on low-voltage ride through (LVRT) of doubly fed induction generator (DFIG) is of great significance.

In order to solve the problem that DFIG cannot operate normally under unbalanced grid voltage, the main research is to adopt the control strategy of double-synchronous rotating coordinate system [6–9]. It is a typical closed-loop control strategy of proportion integration differentiation (PID) that transforms the voltage and current of wind turbine. Although they can effectively control positive-and-negative current under unbalanced voltage, its structure is too complex to regulate with over-abundant parameters, which results in weak robustness and delayed dynamic response. In [10], the state error in active disturbance rejection control is written in the form of proportion + fractional integral, which can better describe the

S. Li (✉) · X. Yao · J. Liu
Hunan University of Technology, Zhuzhou 412007, Hunan, China

L. C. Jain et al. (eds.), *Wireless Technology, Intelligent Network Technologies, Smart Services and Applications*, Smart Innovation, Systems and Technologies 258,
https://doi.org/10.1007/978-981-16-5168-7_10

physical model, and the control effect is considerable, but the fractional integral algorithm is too complex and the calculation is cumbersome. In [11], active crowbar and DC side unloading circuits are used to consume excessive energy caused by stator voltage sag to realize LVRT of wind turbine, but the control of DFIG under unbalanced voltage is not considered. In [12], established the mathematical model of DFIG in the positive and negative sequence synchronous rotating coordinate system under the condition of unbalanced grid voltages. However, the traditional fuzzy control will reduce the control accuracy of systems, and the dynamic quality is also poor, which is not conducive to the long-term operation of wind turbines.

Based on the traditional fuzzy control, a new Fuzzy-ADRC is proposed in vastly improve its efficiency by modifying nonlinear error feedback parameters, so it can not only maintain the advantages of traditional approach, but also overcome the dependency on precise mathematical model, to optimize the control performance. The proposed scheme not only reduces the rotor side transient current, while unbalanced of grid voltages, but also suppresses the fluctuation of the current negative sequence component, torque and power. This will avoid the dependence of the traditional control scheme on the precise parameters of the motor, and effectively improves the system's dynamic response and robustness.

10.2 DFIG Under Unbalanced Power Conditions

Transformer can eliminate zero-sequence voltage from stellar connect to triangle connect, so it only takes positive and negative sequence component into consideration when we analyze the unbalanced grid. The construction of DFIG is shown in Fig. 10.1.

It adopts park transformation to get the positive-and-negative mathematic model by changing ABC coordination into dq axis. The voltage equation in $dq+$ axis and $dq-$ axis is:

$$\begin{cases} U_{sdq+} = U_{sdq+} + \frac{\mathrm{d}\psi_{sdq+}}{\mathrm{d}t} + j\omega_1\psi_{sq+} \\ U_{rdq+} = R_r i_{rdq+} + \frac{\mathrm{d}\psi_{rdq+}}{\mathrm{d}t} + j\omega_{\text{slip}} + \psi_{rdq+} \\ U_{sdq-} = R_s i_{rdq-} + \frac{\mathrm{d}\psi_{rdq-}}{\mathrm{d}t} + j\omega_1\psi_{sdq-} \\ U_{rdq-} = R_r i_{rdq-} + \frac{\mathrm{d}\psi_{rdq-}}{\mathrm{d}t} + j\omega_{\text{slip}} - \psi_{rdq-} \end{cases} \tag{10.1}$$

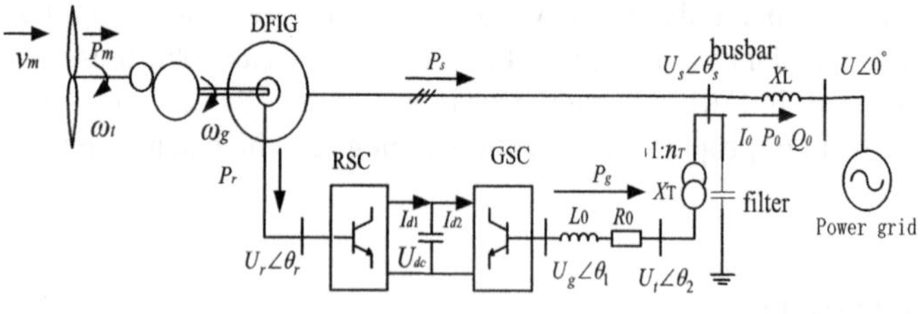

Fig. 10.1 System structure diagram of DFIG

In this equation, U_s and U_r stand for stator voltage and rotor voltage, ψ for flux, ω_1 for angular velocity, ω_{slip} for rotating angular velocity, R_s and R_r for stator and rotor resistance.

Flux linkage equation of DFIG in $dq+$ axis and $dq-$ axis is:

$$\begin{cases} \psi_{sdq+} = L_{s}i_{sdq+} + L_{m}i_{rdq+} \\ \psi_{rdq+} = L_{r}i_{rdq+} + L_{m}i_{sdq+} \\ \psi_{sdq-} = L_{s}i_{sdq-} + L_{m}i_{rdq-} \\ \psi_{rdq-} = L_{r}i_{rdq-} + L_{m}i_{sdq-} \end{cases} \tag{10.2}$$

According to (10.1) and (10.2):

$$\begin{cases} \frac{di_{rd+}}{dt} = \frac{U_{rd+} - R_r i_{rd+}}{\sigma L_r} + w_1 \\ \frac{di_{rq+}}{dt} = \frac{U_{rq+} - R_r i_{rq+}}{\sigma L_r} + w_2 \\ \frac{di_{rd-}}{dt} = \frac{U_{rd-} - R_r i_{rd-}}{\sigma L_r} + w_3 \\ \frac{di_{rq-}}{dt} = \frac{U_{rq-} - R_r i_{rq-}}{\sigma L_r} + w_4 \end{cases} \tag{10.3}$$

In this equation, i_{sd+}, i_{sq+}, i_{sd-}, i_{sq-}, i_{rd+}, i_{rq+}, i_{rd-}, i_{rq-} are $d+$, $q+$, d_-, q_- axial components of stator and rotor currents, U_{sd+}, U_{sq+}, U_{sd-}, U_{sq-}, U_{rd+}, U_{rq+}, U_{rd-}, U_{rq-} are $d+$, $q+$, d_-, q_- axial components of stator and rotor voltages and W_1, W_2, W_3, W_4 are total disturbing momentum, among which $\sigma = 1 - L_m^2/(L_s - L_s)$.

$$\begin{cases} w_1 = Ki_{q+} + \Delta w_{d+} \\ w_2 = Ki_{d+} + \Delta w_{q+} \\ w_3 = Ki_{q-} + \Delta w_{d-} \\ w_4 = Ki_{d-} + \Delta w_{q\_} \end{cases} \tag{10.4}$$

In this equation, ki_{d+}, ki_{d-}, ki_{q+}, ki_{q-} are coupling terms of positive-and-negative sequence current at d with q axial component, and Δw_{d+}, Δw_{d+}, Δw_{d-}, Δw_{q-} are external disturbance value of positive-and-negative current.

10.3 Design of ADRC

10.3.1 ADRC Principle

ADRC improves classical PID control and do not need to measure external interference directly and predict its frequency. Due to the characteristic of ADRC, the coupling effect between the multivariable subsystems can be regarded as an uncertain quantity and can be classified as an "unknown disturbance." In this paper, ADRC is used for decoupling control, and its system structure is shown in Fig. 10.2.

Fig 10.2 Block diagram of ADRC system

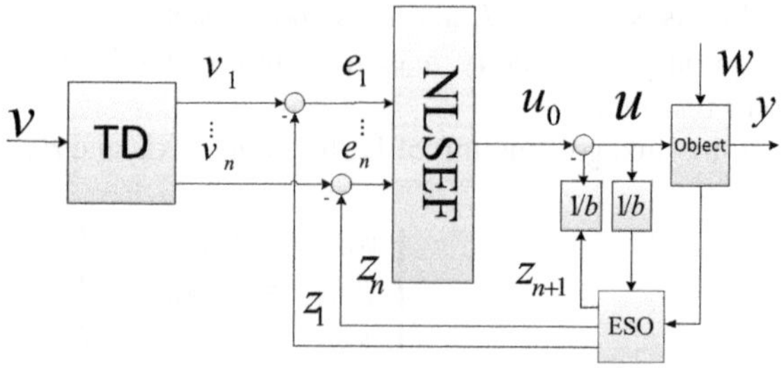

The tracking differentiator in ADRC introduces the generalized differential to realize the fast non overshoot tracking of input signal of the system, and the ESO will estimate by recognizing the disturbance and the unmodeled dynamics as the total disturbance of the system.

10.3.2 Tracking Differentiator Design

The nonlinear uncertain object *x*(*n*) with uncertain interference is indicated as follows:

$$x(n) = f\left(x, x(t), \ldots x^{(n-1)}, t\right) + w(t) + bu(t) \tag{10.5}$$

In the formula, $f(x, X(T), X(n - 1), t)$ is the unknown function; $w(T)$ is the unknown external disturbance; $X(T)$ is the measurable quantity; B is the control input coefficient; $U(T)$ is the control input of the system.

The input of setting and arranging the transition process is reference value i_{*rxy} of the rotor current, and the output is i_{1xy}, i_{2xy}, in which i_{1xy} is current value of tracking rotor, and the i_{2xy} is the "approximate differential" of the rotor current. The TD parameter expression is:

$$\begin{cases} i_{1xy}(t) = i_{1xy}(t-1) + h \cdot i_{2xy}(t-1) \\ i_{2xy}(t) = i_{1xy}(t) \end{cases} \tag{10.6}$$

Among them, x is d axis or q axis, y is positive sequence or negative sequence, i_{1xy} is tracking value of rotor current, z_{1xy} observation value of rotor current, h is step length, and e is error.

In order to debase the overshoot of output and avoid tracking-after discretizing the differentiator, the stable system will arise the phenomenon of high-frequency tremor, and the nonlinear function g in TD is based on:

① Nonlinear function $g(z)$ is continuous and differentiable;

② $g(0) = 0$;

③ Its derivative $g' = \frac{\mathrm{d}g(z)}{\mathrm{d}z} \neq 0$ can be regarded as a function expression fal which is presented as follows

$$\mathrm{fal}(e, a, \delta) = \begin{cases} \frac{e}{\delta^{1-a}}, |e| \leq \delta \\ |e|^a \mathrm{sgn}(e), |e| > \delta \end{cases} \tag{10.7}$$

In formula, a is 0–1, δ is the filtering influence constant (Generally$5T \leq \delta \leq 10T$), and e is the state estimation error.

The parameters of the tracking differentiator are as follow:

$$\begin{cases} e_{1xy} = i_{1xy}(t-1) - z_{1xy} \\ i_{1xy}(t) = i_{1xy}(t-1) + h \cdot i_{2xy}(t-1) \\ i_{2xy}(t) = -h \cdot \mathrm{fal}(e_{1xy}, a_1, \delta_1) \end{cases} \tag{10.8}$$

10.3.3 The ESO Design

The expression of the observation equations that can construct and expand the system are showed as follow:

$$\begin{cases} z_1 = z_2 - g_1(z_1 - x_1(t)) \\ z_2 = z_3 - g_2(z_1 - x_1(t)) + b_0 u(t) \\ z_3 = -g_3(z_1 - x_1(t)) \end{cases} \tag{10.9}$$

Since Z_i separately tracks the expanded state $x_{(i-1)}(t)$, and the parameter b_0 is known, the control quantity can be selected as follows:

Since the nonlinear function is a fal function, the form of ESO is:

$$\begin{cases} e = z_1 - y \\ z_1 = z_2 - \beta_{01}\mathrm{fal}(e, a, \delta) \\ z_2 = z_3 - \beta_{02}\mathrm{fal}(e, a, \delta) + bu(t) \\ z_3 = -\beta_{03}\mathrm{fal}(e, a, \delta) \end{cases} \tag{10.10}$$

Among them β_{0i} is a selected parameter.

In order to achieve ESO observation of the state, the compensation matrix of formula (10.9) $A = \begin{bmatrix} -k_1 & 1 & 0 \\ -k_2 & 0 & 1 \\ -k_3 & 0 & 0 \end{bmatrix}$ is written out. According to the sufficient and necessary condition for the stability of the system, the characteristic roots of a are on the left half of the complex plane and is fully negative, that is, $\lambda^3 + \lambda^2 k_1 + \lambda k_2 + k_3 = 0$. The eigenvalues $\lambda_1, \lambda_2, \lambda_3$ and parameters k_1, k_2, k_3 of A are satisfied by:

$$s^3 + k_1 s^2 + k_2 s + k_3 = (s - \lambda_1)(s - \lambda_2)(s - \lambda_3) \tag{10.11}$$

According to the value k_1, k_2, k_3 obtained by the undetermined coefficient method, an initial value of β_{0i} is obtained by the formula$\beta_{0i} = k_i/\text{fal}(e, a, \delta)$.

If we take i_{rd+}, i_{eq+}, i_{rd-}, i_{rq-} as measuring input; W_1, and W_2, W_3 and W_4 as estimator of disturbance, the extended state observer can be constructed as follows:

$$\begin{cases} e_{2xy} = z_{1xy}(t-1) - i_{xy} \\ z_{2xy}(t) = z_{1xy}(t-1) - \beta_{01}(e_{2xy}, a_2, \delta_2) + bu_{rxy}(t-1) \\ z_{3xy}(t) = -\beta_{02}\text{fal}(e_{2xy}, a_2, \delta_2) \end{cases} \tag{10.12}$$

Among them, x represents d axis or q axis, y indicates positive sequence or negative sequence, and z_{1xy} is the observed value of rotor current disturbance.

10.3.4 Design of Nonlinear State Error of Feedback Control Law

Unlike the traditional linear PID, the improved ADRC proposed in this paper applies fuzzy control to the self-disturbance rejection design. It is input with e_1, e_2 and the output is $\triangle\beta$. Respectively 7 language subsets are defined: {"positive big (PB)", "middle (PM)," "positive small (PS)," "zero (ZO)," "negative big (NB)," "negative middle (NM)," "negative small (NS)".}. It is input with e_1, e_2, and its membership function is gaussmf. It is output with $\triangle\beta_{03}$ and $\triangle\beta_{04}$, and its membership function is trimf. The domain of e_1, e_2 is {− 3, − 2, − 1, 0, 1, 2, 3}. The domain of $\triangle\beta_{03}$ is {− 0.3, − 0.2 and − 0.1, 0, 0.1, 0.2, 0.3}. The domain of $\triangle\beta_{04}$ is {− 0.06, − 0.04, − 0.02, 0.02, 0.04, 0.06}. We take the Mamdani type as fuzzy reasoning theories, and use average weighted methods to defuzzification algorithm [13]. The control table is shown in Table 10.1.

Table 10.1 Delta beta 1, delta beta 2 fuzzy control table

e_1	e_2						
	NB	NM	NS	ZO	PS	PM	PB
NB	PB/NB	PB/NB	PM/NM	PS/NM	PS/NS	ZO/ZO	NS/ZO
NM	PB/NB	PB/NB	PM/NM	PS/NS	PS/NS	ZO/ZO	NS/ZO
NS	PM/NB	PM/NM	PM/NS	PS/NS	ZO/ZO	NS/PS	NS/PS
ZO	PM/NM	PM/NM	PS/NS	ZO/ZO	NS/PS	NM/PM	NM/PM
PS	PS/NM	PS/NS	ZO/ZO	NS/PS	NS/PS	NM/PM	NM/PB
PM	PS/ZO	ZO/ZO	NS/PS	NM/PS	NM/PM	NM/PB	NB/PB
PB	ZO/ZO	ZO/ZO	NM/PS	MN/PM	NM/PM	NB/PB	NB/PB

$$\beta' = \beta + \Delta\beta \tag{10.13}$$

In the form, β is the start value.

$$\begin{cases} u_{0xy}(t) = \beta'_{03} \cdot \mathrm{fal}(e_{1xy}, a_3, \delta_3) + \beta'_{04} \cdot \mathrm{fal}(e_{1xy}, a_3, \delta_3) \\ u(t) = u_{0xy} - z_{3xy}(t)/b \end{cases} \tag{10.14}$$

β_i, a_i, δ are all adjustable.

10.4 Fuzzy-ADRC Design of Rotor Current

From the state equation of the rotor side converter, we can find some coupling terms in the system, and that the components on d axis and the components on q axis are interacted with each other, which is not conducive to the transient performance of the control system. In this paper, the coupling quantity is estimated by Fuzzy-ADRC, which makes it difficult to deviate from the target when the system is disturbed. The DFIG has active P_0 and reactive power Q_0 when the power output characteristics of the grid are unbalanced. They are expressed as follows:

$$P_0 = \frac{3L_m\omega_{\text{slip}}}{2L_s}\left[-\psi_{sq+}\ \psi_{sd+}\ \psi_{sq-} - \psi_{sd-}\right]\left[i_{rd+}\ i_{rq+}\ i_{rd-}\ i_{rq-}\right]^T \tag{10.15}$$

$$\begin{aligned} Q_0 = &\frac{3\omega_{\text{slip}}}{2L_s}\left[-\psi_{sd+} - \psi_{sq+}\psi_{sd-} - \omega_{sq-}\right]\left[\psi_{sd+}\ \psi_{sq+}\psi_{sd-}\psi_{sq-}\right]^T \\ &+ \frac{3L_m\omega_{\text{slip}}}{2L_s}\left[-\psi_{sq+}\psi_{sd+}\psi_{sq-} - \psi_{sd-}\right]\left[i_{rd+}i_{rq+}i_{rd-}\ i_{rq+}\right]^T \end{aligned} \tag{10.16}$$

We take the form $i^*_{rd-} = 0$, $i^*_{rq-} = 0$. From the formula (10.1), (10.2), (10.15) and (10.16), we can know that:

$$\begin{bmatrix} i^*_{rd+} \\ i^*_{rq+} \end{bmatrix} = \frac{2l_s}{3l_m}M_2^{-1}\left\{\begin{bmatrix} P_0 \\ Q_0 \end{bmatrix}\right\} + \frac{3}{2l_s}\begin{bmatrix} U_{sd+}U_{sq+}U_{sd-}U_{sq-} \\ U_{sq+} - U_{sd+}U_{sq-} - U_{sd-} \end{bmatrix}\begin{bmatrix} \psi_{sd+} \\ \psi_{sq+} \\ \psi_{sd-} \\ \psi_{sq-} \end{bmatrix} \tag{10.17}$$

$$M_2 = \begin{bmatrix} U_{sd+}\ U_{sq+} \\ U_{sq+} - U_{sd+} \end{bmatrix}$$

To suppress current negative sequence components and prevent rotor side over-current, we adopt a dual-internal loop Fuzzy-ADRC control block diagram that is shown in Fig. 10.3.

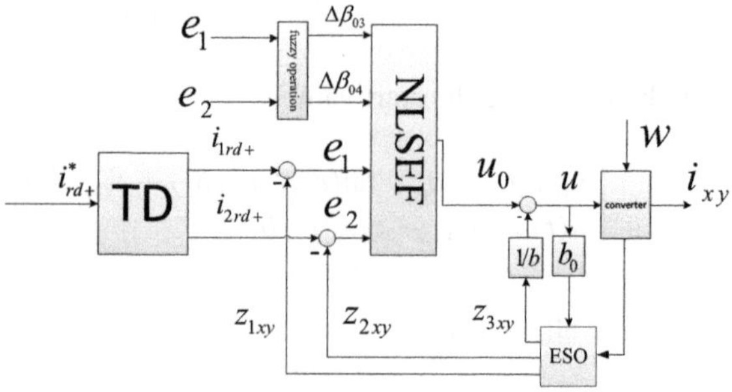

Fig. 10.3 Rotor side converter current fuzzy self-rejection control structure

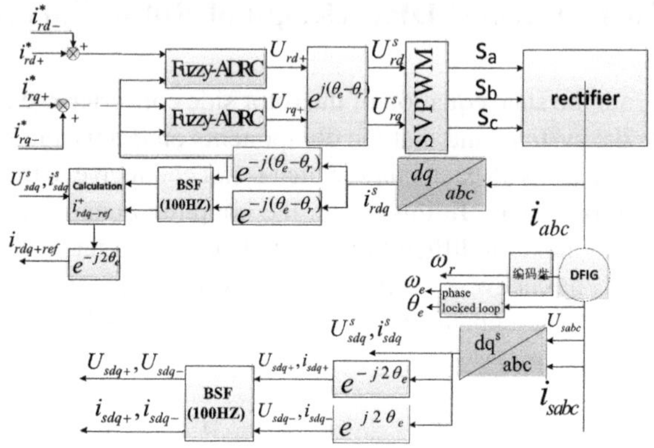

Fig. 10.4 Unbalanced power grid DFIG control block diagram

The block diagram of DFIG system under unbalanced grid is shown in Fig. 10.4, where I_{ird+} is taken as an example, and other current principles such as i_{rd+}.

The control system gives the corresponding differential signal by tracking the transition process of the differentiator, which make the system have no overshoot. The extended observer can receive the observed values of each state variable and the amplitude of system disturbance. By taking the appropriate value of a, the effect of small error and large gain is obtained. The above methods can be merged into a comprehensive method to achieve optimal control within a certain range.

10.5 Simulation Experiment

The results are shown in Fig. 10.5. As illustrated in Fig. (a), the voltage of the grid is unbalanced at 0.24 s about 5%, and it will be restored balance state at 0.80 s. The traditional vector control strategy ignores the negative sequence components of electromagnetic quantities that are caused by the voltage drop in the transient state process, so that less unbalanced voltage will produce a large impact current. This impact current will cause unbalanced heating of wind turbines, and it will seriously damage the normal operation of wind turbines. In addition, when grid

Fig. 10.5 The simulation model of 6 MW doubly fed wind turbine is established in Matlab/Simulink environment, and four 1.5 MW wind farms are simulated. The simulated motor is: the rated stator voltage is 690 V, the stator resistance is 0.0071 (PU), and the stator leakage inductance is 0.175 (PU). The rotor resistance is 0.004 (PU), the rotor leakage inductance is 0.154 (PU), the excitation inductance is 2.7 (PU), the inertia time constant is 5.0 s, and the pole pairs are 3

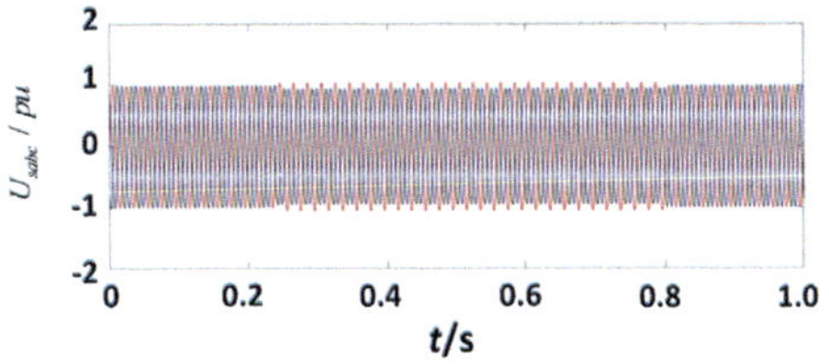

(a) Voltage drop 5%

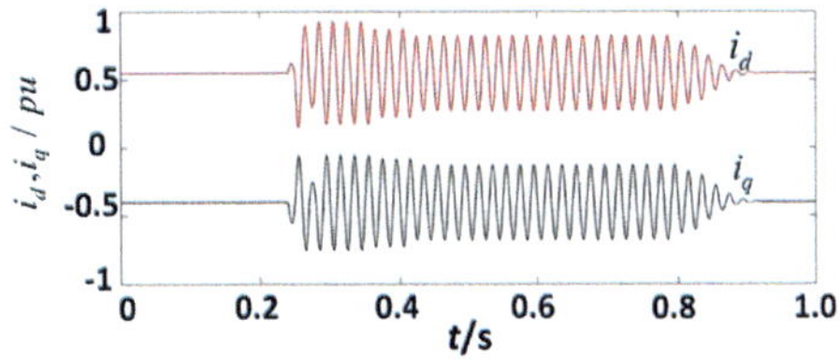

(b) Traditional control strategy dq axis current

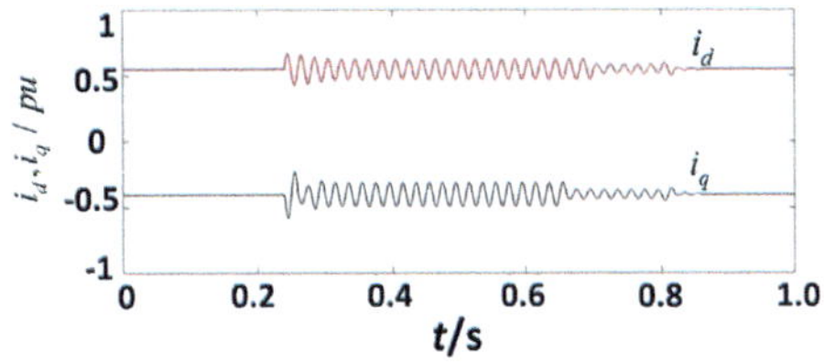

(c) Fuzzy-ADRC control strategy dq axis current

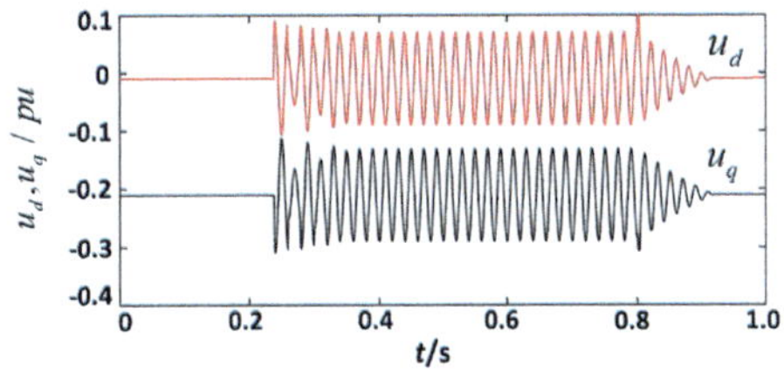

(d) Traditional control strategy dq axis voltage

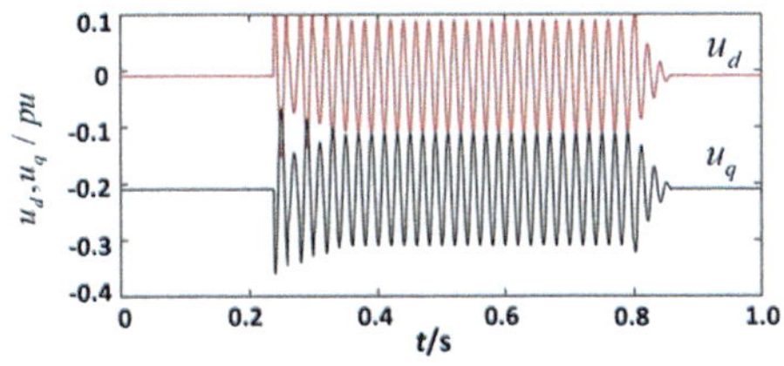

(e) Fuzzy-ADRC control strategy dq axis voltage

Fig. 10.5 (continued)

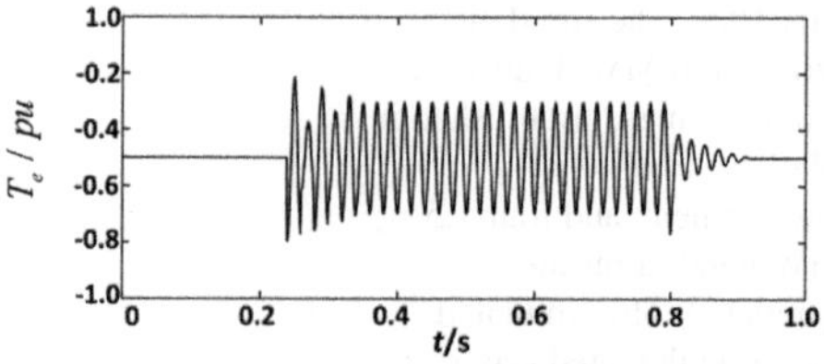

(f) Traditional control strategy torque

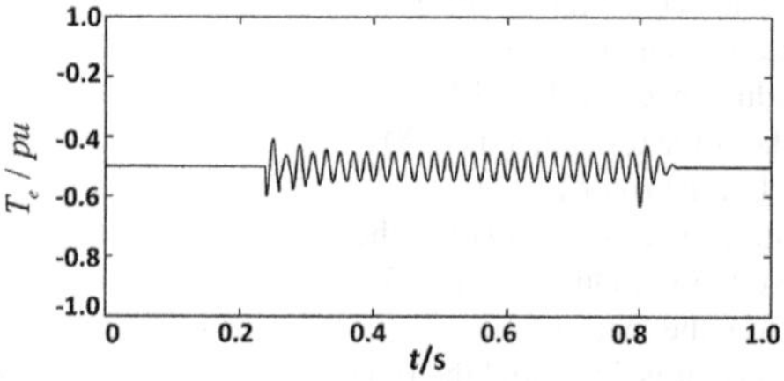

(g) Fuzzy-ADRC control strategy torque

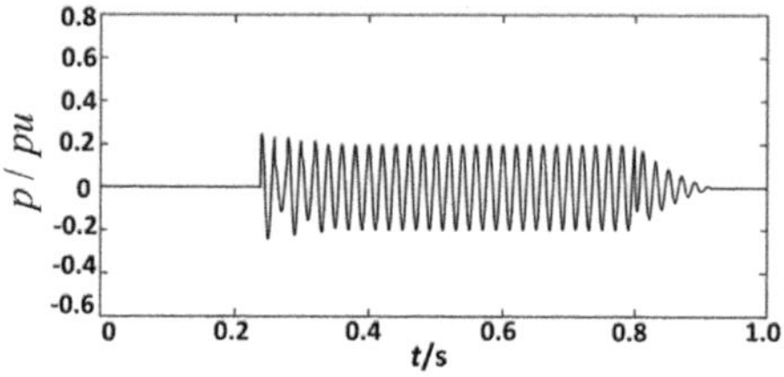

(h) Traditional control strategy power

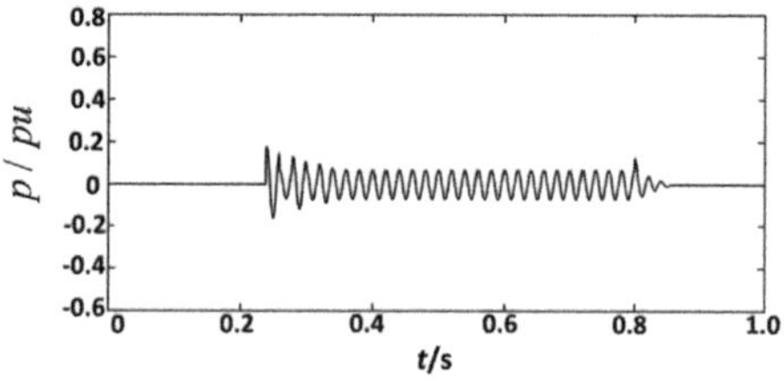

(i) Fuzzy-ADRC control strategy power

voltage returns to normal state at 0.8 s, it will take a long time for the rotor side current to restore to the stable value, and the dynamic response system gets slow. From (b) and (c), we can see that the fuzzy self—disturbance—rejection control is used to modify the controller parameters online to adapt to the system model during the electromagnetic transient process. When the voltage gets unbalanced, the current amplitude of rotor side becomes small, which effectively prevents rotor side overcurrent. From (f) and (g), we can see that the fuzzy self-disturbance—rejection control scheme can restrain the torque jitter better and prolong the operation life of wind turbine when the voltage of power grid fluctuates. From (h) and (i), we can see, it is beneficial to safe operation of the converter when power grid voltage falls, meanwhile the rotor side current, torque and power restore to stability for a shorter time after the gird voltage is restored. It can be seen that the system has stronger robustness and faster dynamic response.

In order to prevent the large impact of the rotor side converter in case of the asymmetrical fault of the power grid, this paper establishes current, voltages and magnetic chain equations of DFIG under the synchronous rotating coordinate system, designs a fuzzy active disturbance rejection control and uses the tracking differentiator to achieve a reasonable transition process and reduce the overshoot of the output; the extended state observer estimates and compensates the total disturbance of system, it calculates the Fuzzy-ADRC control of the sub side. This proposed project controls the negative sequence current, and it slows down the impact of currents and voltages, meanwhile it reduces the adjustable parameters of system and makes the system more robust and dynamic response to the disturbance. However, this project only improves the control strategy and does not change the hardware greatly, and the project needs to be improved when the residual energy of wind turbines becomes abundant because the voltage drop degree of the power grid is too large.

References

1. Xu, H.L., Zhang, W., He, Y.K., Chen, J.S.: A review on low voltage ride-through technologies and prospect for DFIG wind turbines. Autom. Electr. Power Syst. **37**(20), 8–15 (2013)
2. Pannell, G., Zahawi, B., Atkinson, D.J., Missailidis, P.: Evaluation of the performance of a DC-Link Brake Chopper as a DFIG low-voltage fault-ride-through device. IEEE Trans. Energy Convers. **28**(3), 535–542 (2013)
3. Pannell, G., Atkinson, D.J., Zahawi, B.: Minimum-threshold crowbar for a fault-ride-through grid-code-compliant DFIG wind turbine. IEEE Trans. Energy Convers. **25**(3), 750–759 (2010)
4. Nian, H., Cheng, P., He, Y.K.: Review on operation techniques for DFIG-based wind energy conversion systems under network faults. Proc. CSEE **35**(16), 4184–4197 (2015)
5. He, Y.K., Zhou, P.: Overview of the low voltage ride-through technology for variable speed constant frequency doubly fed wind power generation systems. Trans. China Electro Techn. Soc. **24**(9), 140–146 (2009)
6. Yao, J., Chen, Z., Li, Q., Yu, M.T., Zhou, T., Zhao, L.: Control strategy for doubly-fed wind power system under unbalanced and harmonic distorted grid voltage conditions. Autom. Electr. Power Syst. **38**(18), 6–12 (2014)
7. Bollen, M.H.J., Yang, K.: Harmonic aspects of wind power integration. J. Mod. Power Syst. Clean Energy **1**(1), 14–21 (2013)
8. Hu, S., Lin, X., Kang, Y., Zou, X.: An improved control strategy of doubly-fed induction generator under grid voltage unbalance. Trans. China Electrotechn. Soc. **26**(7), 21–29 (2013)
9. Liao, Y., Wu, X.W., Liu. R: Suppression for negative sequence and harmonic current of a doubly-fed induction generator stator in unbalanced grid. Autom. Electr. Power Syst. **37**(19), 26–32 (2013)
10. Wen, J., Huang, Y.: Robust Speed Control of Permanent Magnet Synchronous Motor. Industrial Electronics and Applications (ICIEA), pp. 327–330, IEEE (2013)
11. Zhu, Y., Li, J.L., Zhao, B.: Simulation of LVRT strategy for DFIG wind power system. Electr. Power Autom. Equip. **30**(6), 20–24 (2013)
12. Hu, J.B., Sun, D., He, Y.K., Zhao, R.D.: Modeling and control of DFIG wind energy generation system under grid voltage dip. Autom. Electr. Power Syst. **30**(8), 21–26 (2013)
13. Wang, Y.N., Sun, W.: Intelligent Control Theory. China Machine Press, Beijing (2008)

Chapter 11
Test Methodology and System for 5G Millimeter Wave Terminal RF Conformance

Yu Zhang, Xiang Wu, Xiang Zhang, Yuxin Ren, Chong Pan, and Guiming Wei

Abstract Due to the interface of 5G millimeter wave terminals radio frequency (RF) unit and antennas cannot be separated, the traditional conducted test method cannot be used for the transmission and reception characteristics, 5G millimeter wave devices require over-the-air (OTA) calibration and RF measurement in a plan wave test system. Based on the third generation partner project program technical specification (3GPP TS) 38.521-2 standard, an overview of system design challenges and measurement approaches is provided, the requirements of 5G millimeter wave terminal RF conformance test is analyzed, and the millimeter wave test solution is proposed combined with multi feed compact antenna test range (CATR), and the test procedure and test system for effective isotropically radiated power (EIRP), total radiated power (TRP), spurious emission (SE) and equivalent isotropic sensitivity (EIS) will be introduced in this paper. OTA RF test will run through all stages from research to production of millimeter wave products.

11.1 Introduction

With the development of mobile communication technology, 5th generation mobile networks (5G) is a new network architecture after long-term evolution (LTE), mainly used for enhanced mobile broadband (EMBB) scenario, ultra-reliable and low-latency communications (URLLC) scenario and massive machine type communications (mMTC) scenario, 5G millimeter wave frequency band has attracted extensive attention due to its rich frequency resources. In 2018, the United States and South Korea have successively launched 5G millimeter wave technology for commercial use, and China has also released the frequency band planning for millimeter wave applications. The transmitting and receiving characteristics of wireless communication equipment determine its performance in the network, from 2G, 3G, 4G to today's 5G technology, wireless performance is becoming more and more important. In 5G

Y. Zhang · X. Wu (✉) · X. Zhang · Y. Ren · C. Pan · G. Wei
China Academy of Information and Communications Technology, CAICT, No. 52 Hua Yuan Bei Road, Hai Dian District, Beijing 10091, China
e-mail: wuxiang@caict.ac.cn

L. C. Jain et al. (eds.), *Wireless Technology, Intelligent Network Technologies, Smart Services and Applications*, Smart Innovation, Systems and Technologies 258,
https://doi.org/10.1007/978-981-16-5168-7_11

frequency range 1 (FR1), the characterization of RF performance is evaluated using cabled connections to temporary antenna connectors, bypassing the device's antenna which was assumed to be simple and independent, it would anyway be technically problematic to derive the radiated performance.

In 5G FR2 millimeter wave frequency band, radio frequency unit and antenna interface cannot be separated, all the RF parameters will be tested through OTA test method. In this paper, based on the transmission characteristics of millimeter wave, the frequency band planning and wave propagation characteristics of millimeter wave will be analyzed in Sect. 11.2, the test method of 5G millimeter wave in 3GPP will be introduced in Sect. 11.3. Millimeter wave RF conformance test system will be introduced in Sect. 11.4.

11.2 Characteristics of 5G Millimeter Wave

11.2.1 Transmission Characteristics of Millimeter Wave

Millimeter wave generally refers to the electromagnetic wave with the wavelength from 1 to 10 mm, compared with sub 6 GHz, millimeter wave is less developed and utilized, and larger bandwidth can be allocated. Millimeter wave has a wide frequency band, which can achieve high-speed transmission, but its disadvantage is the free space path loss is really larger than the sub 6 GHz, the propagation distance is short, and it cannot penetrate most solid materials. At the same time, millimeter wave is greatly affected by the rain attenuation, which is caused by the wavelength of millimeter wave that is very close to that of raindrop, the attenuation increases with the rainfall. Due to the influence of diffraction, scattering, high loss and rain attenuation in the propagation process of millimeter wave, it is necessary to adopt some technical methods to ensure the capacity, low delay and high reliability of communication system.

11.2.2 Massive Multiple Input Multiple Output (MIMO) and Beamforming

2 * 2 and 4 * 4 antenna modules have been used in millimeter wave terminals, and multiple antennas are used in both transmitters and receivers. Compared with the single-antenna system, the channel capacity of the communication system is greatly improved. Massive MIMO and beamforming techniques can generate high gain to compensate for the large path loss in the high frequency band. As shown in Fig. 11.1, by adjusting the phase of the array antenna, the amplitude of millimeter wave terminals of some angles could be increased or reduced, so as to form a fixed directional beam. The fixed directional beam is aligned with the direction of the

Fig. 11.1 Beamforming for millimeter wave terminals

incoming wave, could be used not only at the transmitter but also at the receiver, which could obtain higher gain and improve the stability of the system.

11.3 OTA Test Method for 5G Millimeter Wave Terminals

Compared with the device working at sub 6 GHz, 5G millimeter wave products have changed greatly, the RF chips and antenna arrays of the equipment are highly integrated, and there is no test interface, so the device can only be fully characterized by the radiated OTA test methods, which could take the antenna radiated performance in account. There are many solutions for OTA test, including direct far field (DFF), CATR, and near field (NF).

(1) DFF: DFF is the most mature test method for OTA, which can cover all test cases of 3GPP 38.521-2 [1] and 38.521-3 [2]. However, in millimeter wave frequency, the free space loss increases with the test distance. The far field is typically defined as anything beyond the Fraunhofer distance D [3]:

$$D\,(\text{transceiver distance}) > 2d^2/\lambda \quad (11.1)$$

where d is the largest dimension of the antenna under test and λ is the wavelength. At this distance, the phase curvature of the field is 22.5°, taking the device working at 26 GHz as an example, the maximum size of the measured device is 0.2 m, and the far-field distance is nearly 8 m. According to the loss calculation formula of free space:

$$\text{Loss} = 32.4478 + 20 * \log 10(f * 1000) + 20 * \log 10(D/1000) \quad (11.2)$$

where loss refers to the space loss in dB, f is the frequency, D is the transceiver distance. The space loss will reach 80 dB, it is a great challenge for the test system and due to its long transmission distance, and the cost of the test system would be very high.

(2) CATR [4]: the condition of far-field test is achieved through reflector and feed, which is similar to the DFF test, but the test distance is greatly shortened, plane wave can be formed in a small space, and all performance tests of transmitter

and receiver can be executed in the CATR. It is also the RF conformance test system recommended by 3GPP 38.521-2.

(3) NF [5]: NF test system has the advantage of lower space loss, and can be placed in most office space. The NF test system has higher requirements for algorithm, but it cannot directly form far-field conditions and cannot meet the relevant requirements of receiver performance test, so it is not suitable for the receiver performance test.

11.4 Millimeter Wave RF Conformance Test Cases

RF test cases mainly include three aspects: transmitter (Tx) and receiver (Rx) inband test cases, Tx and Rx spurious test cases and Rx blocking test cases. EIRP, TRP, SE and EIS test will be detailed analyzed in this paper.

(1) EIRP and TRP test

For millimeter wave terminals, EIRP and TRP are two key technical indicators to measure the power class of the device under test, higher EIRP or TRP may cause interference to other frequency bands and systems, and lower EIRP or TRP may decrease the coverage. In 3GPP TS 38.521-2, for millimeter wave terminals with power level 3, the EIRP limit in n258 band is set between 22.4 and 43 dBm, and the maximum TRP limit is less than 23 dBm. The diagram of the test system is shown in Fig. 11.2.

3GPP provides two types of spherical sampling points for TRP test, constant step and constant density sampling. Constant step is sampling at a fixed angle in the longitude and latitude, the rotation of the positioner is continuous and the rotation

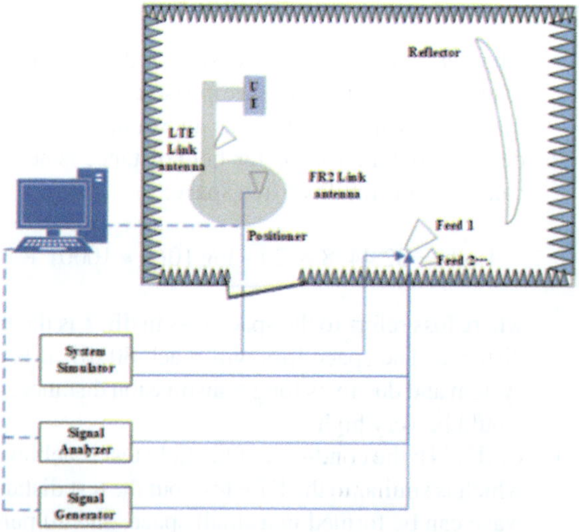

Fig. 11.2 EIRP and TRP test diagram in the anechoic chamber

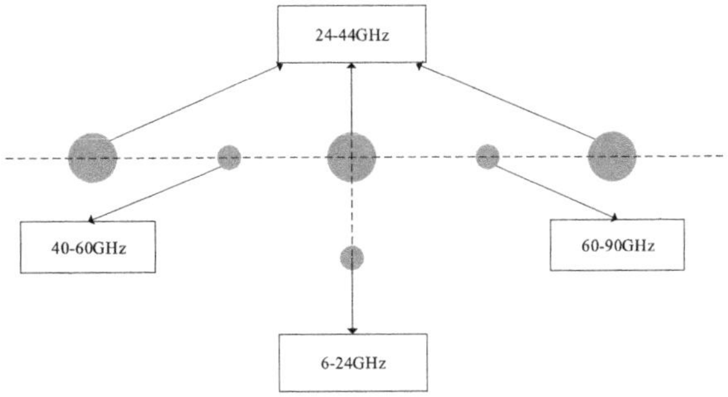

Fig. 11.3 Multi feed design illustration

efficiency is high, but the disadvantage is that there are more sampling points than constant density sampling. Constant density sampling is to discretize the spherical area with fixed points, and the distance between each two points is consistent, the disadvantage the rotation efficiency of the positioner is low.

(2) SE test

Spurious emissions are emissions which are caused by unwanted transmitter effects such as harmonics emission, parasitic emissions, intermodulation products and frequency conversion products. One of the disadvantages of the traditional CATR is that the test efficiency is low when measuring the 3D data of the radiated pattern over a wide frequency, particularly in the spurious emission test. In this paper, we will analyze a multi feed solution for RF conformance test as shown in Fig. 11.3. This multi feed system is composed of three feed working from 24 to 44 GHz, one feed working from 6 to 24 GHz, one feed working from 40 to 60 GHz and one feed working from 60 to 90 GHz, which can meet the requirements of SE test from 6 GHz to 2nd harmonic. This solution can greatly reduce the test time, multiple feed is placed near the parabolic focus instead of the original single feed system, and the offset feed can form a plane wave with a certain angle with the original area in the CATR, so as to realize multi band simultaneous measurement.

(3) EIS test

EIS is a key technical parameter of receiver performance test, which measures the throughput level of terminal equipment in an ideal propagation environment, under low-power level and environment without noise. The throughput of terminal affects the effective coverage of base station. In 3GPP 38.521-2, for millimeter wave terminals with power level 3 supporting 100 MHz bandwidth, the EIS limit in n258 band is lower than −85.3 dbm. In this paper, the terminal is set in the quiet zone of the chamber, and the transmission power of the base station is reduced at the interval of no more than 0.2 dBm. The EIS level is the power of the corresponding base station under the premise of 5% bit error rate.

11.5 Conclusion

In this paper, the characteristics and technical implementation of 5G millimeter wave terminal are introduced, the RF measurement method and requirements of millimeter wave terminal is analyzed, and the RF millimeter wave test solution combined with the multi feed CATR test system is proposed. Base on this OTA test system, the detailed test procedure of EIRP, TRP, SE and EIS is analyzed. In future, low-cost solutions and more efficient system will be proposed continuously in the millimeter wave terminals RF test.

References

1. 3GPP TS 38.521-2: User Equipment (UE) Conformance Specification; Radio Transmission and Reception; Part 2: Range 2 Standalone (Release 16)
2. 3GPP TS 38.521-3: User Equipment (UE) Conformance Specification; Radio Transmission and Reception; Part 3: Range 1 and Range 2 Interworking Operation with Other Radios (Release 16)
3. Selvan, K.T., Janaswamy, R.: Fraunhofer and Fresnel distances: unified derivation for aperture antennas. IEEE Antennas Propag. Mag. 12–15 (2017)
4. Marti-Canales, J., Ligthart, L.P., Roederer, A.G.: Performance analysis of a compact range in the time domain. IEEE Trans. Antennas Propag. **50**(4), 511–516 (2002)
5. Johnson, R.C., Ecke, R.H.A., Hollis, J.S.: Determination of far-field antenna patterns from near-field measurements. Proc. IEEE **61**(12), 1668–1694 (1973)

Chapter 12
A Survey on Energy Saving Technology of Wireless Sensor Network

Huanan Zhang

Abstract Wireless sensor networks (WSNs) have important applications in many fields such as medical treatment and industry. A WSN is typically consists of a large number of sensor nodes that rely on a limited supply of power in many applications. Therefore, improving the energy efficiency of wireless sensor networks has become an important research topic. This paper analyzes the principle of energy consumption of each component unit of wireless sensor node, and studies the energy saving technology and energy saving design idea of MAC protocol from the data link layer and network layer of network model.

12.1 Introduction

Wireless sensor network consists of many sensor nodes. It can sense, monitor and collect environmental information from sensors, such as temperature, heat and the concentration of various gases in the environment. The collected data will be sent to interested observers via wireless communication. Wireless sensor network does not need fixed network equipment, and it has the advantages of rapid expansion, strong damage resistance and so on. In the current application research, the network lifetime is an important index to measure the performance of a wireless sensor network. Traditional wireless sensor nodes use non-rechargeable batteries. Once the electromagnetic energy is used up, the node will die, so the design of wireless sensor network node has become a hot spot. In recent years, with the continuous development of environmental energy acquisition technology, researchers have been able to capture the energy existing in the environment, such as solar energy, mechanical vibration heat energy, etc. This technology is used in wireless sensor networks to design sensors that automatically collect environmental energy. Sensor nodes can convert the solar energy, vibration mechanical energy and heat energy in the environment into usable, and can automatically supplement their own energy, forming a

H. Zhang (✉)
School of Data and Computer Science, Guangdong Peizheng College, Guangzhou, China
e-mail: Topwn2008@163.com

L. C. Jain et al. (eds.), *Wireless Technology, Intelligent Network Technologies, Smart Services and Applications*, Smart Innovation, Systems and Technologies 258,
https://doi.org/10.1007/978-981-16-5168-7_12

wireless sensor network with self-energy characteristics. Combined with this characteristic, this paper aims to improve the performance of wireless sensor network energy storage technology.

Wireless sensors have been widely used in military and civil fields. Extensive deployment has been made in deep-space exploration, deep-sea submersible navigation, environmental monitoring and protection of wetlands, monitoring of rivers and lakes, and urban planning and construction. In remote monitoring, smart city, smart home, smart office and other fields reflect more and more high-practical value. Because of the special application field of wireless sensor network, so many cases, the sensor node is deployed once and is no longer recycled. Some nodes are not easy to recharge. The power of the nodes is depleted, and the entire network fails. Classical algorithms in wired networks and some algorithms in wireless networks with unlimited energy or energy supplementing cannot be directly applied to wireless sensor networks with limited energy. It is necessary to save as much energy as possible to prolong the life cycle of wireless sensor networks (WSNs), so the research on energy saving is always an important branch of WSNs.

In wireless sensor networks (WSNs), the energy capacity of sensor nodes is limited, and the cost of power supply replacement is high and difficult. Therefore, in network design, the use of energy saving technology to extend the life of the network as much as possible has become a crucial issue for wireless sensor networks. At present, the energy conservation of wireless sensor networks is mainly studied from two levels: network level and node level. The network layer can design good clustering algorithm and routing protocol. Data forwarded reduces energy consumption, and thus prolonging the life cycle of the network. Node level can be achieved from both open source and throttling. The open source aspect based on electronic components extends the battery life of wireless sensor power supply or mid-charge. Throttling extends the life cycle of the whole network through the sleep of the node scheduling algorithm.

12.2 Wireless Sensor Node Deployment

Node deployment is random, and the location of each node deployment is not specified in advance. Once all nodes are deployed, and their location does not change. All nodes are homogeneous, that is, all specifications and parameters are the same, especially the sensing radius and communication radius are the same, but not limited to a specific sensor node. A node can perceive all working nodes within its perceived radius. Each node has a unique identity ID, and nodes that are aware of each other can obtain each other's ID. The node communication radius is not less than the perceived radius. It is not necessary for the monitored network area to reach 100% monitoring coverage all the time. During the monitoring process, a small number of areas is allowed to occur that are not monitored. The node does not know the exact location of itself, nor the exact location of its neighbors. In the wireless sensor network, each node can be in a working state or a dormant state. The nodes in a working state can normally carry out various kinds of work, such as perception, monitoring and communication. Such nodes are called working nodes. The nodes in the resting state operate with very low-power consumption, do not perceive and monitor the surrounding area, and do not communicate with other nodes. Such nodes are called sleeping nodes. The working node and the sleeping node can convert to each other. When the working node meets certain conditions, it can enter the dormant period, while the sleeping node will be awakened after a certain period of time and enter the working state. Every wireless sensor network has its coverage area, and some nodes are always located at the boundary of the whole network. Nodes at the boundary cannot be treated as nodes at the center of the network, because the number of neighbors of nodes at the boundary is obviously smaller, so special investigation should be conducted on boundary nodes. Since the network boundary is very large, for each node at the boundary, the boundary can be approximately regarded as a straight line, that is, a diameter. Divided by the diameter, the half that belongs to the network coverage is called the inside of the boundary node, and the other half is called the outside of the boundary node. As for the interior of the boundary node, just like the general node, it has its own neighbor nodes and will be covered by enough neighbors, while for the semicircle outside the boundary node, the number of neighbors and the covered area will be small.

12.3 Analysis of Energy Consumption of Wireless Sensor Network

The energy in wireless sensor network is consumed by the micro-sensor nodes. A typical micro-sensor node consists of four parts, the power, induction unit, processing unit and the wireless communication unit. In addition to the power supply unit, the other three parts all need to consume electric energy to different degrees.

The main function of the induction unit is to sense temperature, humidity, air pressure and other information, and hand it to the processing unit. In the information transformation front end of the induction unit, processing signals and adjusting analog-to-digital conversion and other links, energy consumption of the induction unit mainly depends on the following factors.

According to the different energy consumption, the induction unit can be divided into low-energy consumption, medium energy consumption and high-energy consumption. Three types of low-energy consumption induction units are commonly seen, such as temperature sensor, humidity sensor, photosensitive sensor, speed sensor, etc. Common medium energy consumption induction units include sound sensors and magnetic sensors. Typical high-energy sensor units include image sensors and visual frequency sensors [1]. For example, in scenes with interference, the detection accuracy will be affected to different degrees. The power of the sensor unit and the number of repeated detections need to be increased, and the energy consumption will also increase.

The processing unit is responsible for information fusion and processing in practical application. As the manufacturing process of integrated circuit matures, the power of processor is very small, and its energy consumption mainly depends on the number of wireless sensor nodes and the number of data packets.

The transceiver unit, which is responsible for information forwarding, is the most energy-consuming one of the sensor nodes. There are four working states: idle, receive, send and sleep. In the dormant state, the energy consumption is small, it has a variety of working states, and the energy consumption is almost negligible. The proportion of the three modes of sleep, reception and transmission in energy consumption is 1:1:2.7. In the idle mode of the transceiver unit, listening takes up most of the working time. The study shows that the energy consumed by the listening state accounts for more than 90% of the total energy consumption. There is no data transmission in the idle mode, so the sensor node is in a dormant state, which can greatly reduce the energy consumption. However, if it is turned on too frequently, it will cause certain energy consumption [2].

12.4 Energy Saving Technology of Wireless Sensor Network

The sensor nodes rely on battery power, and the battery capacity they carry is quite limited. Therefore, one of the main goals of the node is how to extend its life as much as possible while completing the task. As a kind of weakened embedded computing equipment, and considering the cost factor, the processor core unit on the node usually chooses the processing chip with weak power, and the processing speed of the node is relatively slow. The storage capacity of the container is small. In addition, the communication module of the node accounts for a large part of the energy consumption of the node. In order to reduce the communication capacity of the energy consumption node, it is often limited to some extent. Communication distance is generally limited to a small range. In the wireless sensor network energy saving problem, there are two kinds of technology: data link layer energy saving technology and network layer energy saving technology.

12.4.1 MAC Protocol Energy Saving Technology

The energy saving technology of the data link layer is realized through the MAC protocol. According to the different energy saving principles, there are mainly four kinds of energy saving technologies in the MAC protocol.

(1) Data transmission conflict control technology

Reduce packet transmission collision rate, reduce the amount of retransmission data. A common node energy path in the data link layer is also one of the basic functions of the protocol implementation to realize the control of data retransmission rate [3]. There are many methods, the common methods are: (1) the use of sleep technology; (2) adoption time multiple access (TDMA) technology; (3) Use data to send preamble.

(2) Dormancy mechanism

Dormancy mechanism refers to the sensor nodes in the free state, the adaptive or program control ways to enter a dormant state, and to control the timing of the dormancy and interval period, in order to reduce the time of the sensor in the idle listening state, from and reduce the energy consumption is enabled after dormancy mechanism, awaken should form a complete set of judgment and decision making mechanism, in order to realize the sensor nodes under the dormant in a timely manner when necessary the state of idle listening and other working condition.

(3) Working mechanism of mixed MAC

It mainly refers to the hybrid mechanism combining TDMA and CSMA. When the number of sensor nodes is fixed and small, the TDMA mechanism of pre-allocated time slot can be adopted. When the number of sensor nodes changes, the TDMA mechanism is immediately transformed into a dynamic slot allocation mechanism. Through this hybrid mechanism, the worker node can send or receive data packets at a fixed time slot, thus avoiding the possibility of collision in the process of data transmission.

(4) Introduction technology of data transmission

The basic idea of the technology is that the sensor node sends a preamble data before sending the data packet to alert the sensor receiving node to be ready for receiving. The introduction data should be sent at least at the time interval of the sampling cycle. Other sensor nodes in the network are sampled to determine the next working state. If the communication channel is found to be busy, continue listening until the correct packet is received or until the channel becomes idle [4]. If the data is received, the control node enters the sleep state after continuing to listen for a certain period of time.

12.4.2 Network Layer Protocol Energy Saving Technology

The common practice of energy saving technology in network layer is to realize the energy saving idea of routing protocol through the energy saving design of routing protocol.

(1) Take the data transmission distance as the route measurement value

In a typical first order radio wireless sensor node, suppose the distance between two sensor nodes is d and the transmitted data packet is K, then data transmission energy consumption $Et(k,d)$:

$$\text{When } d < d_0,\ E(k, d) = E_{\text{el}} \times k + \varepsilon \times k \times d^2.$$

$$\text{When } d \geq d_0,\ E(k, d) = E_{\text{el}} \times k + \varepsilon \times k \times d^4.$$

And d is the threshold value of distance, E_{el} represents a transceiver energy consumption factor required to start a transmission or receiving circuit. ε represents the power coefficient of the transmission amplifier.

According to the above expression, the power consumption of the sensor node is proportional to the square of the distance. Therefore, reducing the transmission distance reduces the energy consumption. However, when transmitting data at nodes, shortening each hop can increase the transmission distance and the hop of the transmission path, and the total energy consumption of the network cannot be reduced enough. At the same time, the transmission delay will also increase. Therefore, the control of the transmission distance should take various factors into comprehensive consideration [5].

(2) Take the number of transmission path hops as the routing measurement value

In the packet transmission path, the fewer hops, the more optimized route. When the transmission power of the sensor node is fixed, the fewer hops, the smaller the total transmission distance and the smaller the total energy consumption of the network, so as to achieve the purpose of energy saving. When the transmission power is variable and the transmission mode is broadcast mode, the packet transmission path hop number is small, which means that the number of nodes participating in the broadcast data forwarding is small, and the total energy consumption of the network also decreases [6].

(3) Dynamically adjust the sending power according to the distance of the routing path

In the wireless sensor network with variable power, adjusting the transmission power according to the distance of routing path can not only reduce the energy consumption of data transmission, but also effectively reduce the radio interference between adjacent nodes and improve the utilization rate of the channel.

(4) Data aggregation technology is adopted

Data aggregation technology means that sensor nodes collect physical information of the monitored area and transmit the data upward in the wireless sensor network. Common data aggregation operation functions are to evaluate the average to evaluate the intermediate value to sum to evaluate the maximum value to evaluate the minimum value to count, etc. Users can also customize the data aggregation operation function according to the application requirements [7].

12.5 Conclusion

The energy consumption of wireless sensor network is limited, and the problem of energy consumption directly affects the life cycle of the entire network. The adoption of energy saving technology has become the core problem in WSNs. This paper analyzes the energy consumption of each component single element of wireless sensor node and its key factors. The research direction in future mainly includes the research on the energy saving technology of data link layer MAC and the network layer routing protocol design based on the energy saving idea and the system design of the energy saving algorithm for each layer of WSN network.

References

1. Zhou, K., Fu, C., Yang, S.: Big data driven smart energy management: from big data to big insights. Renew. Sustain. Energy Rev. **56**, 215–225 (2016)
2. Khan, M., Hussain, S.: Energy efficient direction-based topology control algorithm for WSN. Wirel. Sens. Netw. **12**, 37–47 (2020)
3. Khoa, V., Takayama, S.: Effect of landslide on energy efficiency chain based routing protocol for wireless sensor network. Wirel. Sens. Netw. **12**, 13–36 (2020)
4. Rahimkhani, K., Forouzesh, F.: Improved routing in wireless sensor networks using harmony search algorithm. Wirel. Sens. Netw. **9**, 333–353 (2017)
5. Babber, K., Randhawa, R.: A cross-layer optimization framework for energy efficiency in wireless sensor networks. Wirel. Sens. Netw. **9**, 189–203 (2017)
6. Srour, T., Haggag, A., El-Bendary, M., Eltokhy, M., Abouelazm, A.: Efficient approach for monitoring and controlling water parameters utilizing integrated treatment based on WSNs. Wirel. Sens. Netw. **11**, 47–66 (2019)
7. Alhumud, H., Zohdy, M., Debnath, D., Olawoyin, R., Arefifar, S.: Cooperative spectrum sensing for cognitive radio-wireless sensors network based on OR rule decision to enhance energy consumption in greenhouses. Wirel. Sens. Netw. **11**, 1–11 (2019)

Chapter 13
The Development of Smart Building Under the Background of Internet+

Qing Wei and Qingjuan Liu

Abstract In the development and change of society, the progress of science and technology has played an important role in promoting. In the field of smart building research, with the comprehensive application of different technologies in it, the scope of application of smart building has also been expanded. It began to develop towards the application of wisdom, and formed different application systems, which not only provides a variety of different services for people's production and life, but also makes the building operation process greener and more environmentally friendly, thus fully meeting the needs of sustainable development. Looking back on the development progress of smart buildings and the green buildings, the influences and impacts of new internet technologies on buildings in this information technology era are analyzed and the shapes and features of future buildings and the prospect for smart buildings are explored.

13.1 Introduction

The development of architecture from Earth pit and cave dwelling to brick house and reinforced concrete high-rise building is more perfect in function and quality, including the development of smart building and green building with network and integration of security, fire and protection, the living standard of human beings has been improved and the living environment is more and more safe, comfortable and convenient. With the increase in population and the progress of technology, the negative influence of architecture is increasing day-by-day. Land use, green space reduction, energy consumption increase and environmental damage are urgent problems to be solved. Building scientists around the world are constantly exploring the future direction of building development.

Q. Wei (✉)
Shandong Xiehe University (North Campus), Jinan, China
e-mail: weiqing1986@126.com

Q. Liu
Remote Wall Town, Licheng District, Jinan, China

L. C. Jain et al. (eds.), *Wireless Technology, Intelligent Network Technologies, Smart Services and Applications*, Smart Innovation, Systems and Technologies 258,
https://doi.org/10.1007/978-981-16-5168-7_13

Through summarizing and analyzing the present situation and shortage of smart building in our country, this paper makes a deep study on the characteristics of the development of the times and the influence and function of Internet+ big data technology on the building field, so as to find out the development trend of the future building, this paper studies the necessity of smart architecture and makes a preliminary attempt to explore the future development of architecture [1].

13.2 The Development of Smart Building

Smart building was first put forward to meet people's needs for more functions, solve problems in development, and make people more convenient and comfortable in the process of living, and give full play to the basic advantages of the building, to achieve efficient operation of the building, by building it into a systematic environment, so that different parts of the structure as a subsystem in the system to play a role, so that the building as a whole, in order to facilitate the construction of more scientific management, and reduce the management of the problem [2]. The application of intelligence in smart building refers to the combination of modern technology and traditional technology and equipment through various modern new technologies including big data, cloud technology, Internet of things, intelligent interconnection, modern communication technology, artificial intelligence and so on, to achieve innovative applications (see Fig. 13.1).

After more than 20 years of development, the standard system, technical system and engineering system of smart building in China have taken shape. The intelligent system has become a necessary system for large-scale public buildings, commercial office buildings and high-grade residential buildings, and has greatly enhanced its value and quality, for people to create a safe, comfortable, convenient, efficient, energy-saving work and living environment has played an important role. The size

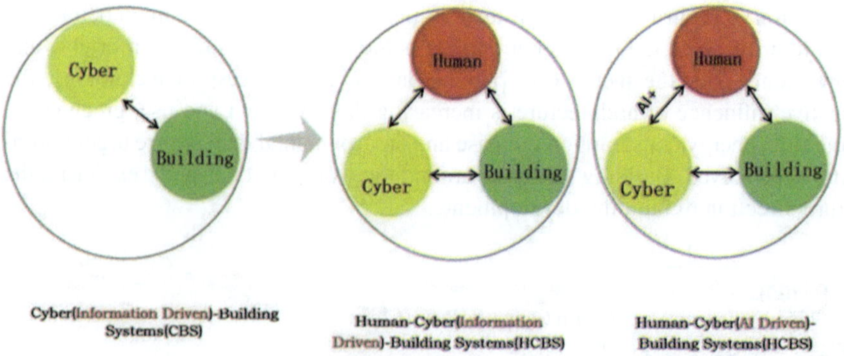

Fig. 13.1 The development of smart building

and output value of the smart building market have also reached the level of 400–500 billion per year.

13.3 The Necessity of Developing Smart Building

13.3.1 Requirements for Application Objects

At present, the application object of architectural design is only the single building itself. For example: Building energy consumption systems, building automation systems, security and fire protection systems, are designed, built and operated in accordance with the building itself, basically do not involve the specific behaviour of the users in the building, will not involve the home, not to mention the surrounding buildings and environment [3]. And the application object of smart building is not only the single building itself, the smart building will involve many factors such as this building and the behaviour habits of the users in this building, it will also involve all surrounding buildings and all possible connections to people, objects and the external environment.

13.3.2 Requirements for Smart City Development

In recent years, under the guidance of the relevant national policies, the construction of smart city has developed rapidly and comprehensively, and smart building will be the concrete application and an important component of smart city, intelligent architecture plays an important role in human life and work. Facing the development of smart city in the future, we should also develop a smart synthesis based on urban Internet of things and cloud architecture [4]. The development of smart city needs more intelligent architecture, and it also needs the implementation of smart architecture and Internet technology.

13.4 Application of Various New Technologies in Smart Buildings Under the Background of Internet+

In the context of the Internet, new Internet technologies emerging in recent years, such as cloud computing technology, Internet of Things technology, big data technology, BIM technology and artificial intelligence technology, further promote the transformation and upgrading of buildings and provide technical support for the realization and development of smart buildings [5]. The architecture of the future will be intelligent architecture, whose development and application cannot be separated from the support of new Internet technologies.

13.4.1 Cloud Computing

Cloud computing is a new computing model based on the Internet. With this model, computers or other devices can share information or hardware and software resources according to demand. The “cloud” in cloud computing is composed of hundreds of millions of servers, which can provide unprecedented computing power and storage capacity, etc. Moreover, cloud computing utilizes virtualization technology to realize centralized management of physical resources, and makes full use of various software and hardware resources to improve the utilization rate of equipment and reduce system energy consumption. In the future, architectural design software suppliers will become architectural design, structural computing and analysis service providers with the development of cloud computing.

13.4.2 The Internet of Things

The Internet of Things is the realization of intelligent association between things, is based on the Internet, traditional telecommunications network and other information carrier. Internet of things is the essence of IT infrastructure can be incorporated into physical infrastructure, and is widely connected, realize something, purpose is to implement dynamic, the wisdom of management and control, the nature and smart building inner intelligence embodied in the various intelligent device embedded in the construction of various objects, and give its run automatically and self-decision-making intelligence work mode. In order to realize effective management of buildings and “smart” thinking of buildings, relevant information of building ecology should be mastered first. Internet of Things technology is an important technical basis for smart buildings to realize comprehensive perception, reliable transmission and intelligent decision-making.

13.4.3 Big Data Technology

Big data technology describes the new generation of IT technology and architecture. Big data is characterized by huge data volume, complex and diverse data types and fast generation speed. With the progress of science and technology and building, the development of smart building of the future will be full of all kinds of sensor and data acquisition and control devices, the smart building based on the technology of Internet of things in real time by building awareness with all kinds of signals, will produce many structural safety, indoor environment, such as professional in the field of building energy consumption data. The massive data accumulated over a long period of time are complex and diverse. Meanwhile the information content of these data is huge, and the capacity will increase from TB to PB, EB and ZB.

13.5 Problems in the Application of New Technology in Smart Building

In the development of intelligent building, there are still some bottleneck problems in the application of many new technologies.

13.5.1 The Limitation of Internet of Things in Intelligent Building

At present, there is no seamless interconnection between the Information Wide Area Network and the TT & C Wide Area Network. In other words, there are many isolated control islands and the Internet of things cannot be extended into them, and the use of the Internet of things by smart buildings is limited to a great extent.

13.5.2 The Problem of the Application of Big Data in the Smart Building

The databases of different industries are separated from each other, and the databases of different search engines cannot be connected seamlessly, which affects the application of big data seriously.

13.5.3 The Problem of the Application of Cloud Computing in the Smart Building

The users of smart building need to use a large amount of external resources, and it is impossible for a concrete smart building entity to carry or possess a huge resource pool, and the resource pool is required to be flexible and up-to-date, so the cloud computing technology needs to be continuously developed. The virtual resource pool with huge data in the cloud can always meet the needs of users.

13.6 Some Suggestions on the Development of the Industry

13.6.1 Pay Attention to the Theoretical Research of Smart Building

Technology enterprises should pay attention to the theoretical research of smart building, the deeper the theoretical research, the deeper the market understanding, the deeper the market excavation, the better you can do it. There are too many practical lessons: the ability to control the market is not strong, the fundamental reason is that the depth of industry support technology research is not enough.

13.6.2 Solve the New Problems in the Development of the Industry in Time

There will be a lot of new situations that have not existed in the past, enterprises must keep pace with the times and make adjustments in technology. For example, air conditioning units, fresh air units and fan coils are used extensively by air conditioning in building equipment monitoring systems, all of which have fresh air supply links, but in recent years air pollution and smog have wreaked havoc, control systems for these devices should be designed to protect against fine particulate matter and other indoor pollutants.

13.6.3 Increase the Content of the Smart Building Subsystem in the BIM Model

In order to prevent many "collisions" in the actual design and construction, resulting in the rework of the change design and construction, it is necessary to reflect and describe more the content of the design and construction of the smart building subsystem in the BIM model, which should be paid attention to by enterprises and experts in the smart building, smart building industry.

13.7 Conclusion

With the advent of the era of information technology and innovation development of various new technologies for the development of industry has brought great opportunities for development, the construction industry in this area is better to promote the development of transformation, realize the breakthrough progress of the industry, should also follow the time development, improve the level of the development of smart buildings, and through the application of new technology make the building's overall function was improved, and achieve better for application development of new technology, make people experience the more wisdom application effect in production and living, to eventually achieve smart city construction play a corresponding role.

References

1. Gao, X., Pishdad-Bozorgi, P., Shelden, D.R., Tang, S.: Internet of things enabled data acquisition framework for smart building applications. J. Constr. Eng. Manag. **147**(2), 0001983 (2020)
2. Zhang, S.J.: Thoughts on smart building and intelligent architecture. Smart Archit. **03** (2019)
3. Wang, S.F.: Model construction and methodological thinking of smart city research. Planners **04** (2012)
4. Cleevely, D.: Why are smart cities proving to be so hard to deliver. IET Smart Cities **2**(4), 165–166 (2020)
5. Liu, F.Y., Shi, Y., Chen, Z.X.: Intelligence quotient test for smart cities in the United States. J. Urban Plann. Dev. **147**(1), 04020053 (2021)

Chapter 14
Paradigm for Urban Safety and Risk Management with Big Data

Nan Jia, Tengfei Zhang, and Siquan Wang

Abstract Risk management of urban public safety, as an important part of urban management, is crucial to the healthy development of cities. In recent years, with the rapid development of big data technologies, urban safety and risk management have also entered the era of big data. Obviously, in this context, it is urgent to reform the traditional risk management model and establish a new paradigm for urban public safety risk management based on big data technology. Firstly, we summarized basic theories related to urban safety and risk management from the aspects of disaster-bearing carrier and risk characteristics. On this basis, it puts forward the connotation of urban safety and risk management and expounds its current situation. Secondly, we analyze and summarize the opportunities and challenges of urban safety and risk management in the era of big data. Finally, the big data-driven paradigm for urban safety and risk management is proposed, and the policy recommendations for urban public security management construction based on this paradigm are given. Obviously, this study will provide theoretical and methodological guidance for the development and implementation of urban safety and risk management of China in the era of big data.

14.1 Introduction

With the continuous advancement of China's industrialization and urbanization, the city's scale and population have expanded rapidly, leading to increased vulnerability of urban buildings and structures, road traffic, lifeline management networks, and personnel gathering places, and the situation of urban safety is grim. How to effectively manage, operate and maintain a complex giant system with huge population size and economic volume is an important issue faced by the urban management in the new era, and achieving a fine urban management has become an important task for municipal governments across the country. The risk management of urban safety, as an important part of fine urban management, is the foundation for ensuring urban

N. Jia · T. Zhang · S. Wang (✉)
School of Informatics and Cyber Security, People's Public Security University of China, Beijing, China

L. C. Jain et al. (eds.), *Wireless Technology, Intelligent Network Technologies, Smart Services and Applications*, Smart Innovation, Systems and Technologies 258,
https://doi.org/10.1007/978-981-16-5168-7_14

residents' personal and property safety. Therefore, it is essential to improve the risk management capability of urban safety.

With the rapid development of big data technology, the risk management driven by big data is a relatively new research field that has emerged with the big data method and has set off a research boom at home and abroad. In 2008, "Nature" launched the "Big Data" research topic to explore the application and challenges of big data in the subjects and fields of the Internet and computer technology. In 2011, "Science" launched the topic of "Dealing with Data," mainly discussed the processing technology and application of big data. In 2018, the United Nations e-government survey report pointed out that building a resilient society needs to have resources and capabilities to predict, reduce, and respond to various risks based on big data and through ICT. At the same time, the arrival of the era of urban management big data has made more and more scholars and research institutions aware of the importance of big data in risk management and conducted related researches. However, the current researches with the application of big data technology to carry out the urban safety and risk management are still very rare. The existing big data-based urban risk management researches are scattered in specific areas or links such as urban risk prevention, urban safety management, and urban safety construction, which haven't been specifically developed for the entire city safety construction and process, and haven't yet formed the universal guiding theory and practice system, which can't meet the theoretical and practical needs of China's urban fine construction which is under a rapid development. It is urgent to conduct an exploration on the systematic model of urban safety and risk management in the era of big data, so as to provide a theoretical basis and practical guidance for the researches on urban safety and risk management of China in the era of big data [1].

14.2 Summary of Risk Management of Urban Safety

14.2.1 Characteristics of Urban Risks

The city has the complexity and intensity of the spatial and temporal overlap of human, buildings, and systems. In a limited area, the accumulation of large population, resources and social and economic activities have formed a dense network, resulting in the urban risks have the characteristics of mobility, diffusivity, emergence, and dynamic uncertainty.

(1) The high mobility of risks, the city as an open and complex system, the interaction of people flow, objects flow and information flow, the frequent flow of various people, and the advanced and convenient urban public transportation, road traffic, rail transit and other public transportation modes have facilitated the transportation of personnel and materials. This can easily lead to high mobility of various types of risks in various areas of the city, such as key personnel of security monitoring, suspicious people, and hazardous chemicals.

(2) The diffusivity of risk emergencies. When the risk evolves into an emergency, due to the spatial and temporal overlap of urban people, buildings and systems, the dense multi-type carriers catalyze and aggravate the spread and diffusion of emergencies. For example, the accidental fire in the residential housing, which is easy to develop into a big fire accident, because the fire can spread quickly with the help of dense urban carriers, which can cause a heavy loss.

(3) Emergence (coupling) of risk accident system. The city is a typically complex system. This kind of urban system is prone to produce a comprehensive disaster that evolves from a single risk to an integrated risk coupled with multiple disasters due to the three-dimensional density of the urban spatial structure and spatial interaction of human, buildings, and systems. Once an accident occurs, it will present an emergent and disastrous effect. Specifically, it will be the occurrence of disaster chain successively in time, spatially dependent on each other. And these catastrophes have cause and effect correlations, interact as both cause and effect, have a chain reaction and occur in turn.

(4) Dynamics and uncertainty in the process of risk evolution. As a typical and composite system, the city collects complicated risk factors and spatial overlaps. And with highly convergent and intricate coupling relationship of the risk factors, it lead to a highly dynamic uncertainty of urban disaster evolution. Or even worse, disaster evolution has an involute and multivariate disaster structure that may cause huge potential risk for disaster state space and disaster variation space.

14.2.2 Present Situation of Urban Safety and Risk Management in China

The urban safety and risk management refer to the monitoring of the whole system and the whole period of human, buildings and systems in the urban scope based on the vulnerability of disaster carriers for urban emergencies. The occurrence time, space, and intensity of possible disasters should be comprehensively calculated analyzed, forecasting and alarm to make positive risk prevention and control response, actively reduce and prevent the disasters, lower the possibility of urban emergencies or mitigate the casualties and losses caused by disaster events. The process of urban safety and risk management is the process of identifying urban risk factors, analyzing risk and giving human intervention to prevent or reduce urban risks and their consequences [2]. At the present stage, the risk management of urban safety in China has the following drawbacks:

(1) Risk perception has the drawbacks of locality and one-sidedness. Urban risks are complex, human, buildings, systems overlap in time and space, and multi-agent risk mobility, and concealment are strong. In the stage of risk discovery and identification of urban safety, most risk response processes are based on risk control and management after the appearance and outbreak, and it is difficult

to truly realize the active perception of the risk of the whole system and whole period within the urban scope.

(2) Risk analysis is mainly appearing in its empirical and subjective characteristics. Urban risks mostly are multi-body convergence and coupling. At the same time, due to the complex and intensive urban spatial structure, the city shifted from the space with low uncertainty to the space with high uncertainty, which increased the difficulty of risk analysis and prediction. At the present stage, risk analysis and prediction are mainly based on expert experience, qualitative and semi-quantitative analysis, and use static data such as case and survey data for calculation and verification. The ability to portray real-time evolution risks is clearly lacking, and the ability of objectivity and generalization is extremely limited.

(3) The risk response has the drawbacks of passiveness and lag. The evolution process of urban risks presents a highly dynamic uncertainty. In the risk response, there is still a state of passive risk management of "no overall plan for a fundamental transformation," and rarely involves early detection and active control of risk factors, and the accurate response after the risk has emerged.

(4) The drawbacks of the existence of coexistence of multiple institutions and lack of synergy in risk management. At present, urban safety risk management belongs to various departmental channels such as public security, health, transportation, water conservancy, electric power, and propaganda. In the actual risk management work, the lack of communication and sharing of risk information is a common problem. Due to the complexity, relevance and comprehensiveness of the risk issues of urban safety, the status quo of "lack of coordination" of various departments and agencies lead to a lack of overall control over comprehensive risk management, making it difficult to comprehensively utilize decentralized data information for intelligent analysis and effective prevention.

14.3 Opportunities and Challenges in Urban Safety and Risk Management in the Era of Big Data

In the wake of the rapid development of mobile Internet, e-commerce, Internet of things, cloud computing and social media, network information flow are inseparably linked with the human behavior in real life. The integration between the real world and the digital world is getting deeper gradually [3]. Simultaneously, data information is more capable of fully reflecting the actual state of urban operation, along with the wide spread and extensive application of new information equipment in urban management such as the construction of digital city and intelligent city, sensors, intelligent terminals, personnel positioning system, identity identification, wireless sensor network and so on [4]. Big data technology makes it attainable to manage the risk of urban safety throughout the whole system and cycle. Informatization, networking and delicacy have gradually become an inevitable direction of risk management of urban safety. The incessant development of Internet of things, radio frequency, sensor, wireless network and other information technologies provide data sources and bases for risk information of urban safety. Meanwhile various computer systems and big data computing platforms have a winged progress in the aspects of computing power, storage capacity, data mining and computing speed and so on. The development of high-performance computing technology has considerably improved the analytical capability for complex problems. Rapid collection of risk data in every corner of the city and deep sharing and extensive analysis of multi-source and massive risk data provides technical support for the risk management and decision-making of normalization, digital, all-weather, seamless urban safety. Big data impel the transformation of the traditional model of risk management from a qualitative analysis to the quantitative management. New scientific research methods and management decision methods can be derived by means of the intersection of data, methods, knowledge and domains, which can provide more accurate means and capabilities of prediction, inference, and deduction and so on for urban public safety and risk management.

Urban public safety and risk management have got a new historical opportunity following the high-speed development of modern information technology, computing system, comprehensive data base and fast data calculation. In the aspect of risk awareness of urban safety, it is achievable to realize the stereoscopic and omnidirectional status detection [5] and the coverage of big data in the whole system and cycle of the city by the fusion of real-time data collected from different forms and channels. Monitoring and analysis of abnormal risk data can help managers to perceive risk changes and identify risk locations in a timely and comprehensive manner, which is favorable to overcome the situation of high-risk concealment caused by the complex interaction of multi-agent such as human, buildings and systems in the city [6]. In terms of risk calculation and analysis of urban safety, real-time information mining and analysis based on physical space and cyberspace can provide the basis for timely and correct research and judgment of risk development and event situation [7]. The event deduction on account of the full-dimension data association can supply the reliable assurance for the risk control strategy of urban safety [8]. The sharing and interaction of real-time GIS data information, personnel trajectory information [8, 9], urban lifeline system, and emergency rescue information can provide identifiable ground for scientific, timely and effective rescue route planning, relief material allocation and personnel evacuation strategy [10]. In view of risk prevention of urban safety, the problems of multi-sector segmentation and complicated data types are overcome to promote multi-sectorial coordination and integrated management in post-disaster reconstruction through the sharing of risk big data and multi-department cooperation [11].

But at the same time, the explosive surge of risk data information due to the widespread coverage of cameras, sensors, intelligent terminals, etc., will increase the difficulty of the risk information analysis of urban safety. The traditional and simple method based on data and experience has been unable to satisfy the need for accurate and rapid risk management of urban safety. Therefore, it is urgent to establish the real-time analysis method of wide-area big data with the guidance of full-system risk analysis. Big data from different sources has great differences in form, structure, attribute, granularity, etc. The convergence, fusion, analysis and calculation of structured, semi-structured, and unstructured multi-source heterogeneous data are the key to risk analysis with the help of big data [12, 13]. The massive increase in risk data of urban safety and multi-agent interweaving of risk factors of a complex convergence also put forward new requirements for the innovation of traditional urban methods of risk prevention. Risk prevention should accumulate thoroughly on the basis of massive data and model analysis to provide more accurate and quantitative support for the exact and active risk prevention scheme and realize the intelligent and active prevention of urban risks.

Under the influence of the technology driven by big data and method changes, the risk management mode of urban safety has a completely new look from the bottom support mode to the upper method application. Obviously, in the context of the construction of urban delicacy management, it is imperative to take full advantage of the new historical opportunity of the urban public safety and risk management brought by the present big data era, convert the traditional experience-oriented mode of risk management and establish the urban public safety and risk management with the support of big data technology.

14.4 The Proposal of Paradigm for Urban Safety and Risk Management with Big Data

Based on the analysis of the basic theories and status quo of urban public safety risks, combined with the core ideas of public safety management, and integrating the new opportunities of urban safety and risk management in the era of big data, the urban safety and risk management paradigm are constructed.

14.4.1 Characteristics of Urban Risks

Various kinds of urban components may make the hidden risk factors obvious. Traditional urban risk management built on experience is partial and one-sided. Simultaneously, risk factors in the city have perfect mobility and urban public emergencies possess the characteristic of significant emergence. As a consequence, the most important task of urban safety and risk management is to carry out the full-cycle and whole system risk monitoring and control from the source in order to avoid changing from unremarkable urban risk to tragic disaster. Timely perception of risk, control of potential urban risk factors, the suppression of occurrence and evolution of risk as early as possible are critical factors to the urban safety and risk management in the stage of risk inoculation.

The state and changes of the monitoring objects are obtained by the full-cycle and whole system monitoring and control on human, buildings, and systems in the city (The status quo of the research objects is obtained through the state monitoring. While the possible change in future is predicted by the object's change or track monitoring). The premise is to obtain the big data of the human, buildings, systems, and environment within the urban coverage by means of monitoring and control. The purpose is to mine the abnormal information in the urban real-time monitoring data with the assistance of big data analysis methods such as data fusion, semantic association analysis, data mining and so on, through the aggregation and fusion of multi-source and heterogeneous data. Meanwhile risk factors and space–time information in the data are extracted, which will break through the bottleneck of the locality and one-sidedness of risk perception of urban safety and actualize the overall perception of urban risks.

14.4.2 Accurate Prediction of Urban Risk Based on Forecasting and Alarm

The urban risks are highly converged and complexly coupled and interrelated. The evolution of urban public emergencies is dynamic and uncertain, with complex and diverse disaster structures, huge potential risk disaster state space and disaster variation space. Therefore, when the risk factors are perceived, make timely and accurate forecasting and alarm about the types of disasters, the consequences of disasters, the locations and the scope of impacts that may be triggered, as well as the prediction of the evolution of disasters and the effects of different rescue measures when disasters occur, so as to realize active response to emergency rescue and targeted.

With the diversity and complexity of the risk issues of urban safety, the demand for risk forecasting and alarm has also developed from qualitative to quantitative. Forecasting and alarm refer to conduct the effective forecast, risk quantitative forecast and risk evolution deduction of urban risks based on the monitoring and control big data of the risk, and through the quantitative analysis of disasters, large-scale and rapid numerical calculation and situational deduction and simulation, which is the key to the prevention and control of urban public emergencies. The goal of forecasting and alarm is to accurately predict the probability and evolution of risk by means of the analysis and calculation of the monitoring and control data, including what disasters may occur, the possibility of disasters, the damage that disasters may bring, the evolution of disasters and the impact of different measures on the evolution of disasters. The purpose is to break through the current risk analysis of urban safety, which is mainly based on the qualitative or semi-quantitative methods in the present situation, reduce the uncertainty of the risks, and the risk of high uncertainty will become low uncertainty risk, in order to provide the basis and guidance for risk prevention and control.

14.4.3 *Collaborative Response to Urban Risks Based on Intelligent Prevention*

The traditional risk management mode of urban safety has the disadvantages such as single method and organization, isolated resources and departments, and lack of multi-agent collaborative response mechanism. Insufficient communication and sharing of risk information lead to the lack of overall and global control of comprehensive urban risk management, making it difficult to use distributed data information for intelligent analysis and effective prevention. Therefore, on the basis of comprehensive risk perception and accurate prediction, overcome the disadvantages of the current risk response passiveness and lag in urban safety management, comprehensively consider the risk factor correlation, business attribute intersection, and realize the information sharing and rational allocation of resources under different actions, fields and permissions, and adapt to multi-agent online collaboration and dynamic coordination of different events, at different stages and different scenarios, as well as the intelligent prevention of the urban risks.

Based on monitoring and control and forecasting and alarm, intelligent prevention integrates urban historical information data, monitoring and control real-time data and the calculation analysis of forecast and early warning, by means of symptom identification and early warning of different types of urban risk emergencies to achieve target-driven and task-driven accurate response plan for urban risks, and to achieve the gateway forward, early warning and intelligent prevention of urban risk management. At the same time, on the basis of data integration analysis, through the construction of integrated urban risk intelligent prevention platform combining police, government affairs and property management, the data-business-service entire process integrated management and interdisciplinary collaborative operation are realized, in order to achieve unified, standardized and multi-level linked integrated information sharing and multi-sector multi-agent collaborative management, so as to conduct comprehensive prevention of urban disasters in advance, make effective rescue in the event and perfect recovery afterward.

14.5 Discussion and Conclusion

China is in a period of economic and social transformation, the foundation of urban public guarantees is relatively weak, the accidents are frequent, and urban safety is facing severe challenges. With the rapid development of big data technology, each city should actively grasp the new opportunities of risk management brought by big data and carry out the construction of safety and risk management in the era of big data.

In terms of risk perception, establishing a sound monitoring network, developing monitoring and control capabilities, and consolidating the city's big data base. The comprehensive coverage of monitoring and control equipment is the basis for obtaining big data on urban risks. In response to the demand for the fine management of urban safety, increasing the investment in monitoring equipment and system, establishing and improving urban safety monitoring networks for public emergencies in different cities, including urban pipe network monitoring systems, rail transit monitoring systems, and meteorological monitoring systems, to eliminate blind spots in monitoring of the perception facilities and sensor networks, realize full period and system coverage of urban data information and the real world, and fully grasp the risk big data of disaster carriers (human, buildings, and systems) in urban. Developing safety monitoring and controlling capabilities, increasing the research and development and investment of monitoring technology, risk spatial and temporal distribution methods to achieve timely and comprehensive awareness and accurate positioning after the risk is revealed.

In terms of risk prediction, improving the ability of big data integration and analysis, breaking through the core technology bottleneck of big data, and strengthening the training, introduction and guarantee system of professional and technical personnel. On the basis of the comprehensive collection of urban risk big data, through the integration and analysis of multi-source heterogeneous data, fully exploit the effective information in the data, so as to truly play the value of risk big data. Focusing on "collecting, communicating, and using" and its security support to break through the core technical bottlenecks such as data collection, aggregation, storage, cleaning, integration and analysis, and provide support for urban public emergency decision-making. At the same time, the smooth operation of urban safety and risk management is inseparable from the scientific research and innovative thinking of big data technology talents, who are the driving force and blood of urban innovation and development. Through the improvement of the training system for big data related professional and technical personnel, the establishment of big data disciplines and related majors, increase training, and reserve big data professional, and technical personnel. Implement practical talent introduction programs, including methods such as simple point system, high-level settlement subsidies, and low-cost housing. And a more targeted selection, use, education, and retention process, a comprehensive talent security system and policies, providing a good development space for professional and technical personnel, providing protection support from the perspectives of medical, pension, housing, and children's education.

In terms of risk response, build and improve the urban safety data resource information sharing mechanism and the deep collaborative and linkage mechanism of multi-department and agent to build a big data sharing service platform. In the event of urban emergencies, it is often necessary to coordinate with various government departments such as public security, transportation, communications, first aid, electric power, and urban management. Therefore, it is necessary to improve and perfect the deep synergistic linkage mechanism of multiple government departments to achieve cross-regional and cross-department, multi-agent unified command, rapid response, coordinated prevention, joint action. At the same time, the multi-departmental high-coordination linkage response must be based on the deep sharing of risk big data. Therefore, it is necessary to establish a resource database and a big data sharing service platform to form a resource complementing and information sharing operation mechanism under the unified data management system. Confirm the boundaries, responsibilities, systems, and rules of data integration and sharing to ensure that urban safety data resources are exchanged and shared reasonably and efficiently, eliminating "data island," maximizing data collection, fully exploiting the added value of data, and comprehensively ensuring urban safety.

Acknowledgements This study is supported by National Key R&D Program of China (No. 2018YFC0809700), National Natural Science Foundation of China (No. 71904095), Beijing Natural Science Foundation (No. 9194027) and Basic Research Fund (2020JKF501).

References

1. Hall, S.W., Carolla, M.A., Deason, J.P.: The use of imagery in environmental disaster preparedness and response. Fedl. Fac. Environ. J. **16**(4), 65–72 (2010)
2. Kiekens, R.M., Maltha, J.C., van't Hof, M.A., et al.: Panel perception of change in facial aesthetics following orthodontic treatment in adolescents. Eur. J. Orthod. **30**(2), 141–146 (2008)
3. Wu, C., Jia, N.: Research on safety humanology connotation and basic principles. J. Saf. Environ. **6**, 153–158 (2016)
4. Comfort, L.K., Springer, C.G.: Emergency management research and practice in public administration: emergence, evolution, expansion, and future directions. Public Adm. Rev. **4**, 539–547 (2016)
5. Wang, Y.: The development of wireless personnel positioning in internet of things based on ZigBee and sensors. Int. J. Digit. Content Technol. Appl. **6**(12), 47–54 (2012)
6. Bouchrika, I., Carter, J.N., Nixon, M.S.: Towards automated visual surveillance using gait for identity recognition and tracking across multiple non-intersecting cameras. Multimed. Tools Appl. **75**(2), 1201–1221 (2016)
7. Mao, G., Fidan, B., Anderson, B.D.O.: Wireless sensor network localization techniques. Comput. Netw. **51**(1), 2529–2553 (2007)
8. Geng, H., Wu, Z., Ji, H., et al.: Risk zonation and assessment to flood disaster in Jiading District of Shanghai. J. Catastrophol. **30**(1), 96–101 (2015)
9. Deng, Q., Liu, Y., Zhang, H., et al.: A new crowdsourcing model to assess disaster using microblog data in typhoon Haiyan. Nat. Hazards J. **84**(2), 1–16 (2016)
10. Zhang, W.Y., Hai, Y.U., Zhang, F., et al.: Personnel positioning system of underground coal mines based on the ZigBee technology. J. Hefei Univ. Technol. **30**(9), 1087–1090 (2007)
11. Hashem, I.A.T., Chang, V., Anuar, N.B., et al.: The role of big data in smart city. Int. J. Inf. Manage. **36**(5), 748–758 (2016)
12. Troy, D.A., Carson, A., Vanderbeek, J., et al.: Enhancing community-based disaster preparedness with information technology. Disasters **32**(1), 149–165 (2010)
13. Zhang, F., Zhou, Z., Xu, W., et al.: Cloud manufacturing resource service platform based on intelligent perception network using fiber optic sensing. Adv. Inf. Sci. Serv. Sci. **4**(4), 366–372 (2012)

Chapter 15
Network Teaching Based on Multimedia Technology: An Example of Ideological and Political Theory Course

Yu Fang, Jing Dai, and Peng Shao

Abstract The cultivation of college students' ideological education mainly depends on the course of ideological and political theory, and the course of ideological and political theory also serves as the main export way to cultivate the scientific, practical and humanistic nature of college students, and promote them to become the spiritual pillar of a new generation. Ideological education follows the form of seeking change and innovation, adapting to development, and teaching students in accordance with their aptitude. Teachers have also changed their teaching methods. Using multimedia technology to update teaching methods can arouse students' learning enthusiasm and stimulate their learning enthusiasm. However, because teachers' thinking teaching methods and curriculum analysis do not pay attention to students from multiple perspectives, the teaching effect is lack, and the ideal state has not been reached after the reform. Therefore, in the ideological and moral classroom based on multimedia technology, students' subjective initiative and creativity should be fully mobilized to avoid the teacher-centered theoretical situation. Based on the multimedia technology platform, teachers combine online teaching with offline teaching to form a new educational concept. Let students have more opportunities to interact with teachers, improve students' cognition and acceptance of ideological and political theory courses, enhance the emotion of teachers and students, and promote the development of ideological and political theory courses. Under the new and old concept, students have a new experience, the corresponding professors and teachers have a new ability to improve. This virtuous circle promoted the development of ideological and theoretical education. That's a nudge.

15.1 Introduction

With the society, schools, teachers, and students paying more and more attention to the newly implemented educational methods, many schools take the ideological and political ethics course as the focus of reform, and there will be friction under the

Y. Fang (✉) · J. Dai · P. Shao
Jiangxi Tourism and Commerce Vocational College, Jiangxi, China

L. C. Jain et al. (eds.), *Wireless Technology, Intelligent Network Technologies, Smart Services and Applications*, Smart Innovation, Systems and Technologies 258,
https://doi.org/10.1007/978-981-16-5168-7_15

alternation of the old educational concepts. The previous education mode was single and inflexible, which resulted in a teacher-centered teaching mode [1]. As a result, the students of streamlined education stagnate or even regress. This is the reason why many colleges and universities can't finish the enrollment rate, teachers can't bring out students, and students' grades are low. Under the tide of contemporary big data Internet, it is urgent for ideological and political education to adapt to the reform of the times. The new ideological and political education attracts students to learn ideological and political education through new teaching methods and colorful teaching thinking. Stimulate students' independent study and teachers' independent research and innovation. Teachers with new ideas and students with strong self-awareness collide with each other. The new teaching mode strives to give full play to students' subjective consciousness and stimulate students' autonomous learning. Through the online education mode of the new media platform, students' thirst for knowledge is opened and the interaction between students and teachers is stimulated. At the same time, it also improves the professional ability of teachers. Under the new education mode and concept, with the joint efforts of teachers and students, it has also delivered talents for this field and laid a solid foundation for research and development projects and field development. Therefore, the ideological and political education under the new media is the key research direction in the future [2].

15.2 Questionnaire Analysis

The project team distributed and recovered 550 valid questionnaires, which constituted the total sample size of this investigation and analysis (see Table 15.1) [3]. To a certain extent, the samples can represent the overall situation of the application of multimedia in ideological and political courses by teachers at different levels of colleges and universities. The results of the investigation are reported as follows, in order to provide reference for relevant universities to better carry out ideological and political courses with the help of the Internet. The project team designed three questions to investigate the current development of ideological and political courses, namely "How much do you think students are interested in ideological and political courses?" "Do you know about online teaching, and do you often use the network to carry out teaching activities?" "How do you evaluate the overall situation of the application of online teaching of ideological and political courses in your university?" Through the exploration of three problems, this paper finds out several problems existing in the ideological and political courses in colleges and universities [4].

Table 15.1 Basic information of samples ($N = 550$)

Project	Category	Number	Percent (%)
Sex	Woman	340	61.82
	Man	210	38.18
Teaching age	1–5 years	140	25.45
	6–10 years	58	10.55
	11–15 years	109	19.82
	16–20 years	51	9.27
	More than 20 years	332	34.91
At the university level	"211" or "985" project colleges and universities	74	14.8
	Main construction, university jointly built by province and ministry	103	18.73
	Common undergraduate course college	176	32
	Independent college, private undergraduate college	31	5.64
	Vocational college	166	28.83

15.2.1 The Survey Found That Most Students Are Not Interested in Ideological and Political Courses

"How much do you think students are interested in ideological and political courses?" 3.64% choose "very interested," 40.18% are interested, 52.73% are not interested, and 3.45% are not interested at all. If the sum of "very interested" and "interested" is used as a measure of interest, then the interest of college students in ideological and political courses is 43.82%, while 56.18% of students are not interested in ideological and political courses (see Table 15.2) [5].

Table 15.2 Students' interest in ideological and political courses

Option	Subtotal	Percent (%)
Be full of interest	20	3.64
Be interested in	221	40.18
Lose interest in	290	52.73
Not at all	19	3.45

Table 15.3 Teachers' understanding of multimedia technology

Option	Subtotal	Percent (%)
Know, often help	157	28.55
Know, not often	342	62.18
Not understand	51	9.27

15.2.2 Teachers Use Multimedia Teaching

The survey found that most teachers didn't understand the use of multimedia, among which 28.55% chose "understand and often use it," 62.18% "understand but not often use it" and 9.27% "don't understand it" [6].

From the results of the questionnaire survey, although the teachers of ideological and political courses in colleges and universities know about online teaching, they do not often use it. To assist the teaching of ideological and political courses see Table 15.3.

15.2.3 The Survey Found That the Ideological and Political Course Teaching in Colleges and Universities Is Not Optimistic

According to the survey, among the teachers of ideological and political courses in colleges and universities, 4.36% of them are "excellent," 26% are "good," 43.28% are "moderate" and 26.36% are "poor." According to the survey results, the overall situation of ideological and political course teaching in colleges and universities is not very good (see Tables 15.4 and 15.5).

Table 15.4 The overall situation of the application of online teaching of ideological and political courses in schools

Option	Subtotal	Percent (%)
Excellent	24	4.36
Good	143	26
Medium	238	43.28
Poor	145	26.36

Table 15.5 Actual use of multimedia teaching

Option	Subtotal	Percent (%)
Network teaching is main, teacher explanation is auxiliary	23	4.18
The teacher explains primarily, the network teaching is auxiliary	401	72.91
Student-led, teachers and network multimedia as auxiliary	126	22.91

Table 15.6 Interactive aspects

Option	Subtotal	Percent (%)
Browse the web, look up and retrieve relevant information	469	85.27
Proficient in PPT courseware design and making	471	85.55
The teaching video can be edited and recorded	181	32.86
Familiar with microblog, WeChat, QQ, news headlines, and other software	357	64.87
Less skilled	90	16.43

Although the vast majority of teachers think that they are skilled in using network information technology, only 32.86% of teachers have made choices in recording and editing multimedia teaching videos assisted by network (see Table 15.6), and 98.02% of teachers have contacted and used multimedia. However, from the results of the questionnaire, teachers' choices for teaching work are mostly simple and boring multimedia presentations such as PPT. In this teaching activity, animation videos are not inserted and students interact with each other, and teachers only serve as "explanations" [7].

From the data analysis of the three tables, it can be concluded that 52.73% of the students are not interested in the subject of ideological and political education, because teachers have not mastered the teaching methods and channels correctly. With the popularity of the Internet, students are no strangers to online education, which proves that students' cognition is acceptable in teaching online courses. But why only 4.36% are satisfied with the development of online courses? 62.18% of teachers know the Internet but lack the ability to apply multimedia technology. Online teaching can't always perceive students' absorption ability like face-to-face, so they can't put teaching first. Teachers should integrate face-to-face experience and multimedia technology into application, develop colorful teaching methods, and interact with each other vividly, so as to drive a lively learning atmosphere, change the dull and dull atmosphere, stimulate students' interest in ideological and political affairs, and cultivate students' autonomous learning ability. Creating a new teaching atmosphere and learning style will naturally make the development of thinking class by going up one flight of stairs [8, 9].

15.3 Conclusion

Ideological and political education needs the joint efforts of teachers and students to be promoted. Teachers should adapt to the new teaching mode, update the teaching methods, adapt measures to local conditions and teach students in accordance with their aptitude. At the same time, "teacher-centered" should be changed into "student-centered" so that students from passive learning to active learning, more conducive to the realization of teaching objectives. Network education is the necessity of developing and reforming traditional education. Teachers play a guiding role in the ideological and political classroom in the new era, and with the popularization of new education, lead students to stimulate new growth, improve and optimize the ideological construction of students, and really apply ideological and political learning resources.

References

1. Zhang, L.: The relationship between self-efficacy and job burnout of ideological and political teacher based on quantitative and qualitative analysis. NeuroQuantology **16**(6) (2018)
2. Qinli, F., Ming, Z.: Building of the college ideological and political education joint forces system based on four identifications. Analyst 2855–2858 (2011)
3. Kemeny, J., Zeitler, B.: Another node on the internet. Comput. Geosci. **27**(9), 1119–1121 (2001)
4. Pujiastuti, H., Fitriah, F.: Design of interactive teaching materials based on a scientific approach to support junior high school students' learning: line and angles. J. Phys. Conf. Ser. **1157**(3), 032087 (6 pp.) (2019)
5. Saddhono, K., Suhartatik, Bagiya, et al.: Learning vocabularies using multimedia-based Teaching Indonesian to Speakers of Other Languages (TISOL). J. Phys. Conf. Ser. **1339**(1), 012108 (5 pp.) (2019)
6. Fan, H., Brief, M.A.: Analysis of the integration scheme of ideological and political teaching and multimedia technology. Wen Yuan (High Sch. Ed.) (002), 213 (2019)
7. Li, L.J.: Exploration and practice of online course construction of ideological and political theory courses in universities based on the cloud space of "world university town". J. Multimed. Netw. Teach. China (03), 22–23 (2019)
8. Wang, J.: Current situation, causes and countermeasures of online teaching of ideological and political theory courses in colleges and universities question and research, No. 1008(33), pp. 28–28 (2019)
9. Yuan, X.: On strategies of multimedia application in ideological and political theory course. Wen Yuan (High Sch. Ed.) (005), 441 (2019)

Chapter 16
Supervised Learning Epidemic Threshold of SIR Model in Complex Networks

Jie Kang and Ming Tang

Abstract Classifying phase transitions of epidemic outbreaks in different kinds of complex networks is a central problem in network dynamics research. Deep learning methods can be used to identify phases and phase transitions in complex networks via supervised machine learning. However, most studies recently published focus on dynamical information of a single node. As a matter of fact, structural features in complex networks also play a significant role in dynamics progress. In this paper, we propose a novel deep learning framework to combine the structural and dynamical information into an image with multiple channels. Then, convolutional neural network (CNN) is used to find phase transition depending on supervised learning labeled image data. By training on regular random network data and scale-free network data, we show our machine learning framework can learn the epidemic threshold of SIR model in a high accuracy and robustness. What's more, complex networks with arbitrary topology and size and real networks can be used universally.

16.1 Introduction

During the last few years, the application of methods and techniques of deep learning (DL) in complex networks has caused a great deal of concern [1–5]. As for the advances in hardware and processing power of computers, which most notably through the use of the GPUs, there were a number of techniques with remarkable abilities to deal with complex networks' high dimensional data. We can use feature extractors to learn features of the complex network automatically, and it has

J. Kang (✉)
School of Communication and Electronic Engineering, East China Normal University, Shanghai 200241, China
e-mail: kangjiechn@163.com

M. Tang
Shanghai Key Laboratory of Multidimensional Information Processing, East China Normal University, Shanghai 200241, China

State Key Laboratory of Precision Spectroscopy and School of Physics and Electronic Sciences, East China Normal University, Shanghai 200241, China

L. C. Jain et al. (eds.), *Wireless Technology, Intelligent Network Technologies, Smart Services and Applications*, Smart Innovation, Systems and Technologies 258,
https://doi.org/10.1007/978-981-16-5168-7_16

been widely used in physics, sociology, economics, biology, etc. [6]. At present, Network representation learning was mostly used for learning the structural data. It can preserve structure information intact in complex networks by network embedding, such as DeepWalk [7], Node2Vec [8], LINE [9], SDNE [4], and so on. However, the applications of these methods are only on nodes such as node classification [10], clustering [11], and link prediction [12].

Many real complex systems can be properly described by complex networks through nodes and links, such as cooperation network, traffic network, social and communication network. Nodes represent individuals and links generally represent the interactions among other nodes [13]. We can fall complex networks into two categories [14], mainly according to the connectivity properties. The first one is exponential networks. A typical example is the random graph model proposed by Erdős and Rényi [15], such as regular random (RR) network. The other one is the scale-free (SF) networks. In particular, we will focus here on the Barabási and Albert (BA) model [16], whose connectivity distribution is $P(k) \sim k^{-3}$ and depict real networks better. Susceptible-infected-removed (SIR) epidemic spreading model is a common infectious disease model, and it has certain reference value to analyze the development trend of Coronavirus Disease 2019 (COVID-19) [17].

Phases and phase transitions are widely present in epidemiological dynamics in complex networks. It is worth noting that propagation dynamics in complex networks are similar with the Ising model. The work of Carrasquilla et al. proposed a standard feed-forward neural network and convolutional neural network (CNN) framework [18], which can detect the Ising model's critical temperature precisely [19]. Especially, imagelike data of Ising model is able to store the topological phases in condensed matter systems. We have already applied deep learning methods to identify the threshold of SIS model dynamics [20, 21]. All these accomplishments were demonstrated that the threshold and the critical point phase transition in the different model can be discovered through these deep learning methods.

In this paper, we exploit a CNN-based deep learning method to learn and identify phases transitions of SIR epidemic spreading in complex networks. There are many differences between the SIR model and the SIS model. The microdynamical state in the complex network of SIR model is more complex because of an extra recovery (R) state. On the other hand, the numerical simulation of SIR model is an irreversible dynamic process with an endpoint. So far, there exist many theoretical studies on its epidemic threshold, but that's not always the case with numerical simulations. Based on the assumption of homogeneous mixing, the numerical threshold of the SIR model can be defined as [22],

$$\lambda_c^{\mathrm{HMF}} = \frac{\langle k \rangle}{\langle k^2 \rangle - 2\langle k \rangle}$$

where the value of $\langle k \rangle$ and $\langle k^2 \rangle$ are the first and second moments of the complex network's degree distribution, respectively.

The rest of the paper is organized as follows. In Sect. 16.2, we introduce the epidemic threshold of SIR model and the proposed CNN-based deep learning framework. Section 16.3 presents the results and performance of the proposed model by supervised learning method. Finally, Sect. 16.4 is devoted to the analysis of robustness and Sect. 16.5 gives our conclusion in this work and discussion for further research studies.

16.2 Basic Definitions and Models

In this section, we will introduce the standard SIR spreading model and deep learning method that are used in our work, as well as the structure of the training imagelike dataset.

16.2.1 SIR Epidemic Model and Threshold

The standard SIR model [23] is suitable for two situations: the first is that patients can obtain lifelong immunity after cure, such as mumps, measles, and smallpox. The second is death is almost inevitable, such as AIDS. In the first spreading process, an infected node can transmit a disease to its neighbors whose state is susceptible with the infection rate β. Then, in the next moment, suppose that the infected nodes become removed with a steady rate γ. The basic reproduction number is defined as the average number of other susceptible neighbors that an infected node can infect before being removed $\lambda = \frac{\beta}{\gamma}$. In this work, we suppose $\gamma = 1$ which means an infected node will become removed in the next time step, and the threshold value of λ, denoted as λ_c. The active state and absorbing state are two phases in the epidemic-spreading dynamic. When the basic reproduction number $\lambda < \lambda_c$, the system tends to approach the absorbing state in due time and an epidemic will be limited in very few nodes. When $\lambda > \lambda_c$, the disease in the system will outbreak and be in the active phase. A finite percentage of nodes on the network will eventually infect an epidemic. The SIR epidemic model is an irreversible spreading dynamics, so the system in the final state has no infected node. The differential equation is described as follows:

$$\begin{cases} \frac{ds}{dt} = -\beta si \\ \frac{di}{dt} = \beta si - \gamma i \\ \frac{dr}{dt} = \gamma i \end{cases}$$

Numerical simulations are typically employed in the study of phase transitions. Variability measure is a standard measure to determine the threshold of SIR model [24]:

$$\Delta = \frac{\sqrt{\langle \rho^2 \rangle - \langle \rho \rangle^2}}{\langle \rho \rangle}$$

where ρ is denoted as the outbreak size, $\langle \rho \rangle$ and $\langle \rho \rangle^2$ denote the first and second moments of ρ in a large number of realizations, respectively. The peak of the variability is an indication of the SIR epidemic threshold λ_p^Δ, which gives the theoretical threshold. The training data sets are obtained by Monte Carlo simulation of SIR model.

16.2.2 Proposed Deep Learning Framework

We propose a CNN-based machine learning framework to identify the threshold of SIR dynamics model by supervised learning method. First of all, we embed the high dimensional network into 2-dimensional space. Our strategy to use the CNN model is to convert the network into an image with epidemic dynamical information (S state and R state) added by 2 channels. After that, we feed the representative images into the CNN to indicate how possible the input image is in one specific phase.

In the following text, specific content in detail of our proposed framework is expounded and why it is powerful in the model.

a. Network Representation Learning (Network Embedding): Due to the development of network representational learning, we have a wide selection of network embedding algorithms. We select Poincaré embedding [25], which is suitable to model hierarchical structure, and which is existing in real complex networks widely [26, 27]. In the Poincaré ball model, the information of hubs is able to maximize and the complex network is represented with a hierarchical structure.
b. Discretize network embedding and add nodes' dynamical information: Since the bivector of nodes in the complex network is a point cloud in a 2-dimensional space and cannot be processed by CNN, we then take forward to rasterize the 2-dimensional scatter into an image [28, 29]. We can divide the areas into 48 * 48 grids and then make the pixel grayscale in the image as the number of nodes which is S/R state in each grid in 2 channels, respectively. In this manner, a grayscale image with 2 channels obtained contents both the structural information of the overall complex network and nodal dynamical states. We consider the image data sets with a certain value of λ consists 1000 SIR dynamical simulations. As is represented in Fig. 16.1, the Zachary's Karate Club network [30] after a spreading dynamical simulation is converted into a grayscale image.
c. CNN architecture: The proposed CNN model is able to learn the structural and dynamical information from the training image data. And in the test image data, the model can predict the probability of test images which phase it belongs to. In this work, the size of input images is (48, 48, 2), where 2 is the number of channels. The CNN architecture contains 2 convolution layers, 2 max-pooling layers. The size of the convolutional filter is 3 * 3, and the size of the pooling

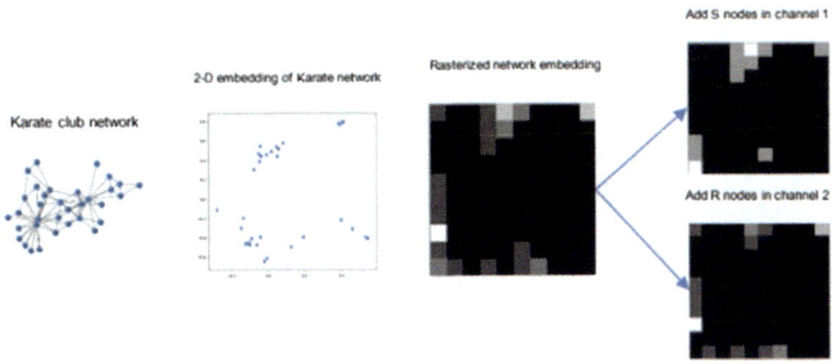

Fig. 16.1 The pipeline of embedding Karate Club network into 2-D image with 2 channels

kernel is 2 * 2. The first convolutional layer has 32 convolutional filters, and the second convolutional layer has 64 convolutional filters, and the fully connected layer has 128 units. ReLU activation function map data in a nonlinear way to prevent gradient vanishing. In addition, we use Dropout [31] and L2 regularization to avoid overfitting. The ratio of dropout is rare $= 0.5$, and regularization coefficient is $l2 = 0.01$. Adam optimizer [32] is used to improve the learning efficiency. The output layer has only one unit which gives the probability of the phase constrained from 0 to 1 by sigmoid function. Specifically, the closer the output is to 0 (or 1), the more likely the neural network thinks that the current input system state is in the absorbing (or active) phase. Other hyper-parameters are set as: batch size $N_b = 128$, learning rate $\alpha = 0.001$. This architecture can be easily built from Tensorflow [33].

16.3 Learning by Supervised Method

We conduct a large number of synchronous updating Monte Carlo simulations of SIR model to obtain the training and test image data set, and the CNN's predicting results of a RR network and a BA network are present in this section. Provided that we have the image set is determined along a tuning parameter λ, which also controls the image's true label (less than the threshold is denoted as 0, greater than the threshold is denoted as 1). Thus, a supervised learning method is used to learn the relationship with training data and labels. After multiple training the parameters in the CNN, the test image set will obtain the probability of its label. When λ is far less than λ_c, the neural network predicts the labels' results of 0. When λ is far more than λ_c, the neural network predicts the labels' results of 1 accurately. However, when λ is close to the true threshold λ_c, the accuracy will be significantly reduced. The predicted output of 50% represents the threshold of the SIR model of the complex network.

The first task is applied in a homogeneous complex network. We generate a RR network with $N = 1000$, $\overline{k} = 10$. As is shown in Fig. 16.2a, when λ is far less than λ_c, the result of output layer is 0, and when λ is far more than λ_c, the result of

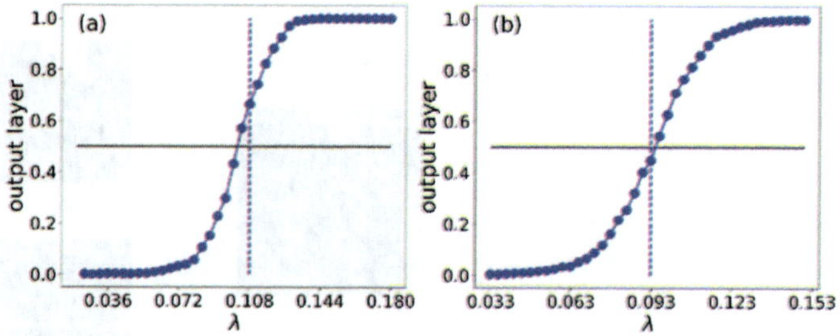

Fig. 16.2 Learning phase transition on a homogeneous and heterogeneous network by supervised method. **a** For a RR network of size $N = 1000$ and average degree $\bar{k} = 10$. **b** For a BA network of the same size $N = 1000$ and $m = m_0 = 3$. The function of the average results of output layer and infection rate λ for the SIR spreading dynamics is as shown. The intersection of the dotted blue line and the output layer's predicted probability of 0.5 is basically no difference

output layer is 1. The CNN can predict the system dynamical state accurately. The solid black line represents the predicted probability of 0.5, and the dotted blue line represents the variability index λ_p^{Δ} of the random network, $\lambda_c = \lambda_p^{\Delta} = 0.108$. At the average predicted probability of 0.5, the result of the supervised learning method is $\lambda_c^{'} = 0.104$, which is very close to the actual threshold value.

The BA network is a typical heterogeneous network, the result is shown in Fig. 16.2b. For the BA network with the same size ($N = 1000$) and $m = m_0 = 3$, the peak of the variability is $\lambda_c = \lambda_p^{\Delta} = 0.093$. The result of supervised learning shown in solid black lines is $\lambda_c^{'} = 0.093$, which is indistinguishable with the theoretical threshold. In general, we can say the proposed method can detect phase transition accurately for homogeneous and heterogeneous complex networks. Theoretically, this method can be applied to data of various network structures.

16.4 Robustness Analysis

In order to test the robustness of the supervised learning method against noises, we generate a proportion of labels inverted for the training image set, and then we observe the CNN's predicted output layer and the theoretical threshold λ_p^{Δ}. We systematically use three error rates of labels, which are 2, 10, and 20%, which are visualized in Fig. 16.3. The results of the experiments are shown in Fig. 16.3. This CNN framework has good robustness. The results of output layer change very little, as the error rate of labels increases. Experimental results show that the proposed model is effective and has a good performance.

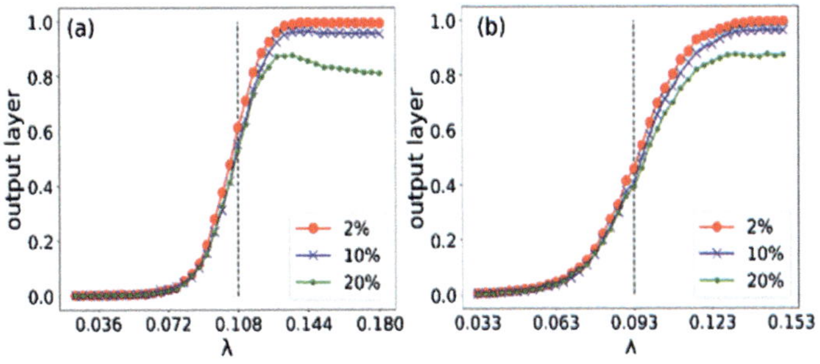

Fig. 16.3 Robustness analysis of supervised learning method. The dashed black line represents the theoretical threshold λ_p^{Δ}. Circles, triangles, and crosses represent the results for 2, 10, and 20% error rates

16.5 Conclusion and Discussion

In this paper, we propose a CNN-based framework, which mainly incorporates complex network and network representation learning, to learn and identify the phases transition and the threshold of SIR model's epidemic dynamics. With Poincaré embedding, we obtain an image for each SIR dynamical simulations in a complex network, and then CNN framework is used to implement the binary classification task of SIR epidemic dynamic's two phases. Our method performs well on homogeneous and heterogeneous networks. The method is robust, well-behaved, and computationally efficient.

References

1. Perozzi, B., Al-Rfou, R., Skiena, S.: In: Proceedings of the 20th ACM SIGKDD International Conference on Knowledge Discovery and Data Mining, pp. 701–710. ACM (2014)
2. Grover, A., Leskovec, J.: In: Proceedings of the 22nd ACM SIGKDD International Conference on Knowledge Discovery and Data Mining, pp. 855–864. ACM (2016)
3. Kipf, T.N., Welling, M.: Semi-supervised classification with graph convolutional networks. e-print arXiv: pp. 1609.02907 (2016)
4. Wang, D., Cui, P., Zhu, W.: Structural deep network embedding. In: Proceedings of the 22nd ACM SIGKDD International Conference on Knowledge Discovery and Data Mining, pp. 1225–1234. ACM (2016)
5. Hamilton, W.L., Ying, R., Leskovec, J.: IEEE Data Eng. Bull. **40**(3), 52–74 (2017)
6. Newman, M.E.J.: Networks: An Introduction. Oxford University Press, Oxford (2010)
7. Perozzi, B., Alrfou, R., Skiena, S.: DeepWalk: online learning of social representations, pp. 701–710 (2014)
8. Grover, A., Leskovec, J.: node2vec: scalable feature learning for networks, p. 855 (2016)
9. Tang, J., Qu, M., Wang, M., Zhang, M., Yan, J., Mei, Q.: LINE: Large-Scale Information Network Embedding, 2, pp. 1067–1077 (2015)
10. Bhagat, S., Cormode, G., Muthukrishnan, S.: Social Network Data Analytics, pp.115–148. Springer (2011)

11. Ding, C.H., He, X., Zha, H., Gu, M., Simon, H. D.: In: Proceedings IEEE International Conference on Data Mining, ICDM 2001, pp. 107–114. IEEE (2001)
12. Liben-Nowell, D., Kleinberg, J.: The link-prediction problem for social networks. J. Am. Soc. Inf. Sci. Technol. **58**, 1019 (2007)
13. Jordan, M.I., Mitchell, T.M.: Machine learning: trends, perspectives, and prospects. Science **349**, 255 (2015)
14. Romuado, P.-S., Alessandro, V.: Epidemic dynamics and endemic states in complex networks. Phys. Rev. E **63**, 066117 (2001)
15. Erdős, P., Rényi, A.: On the evolution of random graphs. Publ. Math. Inst. Acad. Sci **5**, 17 (1960)
16. Barabási, A.-L., Albert, R.: Emergence of scaling in random networks. Science **286**, 509 (1999)
17. National Health Commission of the People's Republic of China. Accessed 2 Mar 2020
18. Krizhevsky, A., Sutskever, I., Hinton, G.E.: ImageNet classification with deep convolutional neural networks. Commun. ACM **60**, 2 (2012)
19. Carrasquilla, J., Melko, R.G.: Machine learning phases of matter. Nat. Phys. **13**, 431 (2017)
20. Ni, Q., Tang, M., Liu, Y., Lai, Y.-C.: Machine learning dynamical phase transitions in complex networks. Phys. Rev. E **100**, 052312 (2019)
21. Ni, Q., Kang, J., Tang, M., Liu, Y., Yong, Z.: Learning epidemic threshold in complex networks by convolutional neural network. Chaos **29**, 113106 (2019)
22. Barthélemy, M., Barrat, A., Pastor-Satorras, R., Vespignani, A.: Velocity and hierarchical spread of epidemic outbreaks in scale-free networks. Phys. Rev. Lett. **92**, 178701 (2004)
23. Kermack, W.O., McKendrick, A.G.: A contribution to the mathematical theory of epidemics. Proc. R. Soc. Lond. A **115**, 700–721 (1972)
24. Shu, P., Wang, W., Tang, M., Do, Y.: Numerical identification of epidemic thresholds for susceptible-infected-recovered model on finite-size networks. Chaos **25**, 063104 (2015)
25. Nickel, M., Kiela, D.: In: Advances in Neural Information Processing Systems, pp. 6338–6347 (2017)
26. Muscoloni, A., Thomas, J.M., Ciucci, S., Bianconi, G., Cannistraci, C.V.: Machine learning meets complex networks via coalescent embedding in the hyperbolic space. Nat. Commun. **8**, 1615 (2017)
27. Papadopoulos, F., Kitsak, M., Serrano, M.Á., Boguná, M., Krioukov, D.: Popularity versus similarity in growing networks. Nature **489**, 537 (2012)
28. Tixier, A.J.P., Nikolentzos, G., Meladianos, P., Vazirgiannis, M.: Graph classification with 2D convolutional neural networks. In: International Conference on Artificial Neural Networks, pp. 578–593. Springer, Cham (2019)
29. Xin, R., Zhang, J., Shao, Y.: Complex network classification with convolutional neural network. Tsinghua Sci. Technol. **25**, 447–457 (2020)
30. Rozemberczki, B., Kiss, O., Sarkar, R.: Karate Club: an API oriented open-source python framework for unsupervised learning on graphs. In: Proceedings of the 29th ACM International Conference on Information & Knowledge Management, vol. 20, pp. 3125–3132 (2020)
31. Srivastava, N., Hinton, G., Krizhevsky, A., et al.: Dropout: a simple way to prevent neural networks from overfitting. J. Mach. Learn. Res. **15**(1), 1929–1958 (2014)
32. Kingma, D.P., Ba, J.: Adam: a method for stochastic optimization. arXiv, preprint arXiv: pp. 1412.6980 (2014)
33. Abadi, M., Barham, P., Chen, J., Chen, Z., Davis, A., Dean, J., Devin, M., Ghemawat, S., Irving, G., Isard, M., et al.: Tensorflow: a system for large-scale machine learning. OSDI **16**, 265–283 (2016)

Chapter 17
Research on the Method of Alleviating Depression in Pregnant Women Based on Social Mobile Phone Application

Xiaoyan Wei

Abstract The use of social mobile apps can significantly improve the health of pregnant women, thus reducing their likelihood of developing depressive symptoms. For this reason, the relevant research contents will be summarized first in the study of this paper, and then the analysis of specific results will be completed by collecting experiments, data screening, and correlation analysis, thus completing the study of this paper.

17.1 Depressive Symptoms in Pregnant Women

The symptoms of depression in pregnant women are first reflected in the change of mood, among which the typical mood will be persistent low, lack of energy, easy to sleepiness, and some pregnant women will often cry. Most patients experience a decrease in self-evaluation during pregnancy and even a loss of confidence in later life. At the same time, some first-time pregnant women will also have a discordant relationship with their husbands, resulting in a loss of enthusiasm for life. After reviewing the relevant literature, it can be found that the severity and occurrence probability of depression symptoms of pregnant women are related to their educational level, marital status, family income, and social situation of pregnant women. Therefore, this paper will choose social mobile application to study it [1–3].

17.2 Research on How Social Mobile Apps Can Alleviate Depressive Symptoms in Pregnant Women

Social mobile apps are not only accurate and efficient, but also non-invasive to study maternal health and depression. Due to the characteristics of its already exists

X. Wei (✉)
School of Information Technology and Electrical Engineering, The University of Queensland, Brisbane, QLD 4072, Australia
e-mail: lianchao2019@163.com

L. C. Jain et al. (eds.), *Wireless Technology, Intelligent Network Technologies, Smart Services and Applications*, Smart Innovation, Systems and Technologies 258,
https://doi.org/10.1007/978-981-16-5168-7_17

Table 17.1 Statistics on depression among users of social mobility apps

Social media types/user types	Normal users	Mildly depressed user	Major depressive disorder users
QQ zone	461 (60.5%)	280 (36.7%)	21 (2.8%)
WeChat moments	131 (36.8%)	218 (61.3%)	7 (1.9%)

as a kind of mainstream psychology research way, after a lot of research to find pregnant women suffer from depression or with depression tendency compared with the ordinary people in social mobile applications shows behavior is very different, its social activities is relatively small, not only emotionally and more negative. After analyzing the content of his posts, it can be found that he also uses more angry words and uploads some private information to express his depression. At present, this paper starts with several mainstream social mobile applications in China. However, compared with normal users, the number of pregnant women with depression is relatively small, so the sampling is unbalanced. In the future studies, a large amount of research work will be used to complete the training samples.

17.3 Research Methods

17.3.1 Data Acquisition

In this study, we invited users to fill out the self-rated depression questionnaire to obtain usage content and complete data analysis. The results were compared with those of ordinary users. Meanwhile, the social account information of some people was also collected, and the data of those who fully participated in the experiment were kept strictly confidential. At the same time, in order to obtain the safety of this paper, the quality of the questionnaire data can be used Beck Depression Inventory (BDI) and self-rated depression scale, which is the most useful in measuring the scale of clinical depression, as shown in Table 17.1.

17.3.2 Data Feature Screening

The above data is obtained after data feature screening, and in the process of feature screening, we should first analyze its behavior characteristics. Its behavior characteristics include original post vocabulary content, interactive content between users, and the number of posts from 0 to 6. Secondly, it contains language features TF. The content of this study mainly comes from linguistic inquiry and word count (LIWC), including emoticons d_i and first person singular and plural words ω. At the same time, frequency is also an important reference point in the process of research. For language features, it is the total amount of category J feature words published by users in their posts

$$\begin{aligned}\mathrm{TF}_j &= \sum_{i=1}^{n} \text{include} \quad (d_i, \omega_j) \\ \text{include} \quad (d_i, \omega_j) &= \begin{cases} 0, \text{ other} \\ 1, \text{ If post } d_i \text{ contains feature } \omega_j \end{cases}\end{aligned} \tag{17.1}$$

In the normalized frequency NTF (An acronym or abbreviation for a simple language definition), the JTH (An acronym or abbreviation for a simple language definition) feature of the user is converted into frequency, and the mapping of [0, 1] is completed at the same time. In this process, it must satisfy:

$$\mathrm{NTF}_j = \frac{\mathrm{TF}_j}{n} \tag{17.2}$$

Finally, in the Z-score standardization frequency, a certain feature ZTF of all users will be normalized by NTFμ:

$$\mathrm{ZTF}_j = \frac{\mathrm{NTF}\mathbf{j} - \mu}{\sigma} \tag{17.3}$$

17.3.3 Data Correlation Analysis

The correlation analysis carried out in this study was based on the subject vector and data analysis results, so it does not only rely on individual words or parts of speech. In two different mobile applications, a total of 74 word types of mobile phones were completed, but not all of the mobile phone contents were related to depression [4–6]. Therefore, the research center in this paper can select the classification model through the relationship between the eigenvalues and the self-rating scale. It is assumed that all the eigenvalues in this paper show a normal distribution relationship with the eigenvalues; that is, the Pearson correlation coefficient can be used to calculate the correlation. In order to display the selection results of correlation characteristics, the details are shown in Table 17.2.

Table 17.2 Relevant feature details

Characteristic	Sense	Pearson coefficient	Significant level
Mention	Symbol	−0.24	0.035
Shehe	Third person singular pronoun	−0.247	0.030
Verb	Verb	−0.228	0.039
Preps	Prepositions	−0.258	0.018
Interjunction	Particles	−0.218	0.036

17.4 Data Result Analysis

According to the above analysis method and experimental results, there are 25 subjects set in the experiment, and the details of their influence on the data results are shown in Fig. 17.1.

By comparing the experimental results with other results such as bag of words model (BOW) and Word2Vect (bag of words model), it can be found that the characteristic value of LIWC is relatively good, which is mainly because it is more consistent with the statistical logic of this paper from the perspective of psychology. At the same time, according to the related literature statistics, it is found that other studies have used Weibo database as a source of research and add the analysis of the characteristics, and the related conclusions can find it not only can use the user-rating scale approach to statistical results; dialects can also cooperate with relevant content further complete the user depression tendency of confirmation. Not only will the data collected be of better quality, but will also make a clearer distinction between depressed and non-depressed users. If conditions permit, statistics should be carried out in more types of software programs, so as to enrich the emotional and social relationship characteristics of users [7, 8].

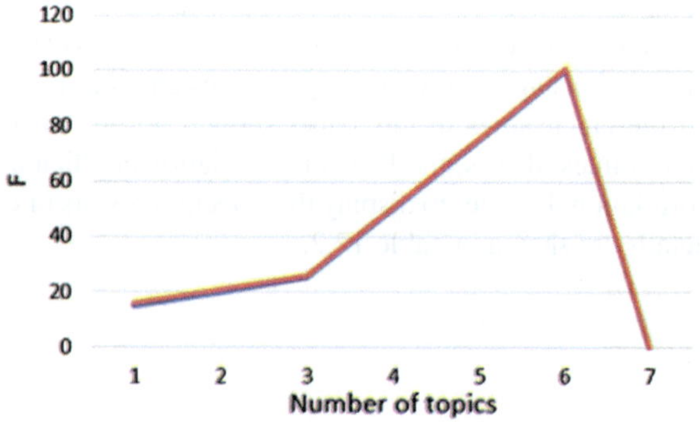

Fig. 17.1 Influence of the number of topics on the detection model

17.5 Conclusion

Through the study of this paper, it can be found that social mobile applications have an obvious effect on alleviating the depressive symptoms of pregnant women. Through the analysis of data results, the specific impact was discussed in terms of sample selection results and characteristic quantification. It is hoped that the content of this article will arouse the attention of all sectors of society and help pregnant women relieve the symptoms of depression during pregnancy in an all-round way. At the same time, limited by the length of the research content is not comprehensive, I hope that the research in this aspect can be further carried out in the future.

References

1. Wang, Y.: Research on Social Support of Pregnant Women Schools from the Perspective of Health Communication. Southwest University of Political Science and Law (2019)
2. Ni, X.X.: Construction and Evaluation of Self-Management Support Module for Mobile Health Lung Cancer Patients. Hangzhou Normal University (2019)
3. Shen, R.: Investigation and Analysis of Current Situation and Influencing Factors of Postpartum Depression and Research on Psychological Intervention Effect. Shanxi Medical University (2011)
4. Yue Yun, W., Yan Ru, Z., Ming Zhen, W., et al.: Effect of 'music chair' and 'cranial electrotherapy stimulator' in relieving pregnant women's anxiety and depression. Health Res. (2017)
5. Shiwen, S., Junqin, L., Huilian, P., et al.: Survey and analysis on present situation of social support and family care for pregnant women with fetal abnormality. J. Nurs. Rehabil. (2018)
6. Nazik, E., Eryilmaz, G.: Incidence of pregnancy-related discomforts and management approaches to relieve them among pregnant women. J. Clin. Nurs. **23**(11–12), 1736–1750 (2013)
7. Clemente-Fuentes, R.J.W., Pickett, H., Carney, M.: Clinical inquiry. How can pregnant women safely relieve low-back pain? J. Fam. Pract. **62**(5), 260–268 (2013)
8. Podungge, Y.: Endhorpin massage and pregnancy exercise as a method to relieve lower back pain in trimester III pregnant women. Health Notions **3**(4), 166–172 (2019)

Chapter 18
Research and Development of Virtual Simulation Software Based on Multimodal Sensory Theory

Guan Huang and Dan Chen

Abstract The presentation of multimodal sensory theory and the continuous development of virtual reality technology provide new ideas and methods for the teaching reform of experimental education in colleges and universities. On the basis of defining the core concepts of multimodal sensory theory and virtual reality technology, this paper constructs a virtual simulation software development framework based on multimodal sensory theory, expounds its key technologies, and tries to develop a series of visual, operable, understandable, and interactive virtual simulation software under the guidance of multimodal sensory theory and uses deep learning algorithm to quantify and visually analyze the data of users' behavior, emotion and cognitive engagement, and diagnose and evaluate users' learning engagement and concentration, which effectively promote the reform of virtual reality education.

18.1 Research Background

With the rapid development of the information age of "Internet + education," our country put forward "selecting and developing 1500 sets of virtual simulation and training experimental systems" and "innovating the application mode of simulation and training resources to improve the use efficiency" in the Ten-year Development Plan of Educational Informatization (2011–2020) [1]. The remarkable breakthrough in the construction and application of virtual simulation educational resources has put forward a new direction for the development and reform of experimental teaching construction and educational informatization in colleges and universities. In higher education form, meaningful learning and actively constructing knowledge representation encourage learners to real learning and working system memory theory, dual coding theory, cognitive load theory such as multimedia general cognitive psychological theory, and the depth of the education teaching, learners perception intentions give full play to the learners' subject-learning initiative, enthusiasm, and creativity.

G. Huang (✉) · D. Chen
China West Normal University, Sichuan, China
e-mail: helen1983226@aliyun.com

L. C. Jain et al. (eds.), *Wireless Technology, Intelligent Network Technologies, Smart Services and Applications*, Smart Innovation, Systems and Technologies 258,
https://doi.org/10.1007/978-981-16-5168-7_18

Single-modal information of sources can cause errors in learners' perception or intention understanding. The single information sources to be integrated is weighted according to their reliability, and the combined estimation obtained is more reliable than any information source itself [2]. Under the guidance of multimodal sensory theory, a new generation of virtual reality technology is integrated to seamlessly transform the physical space where experimental teaching learners live into virtual space, which is conducive to the development of virtual reality educational resources, enhancing learners' learning experience, and improving the quality of virtual simulation experiment teaching in colleges and universities.

18.2 The Definition of Core Concepts

18.2.1 Multimodal Sensory Theory

Multimodality refers to multiple senses. Kolb and Whishaw [3: 174] pointed out that the five senses of sights, hearings, touches, smells, and tastes should be more accurately referred to as "five modality" or "five groups of SCDSeS." Each modality has its own sub-modalities [3]. In the process of constructing information learning and cognition, human sensory systems are interrelated and work together to integrate the information in different sensory channels to form a coherent perceptual process. Therefore, the multimodal sensory theory can interconnect the sensory subjects inside the body with the outside world and interact with each other. In virtual reality, the innovation of entering multichannel presentation mode through single-channel presentation mode points to a new reform mode of education and teaching. The emergence of multimodal sensory theory provides new ideas and guidance for the design of virtual simulation educational resources, which can not only improve the teaching quality, but also have important significance for training learners to learn efficiently. In the book Virtual Reality Technology (Second Edition), the definition of virtual simulation is: "Virtual reality is a high-end man–machine interface, including real-time simulation and real-time interaction of various sensory channels such as touches, hearings, smells, and tastes [4]."

18.2.2 Virtual Simulation Software

Virtual reality (VR) is a comprehensive integration technology with computer technology as its core. It uses three-dimensional display technology, environment modeling technology, realistic real-time rendering technology, etc., to generate a three-dimensional realistic virtual environment. With the help of wearable helmets, handles, and other sensing devices, users can experience the immersive virtual realm,

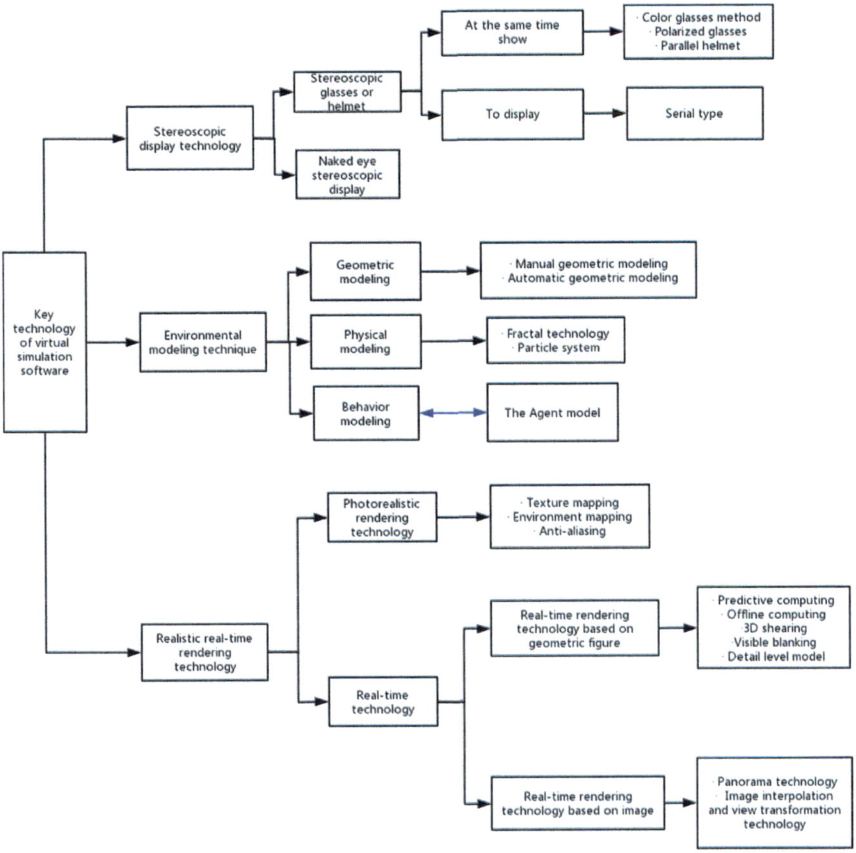

Fig. 18.1 Key technologies of virtual simulation software development

perceive and operate various elements in the virtual environment in real time, and generate immersive deep interaction.

18.3 The Key Supporting Technologies of Virtual Simulation Software

The key technology of virtual simulation software mainly revolves around the accuracy of the virtual environment, sensory information synthesis of the authenticity of the virtual environment, and human and virtual environment interaction of naturalness, real-time display, and graphics generation, for solving the problems of the intelligent technology enables users to immersive virtual environment perception, so as to achieve the aim of exploring and understanding objective things [5] (Fig. 18.1).

18.3.1 Stereoscopic Display Technology

Stereo display technology can make people feel more immersed in the virtual world, and the introduction of stereo display can make the simulation of various simulators more realistic. Research stereo imaging technology, and use the existing microcomputer platform, combined with the corresponding software and hardware system is to display the stereo scene on the flat-panel display. Computer and projection system use four stereo display technologies (color separation technology, light splitting technology, time-sharing technology, and grating technology) to deliver the film source to both eyes, produce appropriate image offset, and form observable stereo images.

18.3.2 Environmental Modeling Technology

Obtain the 3D data of the actual 3D environment and establish the acquired 3D data and the corresponding virtual environment model according to the application requirements. Imitate the environment in the real world, the environment is constructed subjectively by human beings and the invisible environment of human beings in the real world. The main research object of geometric modeling technology is the representation and processing of geometric information of objects. It is an algorithmic modeling method involving the representation of geometric information data structure and the related construction and manipulation of data structure, which measures the absolute distance between two points in multidimensional space.

18.3.3 Realistic Real-Time Rendering Technology

Realistic real-time rendering technology is divided into realistic rendering technology and real-time rendering technology. Realistic rendering technology refers to the process of reproducing real-world scenes in computers. Real-time rendering means that when a user's viewpoint changes, the scene he sees is updated in time, so that the speed of graphic display update must keep up with the speed of viewpoint change.

18.4 Research and Development of Virtual Simulation Software Based on Multimodal Sensory Theory

Virtual reality technology produces the experience of "existence" by mixing visual, auditory, tactile, even olfactory, and gustatory elements. By the end of the twentieth century, Burdea and Coiffet mentioned three basic characteristics of virtual reality technology, namely immersion, interaction, and imagination [6]. From the perspective of communication theory, the more senses are mobilized when receiving information, the more the psychological image of things tends to be true, and the sensory experience is maximized, while the communication utility is constantly increasing.

The Action Plan of Educational Informatization 2.0 requires that the construction of demonstration virtual simulation experimental teaching projects should be taken as the carrier to strengthen the construction of large-capacity intelligent teaching resources and promote the mode change and ecological reconstruction of education supported by new technologies [7]. After investigating a large number of virtual simulation software technologies and products, the technologies and products that can solve the above problems are integrated, and the key points of multichannel and immersive interaction are grasped. After repeated technical demonstration and practice, the virtual simulation software development framework based on multimodal sensory theory is finally constructed (as shown in Fig. 18.2), and teaching practice activities are carried out in schools, and the practical application of virtual simulation teaching resources covering the whole province is gradually built through integration and optimization.

The framework integrates the new concept of "multimodal + VR" to meet the needs of learning, experience, and interaction. It takes computer technology as the soul, takes Idea VR resource base as the center, takes virtual reality tool software as the foundation, and uses nature-oriented human–computer interaction equipment to independently develop the model library and the action library to realize the whole

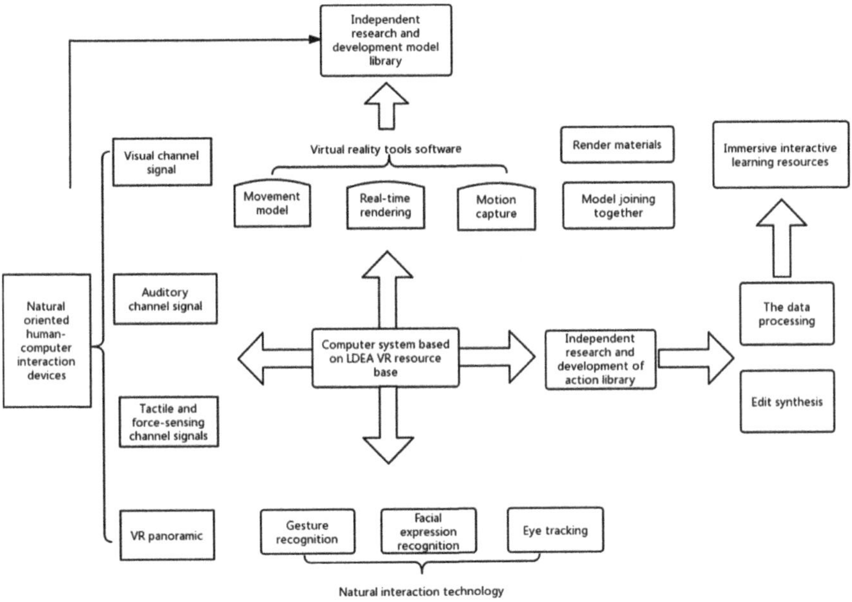

Fig. 18.2 Virtual simulation software development framework based on multimodal sensory theory

process immersion teaching, deeply understand learners' emotions, innovate learning methods, and build an intelligent, adaptive and innovative learning environment in the future.

18.4.1 Based on IdeaVR Resource Pool as the Center

IdeaVR is a virtual reality engine platform independently developed by Shanghai Mangan Digital Co. LTD [8], which supports multiple people in different places. With the help of the built-in powerful distributed multiperson collaborative system, the multiperson collaborative work environment can be quickly constructed, so that all users can be placed in the same real scene. In the new functional modules of IdeaVR, such as voice communication, whiteboard, PPT play, and answering exams, the efficient communication between users and the full coverage of "teaching, learning, testing, and training" is realized under the multiperson collaboration mode [3]. The 3D model of related scenes is established based on IdeaVR resource library, and the 3DMAX software self-developed model is imported into IdeaVR 2019 software. After IdeaVR performs spatial orientation conversion, position coordinate adjustment, model size scaling and material setting, and PPT and animation production are established according to the actual needs of the experiment, thus to understand the generation of virtual simulation resources [9].

18.4.2 Natural Interaction Technology

Various behaviors of learners are recorded through intelligent video terminals and natural data such as identities, emotions, locations, and actions are identified. Speech recognition is used to convert learning content into words and analyze their emotional data. Content recognition and emotion recognition through written words record user operation behavior data of learning terminal.

In the design of virtual simulation software, the use of multimodal sensory theory has three levels of connotation: One is to mobilize the sensory function of the audience through multisensory integration, making the information receiving process more active, interesting, and intuitive. Second, the image language can be extended from a single visual communication information, using the physiological and psychological characteristics of human senses to broaden the artistic expression form from the perspective of cognition. The third is to explore the influence of multisensory association on the information expression of virtual simulation software to better evaluate the effectiveness of information communication and the audience's internal and external experience.

18.4.3 The Advantages of Immersive Interactive Learning Resources

① Facing education

Virtual simulation software based on multimodal sensory theory has very good compatibility, requires people-oriented and self-learning, and can be applied to students of basic education and higher education. Learners can choose their own learning resources, automatically match related models and actions, and have interactive and immersive feelings. At the same time, learners can choose the story plot to decide the final story of outcome and can also choose the perspective of a certain character in the story to watch the development of the story. It breaks the limitation of students' contact with knowledge and makes interest, teaching, and experience become immediate and real personal experience.

② Rich in content

The software develops a series of VR education products aiming at the basic education and higher education courses in primary and secondary schools, which covers all school sections and disciplines, and can be quickly customized according to special needs. It adopts deep integration of various technologies and presents and restores teaching knowledge points 1:1, which can fully meet the individualized needs of users, increase learners' fun, combine theory with practice, promote students' acceptance and understanding of knowledge, and help students develop in an all-round way.

③ Strong pertinence

Through the project research, we fully understand the characteristics of the school's characteristic courses, complete the professional excellent course resource database with each key specialty, realize the unified teaching resource information sharing, promote the digitalization and networking process of the school's teach, and provide new convenient support for teaching of information. And, form a teaching environment in which various multimedia teaching forms coexist, such as teaching and learning, teaching and training, physical projection, micro-display, and simulation, which realizes the sharing of data resources and improves the teaching environment.

Acknowledgements Project Name and NO: 2017 Outstanding Talents Research Fund Project "Construction and Application of Innovative Learning Mode in Intelligent Classroom under information Environment" (Project No. 17YC203) Social Science Research of Nanchong city; Practical Research on Flipped Classroom Based on the concept of gamification—Taking the course of TV Textbook Writing and Directing and Making as an example (paper), number: NC2019B165; Research on the construction and application of intelligent classroom innovative learning mode in the information environment (Project No. 17yc203).

References

1. Wang, J.J., Wei, X.F.: Hot status and cold thinking of virtual experiment. Audio-Vis. Educ. China **04**, 126–129 (2011)
2. Ernst, M.O.: Multisensory integration: a late bloomer. Curr. Biol. **18**(12), R519–R521 (2008)
3. Kolb, B., Whishaw, I.Q.: Fundamentals of Human Neuropsychology. Worth Publishers, New York (2003)
4. Yuan, H.P.: Design and development of simulation mechanics experiment teaching software. MA thesis, Sichuan Normal University (2009)
5. Yang, P.P.: Design and implementation of thematic learning activities based on 3D virtual environment. MA thesis, Southwest University (2019)
6. Han, R.P.: Intention perception method based on multimodal fusion and its application in VR 2020. MA thesis, Jinan University (2019)
7. Education informatization 2.0 action plan: West. Qual. Educ. **4**(10), 123 (2018). https://doi.org/CNKI:SUN:XBSJ.0.2018-10-079
8. Chen, M.Y., et al.: Research on the construction and practice of mixed learning space from the perspective of wisdom education. China Dist. Educ. **11**, 79–85 (2019). https://doi.org/10.13541/j.cnki.chinade.2019.11.007
9. Xiong, L., et al.: Multi-modal virtual reality teaching experiment system based on HoloLens. Softw. Guide **19**(09), 152–154 (2020). https://doi.org/CNKI:SUN:RJDK.0.2020-09-035

Chapter 19
Construction of Intelligent Information Platform for Preschool Education Based on Big Data

Xia Yu

Abstract Preschool education has always been an important teaching field in China. The state has issued many relevant policies to support the development of public and private kindergartens and invested a lot of money to promote the quality of preschool education and ensure the popularization of preschool education and the diversified development of schools. At present, there are a large number of kindergartens in China, and the educational level of kindergartens in urban and rural areas is polarized. The main reason lies in the imbalance of economic development between urban and rural areas in China, which leads to the uneven distribution of all teachers and hardware facilities owned by kindergartens. Therefore, how to use administrative policies to eliminate the polarization of kindergarten education resources and improve kindergarten education level is the key issue at present. With the help of big data analysis, we can optimize the allocation of preschool education resources, establish a kindergarten information collection system and obtain data, provide basic data for formulating relevant education policies, and promote the optimal allocation of preschool education resources.

19.1 Big Data Infrastructure Platform

Big data core platform has the ability to provide basic big data information. It is divided into three subsystems: big data cloud extract-transform-load (ETL), big data basic kernel, and big data analysis mining. Among them, the main functions of big data cloud ETL subsystem are data extraction, data cleaning and transformation, and data loading. With the help of big data platform, the relevant data from the education evaluation system are processed concretely. The main function of big data basic kernel subsystem is to store, calculate, and analyze data. MapReduce batch computing, Hadoop distributed file system (HDFS) distributed file storage, Hadoop database (HBASE) semi-structured data storage, and other related technologies are usually adopted. MapReduce is a programming model, which is mainly used to

X. Yu (✉)
Shangrao Preschool Education College, Shangrao 334000, Jiangxi, China
e-mail: yueyin019@163.com

L. C. Jain et al. (eds.), *Wireless Technology, Intelligent Network Technologies, Smart Services and Applications*, Smart Innovation, Systems and Technologies 258,
https://doi.org/10.1007/978-981-16-5168-7_19

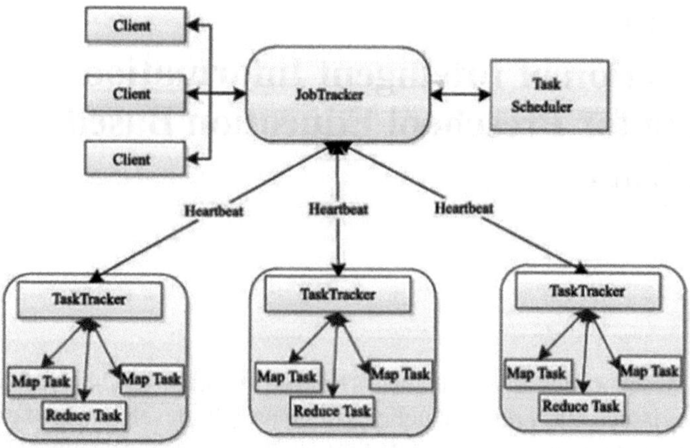

Fig. 19.1 MapReduce model

simplify parallel computing and deal with a large number of computing problems. Processing and generation of large data sets are often realized through algorithm models [1]. Figure 19.1 shows the correlation model. The main function of big data analysis and mining subsystem is to assist users in business and management activities. It is mainly divided into data exploration, mining modeling, and model evaluation. Its main processing object is the data stored in the platform. By obtaining data in advance, identifying data characteristics, and selecting suitable processing and analysis technologies, the analysis and mining of structured and unstructured data in the platform can be realized [2].

19.2 Preschool Education Information Data Collection Part

In this paper, the effective school-running efficiency is the main index of collecting data. In 2020, the number of effective preschool education provinces in China is 13, accounting for 41.94% of all decision-making units. This shows that there are regional differences in effective preschool education in China. In addition, the number of effective provinces with pure technical efficiency is 20, accounting for 64.52%. The number of provinces with effective scale efficiency is 13, accounting for 41.94%. Among them, the effective school-running rate and effective scale rate are 13, accounting for 41.94% of all decision-making units. It shows that in terms of pure technical efficiency and scale efficiency, preschool education in about half of the provinces has failed to meet 1. It means that the educational resources invested by the state do not play its due role in management and technical level [3] (Table 19.1).

According to the data displayed in 2020, the preschool education in about half of the provinces failed to meet 1. It means that the educational resources invested by

Table 19.1 Evaluation statistics of preschool education efficiency in China in 2020

	Province	Minimum value	Maximum value	Mean	Number of effective provinces	Effective provincial specific gravity (%)
Efficiency of running	31	0.22	One	0.70	13	41.95
Pure technical of efficiency	31	0.63	One	0.92	20	64.52
Scale of efficiency	31	0.22	One	0.73	13	41.94

Table 19.2 2020 China's preschool education adjustment evaluation statistics

	Province	Minimum value	Maximum value	Mean	Number of effective provinces	Effective provincial specific gravity (%)
Efficiency of running	31	0.34	One	0.71	10	32.26
Pure technical of efficiency	31	0.70	One	0.95	23	74.19
Scale of efficiency	31	0.41	One	0.74	10	32.26

the state have not played their due role in the management and technical level and need to be adjusted in a targeted way (Table 19.2).

Slacks-based model (SBM) is a non-parametric method, and its model is a kind of data envelopment analysis (DEA) extended model, which is perfect. SBM model can comprehensively consider the slack problems in input and output of relevant decision-making units. Furthermore, it is not necessary to give the function forms involved in production in advance, and it is not necessary to manually select the weights of relevant indexes. Therefore, it has an extremely objective evaluation effect. In addition, it can carry out effective treatment evaluation under the condition of more input and output. Because SBM has the characteristics of dealing with multiple inputs and outputs, it is widely used in the field of educational efficiency evaluation [4].

Let n be the number of decision-making unit (DMUs), and any DMU has corresponding input vectors and output vectors. Based on SBM model, its condition is based on relaxation measure.

$$p = \min \frac{1 \pm (1/m) \sum_{i=1}^{m} s_i / x_{iv}}{1 \pm (1/h) \sum_{r=1}^{h} s_r / y_{iv}}$$

In the above formula, the efficiency evaluation standard is ρ. The type of input factor is m. The type of output element is H. Overall output of DMU is X. Overall investment of DMU is Y. X is the input vector of each DMUv to be evaluated. The input vector of each DMUv to be evaluated is y_{iv}. The ith index value of vector xv is x_{iv}.

$$\min_{\delta} = t - (1/m)\sum_{i=1}^{m} s_i/x_{iv}$$

$$\text{s.t. } 1 = t + (1/h)\sum_{r=1}^{h} s_r/y_{rv}$$

After conversion, the measure value of efficiency is expressed by δ. Input redundancy of the vth decision unit can be obtained by s_i/x_{iv}. Output shortage can be obtained by s_r/y_{rv}.

SBM model evaluates efficiency through input-oriented model and output-oriented model. Input-oriented model refers to the assumption that the estimation efficiency is to minimize the input based on a given DMU output level. Output-oriented model refers to the assumption that the estimation efficiency is to maximize the output based on the given DMU input. This can be analogized to the preschool education field, based on the limited input of educational resources by the government, in exchange for the maximum output. Based on this, this study decided to use SBM model to study the output orientation of preschool education resources.

Technical efficiency (TE) is expressed as pure technical efficiency (PTE) $\times$ scale efficiency. The expression of scale of efficiency (SE) of DMU is technical efficiency/pure technical efficiency. Therefore, the efficiency of running a kindergarten can be calculated. When the pure technical efficiency is 1 and the scale efficiency is 1, it shows that the state is optimal; that is, the technical efficiency is effective. It is an important index to describe the efficiency of running a kindergarten. Management and technical level can affect the production efficiency of production units, that is, pure technical efficiency. It is usually used as an important indicator to measure and evaluate the efficiency of school education (e.g., school management, teachers' level, and teaching ability). The index that production efficiency is affected by unit scale is scale efficiency. That is, the greater the productivity, the higher the scale efficiency, and the more appropriate the production scale [5].

19.3 System Architecture

The system consists of two parts: the big data core platform and the baby platform. Big data core platform has the characteristics of big data system operation and resource management and scheduling. It can fulfill the main related requirements of big data system, such as deployment, scheduling, and cluster management. Big data core platform can fully meet the needs of running early childhood education and processing data. Big data core platform can manage data in the whole process. Grab,

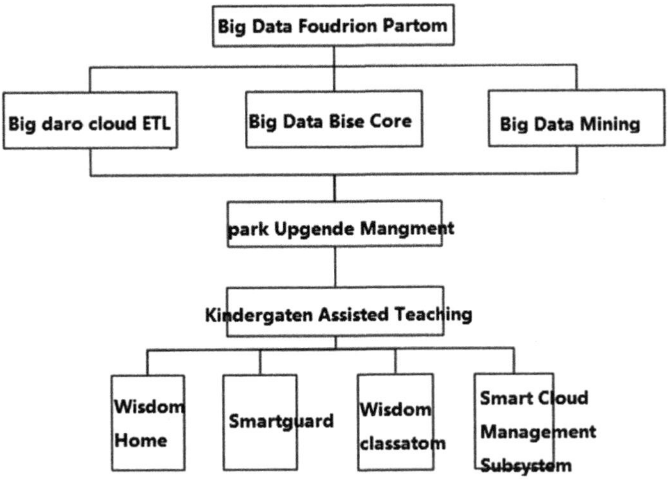

Fig. 19.2 System architecture

clean, and import data by ETL and save the data separately according to functions or characteristics (such as original number, and calculated data). If you want to deal with a large amount of structured data, you can use the spark data engine and batch computing (M/R) data engine. Its main advantage is that it is easy to import and effectively reduce the difficulty of developing and migrating applications. Big data core platform has the ability to provide basic big data information. It is divided into three subsystems: big data cloud ETL, big data basic kernel, and big data analysis mining. With the help of big data platform and big data technology, it can effectively face different groups with multiple attributes, meet the specific needs and communication of different groups for related data, and make resource utilization more effective. By building an early childhood education platform, we can have more effective and intelligent teaching methods and learning methods. At present, the infant system platforms in use include intelligent guardian and intelligent classroom. As shown in Fig. 19.2 [6].

19.4 Conclusion

At present, under the great environment of vigorously developing education for all and education informationization in China, the development of early childhood education has become a short board in the development of China's education industry. With the continuous development of China's network, a variety of information platforms are constantly emerging, which also drives the development of online education platform for children in a certain process. Through big data, we can get the relevant data in online early childhood education and keep the learning situation of early childhood in time, which reflects the great development potential of early childhood education. Taking this as the research background, this paper uses big data technology to design the infant platform and put forward the corresponding construction methods. It is hoped that the application of this platform will promote the realization of personalized early childhood education.

References

1. Tao, F.M., Xi, Y.Y.: Value-added evaluation: big data fine helps children's scientific development. Aid Chongqing World Teach. Dev. **9**, 48–49 (2015)
2. Cao, R.: Smart kindergarten construction based on big data integration. Invent. Innov. Educ. Inform. **10**, 21–24 (2015)
3. Yan, Y.: An analysis of kindergarten management in the "Internet plus" era. Urban Fam. Educ. Mon. **5**, 34–37 (2016)
4. Department of Education, Employment and Workplace Relations, Australia: Universal Access to Early Childhood Education [EB/OL]. http://www.deewr.gov.au/, 9.14 (2009)
5. Dalli, C., Te One, S.: Early childhood education in 2002: pathways to the future. N. Zeal. Rev. Educ. **12**, 177–202 (2003)
6. Karila, K.: A Nordic perspective on early childhood education and care policy. Eur. J. Educ. **47**(4), 584–595 (2012)

Chapter 20
Development and Analysis of Solenoid Valve Controller Based on LoRa Technology

Wufen Chen, Xianqun Jiang, Mingmin Chen, Qibin Liang, and Bo Zhang

Abstract In order to solve the shortcomings of wireless controllers such as short communication distance and high power consumption in current automatic agricultural irrigation, a solenoid valve controller based on long range (LoRa) technology was developed. The controller is based on MSP430 microprocessor, supplemented by charge and discharge control module, LoRa wireless communication module, solenoid valve control module, data acquisition module, and other peripheral structure circuits. It also uses solar energy as the energy source to achieve long-term work without battery replacement. The test results of the controller show that the transmission distance of the controller can reach 2.5 km on the premise of less than 10% packet loss rate, which can meet the requirements of automatic irrigation control.

20.1 Introduction

With the urgent demand for water saving, automatic controllers are widely used in agricultural irrigation [1–3]. The traditional wired way inevitably presents difficulties in field installation and maintenance, so more and more farmland adopts wireless way to control solenoid valve [4–6]. At present, the traditional wireless controller mainly adopts general packet radio service (GPRS) or global system for mobile communications (GSM) mode, and the disadvantage is that it cannot work normally without public network signal. The controller based on ZigBee technology has limited propagation distance due to poor diffraction performance of electromagnetic wave in 2.4 GHz frequency band [7, 8]. In 2011, Wang and Li et al. designed and implemented an automatic irrigation controller based on wireless fidelity (Wi-Fi), which overcomes the problems such as difficult wiring of traditional wired controller and unfriendly operation interface [9]. In 2014, Ping and Guo who come from Wuxi Institute of Technology designed an automatic irrigation controller based on low-power microcontroller Launchpad, which realized effective switch control for irrigation equipment pumps [5].

W. Chen (✉) · X. Jiang · M. Chen · Q. Liang · B. Zhang
Pearl River Water Resources Commission of the Ministry of Water Resources, Guangzhou 510610, China

L. C. Jain et al. (eds.), *Wireless Technology, Intelligent Network Technologies, Smart Services and Applications*, Smart Innovation, Systems and Technologies 258,
https://doi.org/10.1007/978-981-16-5168-7_20

Based on the actual demand of agricultural irrigation system, a kind of solenoid valve controller based on LoRa technology is developed in this paper. The solenoid valve controller can collect data and control the irrigation solenoid valve and has the advantages of large monitoring coverage, strong stability, and flexible node setting.

20.2 General Structure of Irrigation Decision Control System

Irrigation decision control system is mainly composed of field monitoring unit and cloud monitoring center unit. According to the real-time information such as soil moisture, pipeline flow, and meteorology, the irrigation decision-making control system accurately analyzes the information of crop water demand and makes the precise water control of the irrigation system, so as to realize the automation and intelligence of the irrigation system.

The decision-making and control system of irrigation takes crops as the research carrier. It detects the soil moisture content of each layer of crops and the micro meteorological information in the field in real time and makes appropriate and timely precise irrigation strategy for crops according to the water demand model of crops. After receiving the control command, the control node (solenoid valve controller) will analyze the command frame and data frame according to the specified communication protocol, complete the corresponding action that is opening or closing the solenoid valve, and complete the irrigation task. At the same time, the working state of the solenoid valve is fed back to the monitoring center through the joint point of the network. The network topology of the system is shown in Fig. 20.1.

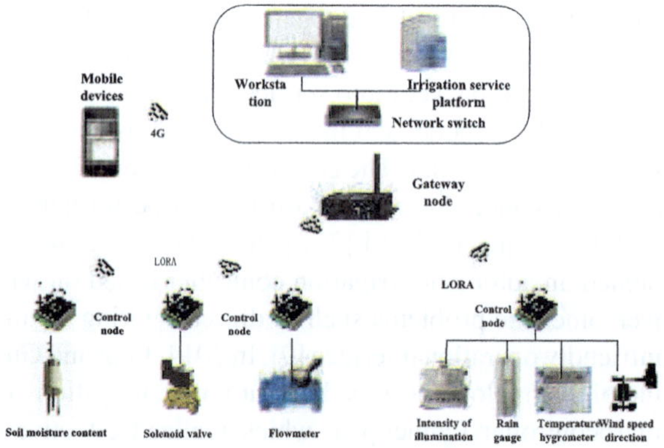

Fig. 20.1 Network topology of system

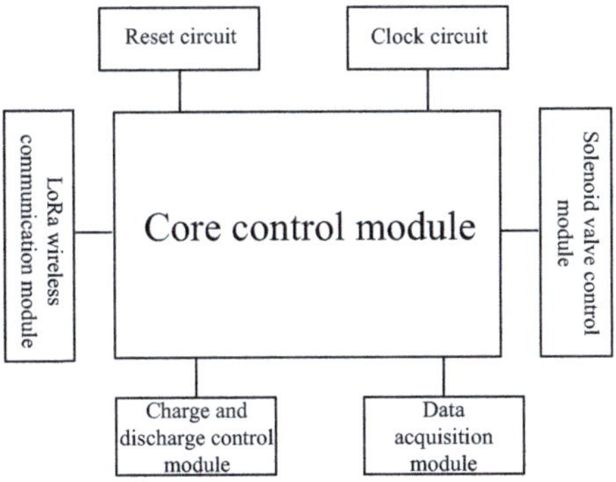

Fig. 20.2 Overall hardware structure of controller

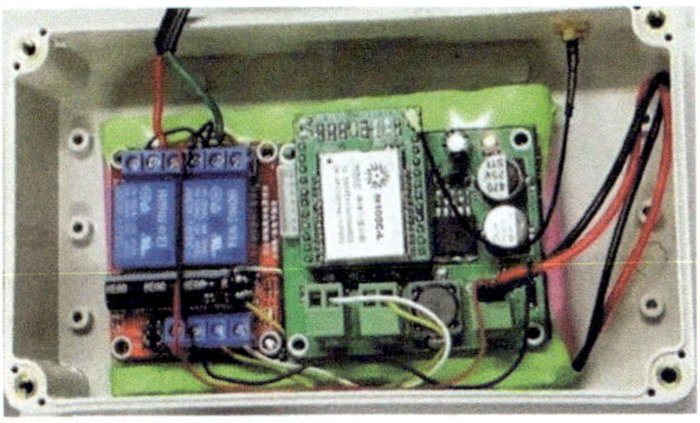

Fig. 20.3 Physical object of controller

20.2.1 Hardware Design of Solenoid Valve Controller

The solenoid valve controller has a very low standby current and can work reliably for a long time in continuous rainy weather. The controller adopts modular design scheme, which is designed into different hardware modules according to the different functions. It is convenient for upgrading and function expansion. It can be divided into charge and discharge control module, core control module, data acquisition module, solenoid valve control module, and LoRa wireless communication module. The overall hardware structure of the controller is shown in Fig. 20.2. The physical object of the controller is shown in Fig. 20.3.

20.2.2 Software Design of Solenoid Valve Controller

The software of the solenoid valve controller must have the following functions, such as being able to accept the command frame sent from the monitoring center, organize and analyze the command frame message according to the specified communication protocol, open or close the solenoid valve according to the specific message content, complete the filling action within the specified time, and feedback the working status of the solenoid valve to the monitoring center in real time. Therefore, the system uses a gateway node to communicate with multiple controllers. The communication process is as follows: Firstly, after the gateway node starts or joins a wireless network, it is placed in the allow binding mode to respond to the binding request sent from the controller. Then after, the controller successfully joins the network established by the gateway node, the controller will automatically discover and bind to the gateway device because the gateway is in the binding-allowed mode, and then it will begin to report the node status data to the gateway node. The switch of the solenoid valve transmits the state of the controller and the state of the solenoid valve to the gateway node through the LoRa wireless communication technology.

20.3 Analysis of Joint Adjustment of Solenoid Valve Controller

20.3.1 Distance Test of Controller

Communication distance is one of the most important indicators to measure the performance of the controller. The effective communication distance of wireless LoRa is the premise to ensure the normal communication. In the test, we adopt the point-to-point communication mode and fix the gateway node, then move the controller. The controller sends data to the network joint point, and the node sends data packets at a fixed time (5S). The frequency of test LoRa module radio frequency (RF) center is set to 433 MHz, and the transmission power is 20 dBm. Fewer than two conditions, the test carries out point-to-point test on the received signal strength indicator (RSSI)-received signal strength and communication success rate. The test results are shown in Table 20.1, and the broken line Fig. 20.4 is drawn according to the test results.

It can be seen from Table 20.1 that under the two test environments, the received signal strength and communication success rate decrease with the increase of distance. The communication success rate of unsheltered open land is better than that of sheltered farmland. The communication success rate can reach more than 90% within the range of 2500 m. It can be seen from Fig. 20.4 that in the two test environments, with the increase of communication distance, the received signal strength and communication success rate are continuously reduced, and the test results in the same test distance without shelter environment are better than those in the presence of shelter, and the test results are in line with the pre-determined.

20.3.2 Packet Loss Rate Test of LoRa Module

The packet loss rate test environment of LoRa module is carried out under the condition that there are obstructions within 3 km. The test is divided into single-point

Table 20.1 LoRa point-to-point communication distance test

Testing environment	Unsheltered open land		Sheltered farmland	
Communication distance (m)	RSSI (dBm)	Communication success rate (%)	RSSI (dBm)	Communication success rate (%)
500	−53	100	−59	99
1000	−56	99	−65	96
1500	−62	98	−77	95
2000	−66	97	−85	92
2500	−73	96	−82	91
3000	−70	95	−98	89
3500	−75	93	−109	84
4000	−83	93	−115	78
4500	−92	88	−127	75
5000	−99	82	−139	69

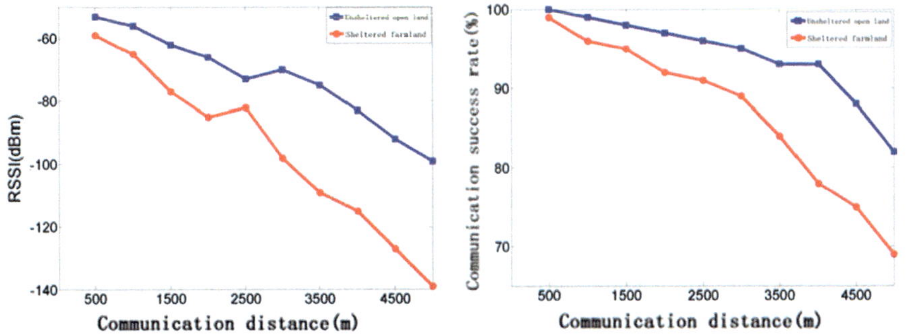

Fig. 20.4 Relationship between LoRa communication distance, RSSI, and communication success rate

communication and multi-point communication. The packet size is divided into 15 and 30 bytes, and each group of tests receives and sends 1000 packets.

In the single-point communication test environment, we select a controller to transmit data to the gateway node in a fixed time and one-way manner. The packet size is divided into 15 and 30 bytes. The experimental results are shown in Table 20.2, and the change of packet loss rate under different packet sizes is plotted in Fig. 20.5.

It can be seen from Table 20.2 and Fig. 20.5 that packet size has a significant impact on packet loss rate. Under the same communication distance, the larger the packet is, the higher the packet loss rate is. The reason is that the sending time interval of 30 byte packet is longer than that of 15 byte packet, so the packet loss rate will increase.

In the multi-point communication test environment, three controllers and one gateway node are selected to form a wireless sensor network. The controller sends

Table 20.2 Test results of packet loss rate of single-point communication

Communication distance (m)	15 byte packets		30 byte packets	
	Packets received	Packet loss rate (%)	Packets received	Packet loss rate (%)
500	996	0.40	981	1.90
1000	985	1.50	953	4.70
1500	972	2.80	938	6.20
2000	947	5.30	914	8.60
2500	933	6.70	905	9.50
3000	878	12.20	864	13.60

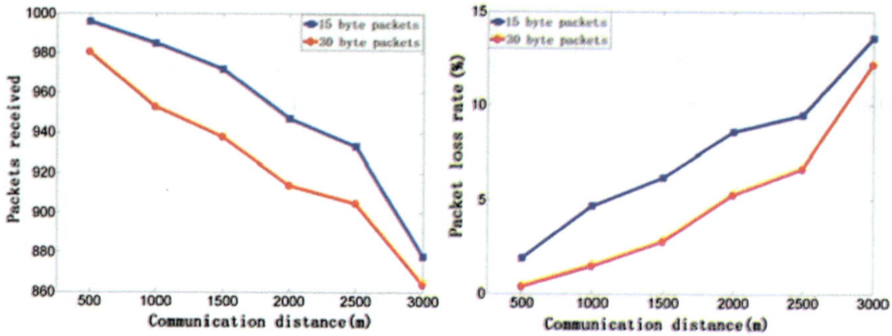

Fig. 20.5 Test results of packet loss rate under different packet sizes

data to the network joints at the same time. The packet size is set to 15 bytes, and each node sends 1000 bytes. The test results are shown in Table 20.3, and the change trend of packet loss rate of multi-point communication is plotted accordingly in Fig. 20.6.

The test results are analyzed as follows: From the test data in Table 20.3, it can be estimated that in the actual application scenario, when the communication distance is less than 1.5 km, the packet loss rate is basically maintained within 5%, and when

Table 20.3 Packet loss rate test of multi-point communication

Communication distance (m)	Packets received			Packet loss rate		
	Node 1	Node 2	Node 3	Node 1 (%)	Node 2 (%)	Node 3 (%)
500	991	988	990	0.90	1.20	1
1000	976	981	979	2.40	1.90	2.10
1500	950	953	964	5	4.70	3.60
2000	919	922	917	8.10	7.80	8.30
2500	881	887	892	11.90	11.30	10.80
3000	862	871	873	13.80	12.90	12.70

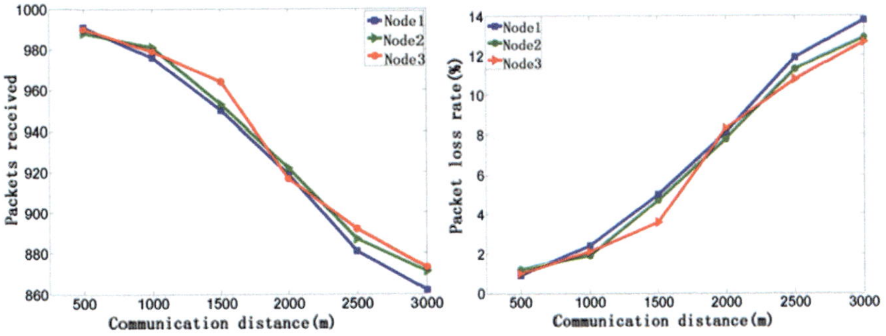

Fig. 20.6 Packet loss rate test of multi-point communication

the communication distance is more than 2.5 km, the packet loss rate is more than 10%, which exceeds the maximum packet loss rate required by the system, so it is not considered. Figure 20.6 shows that the packet loss rate will increase with the increase of communication distance, and the packet loss rate of communication between the three controllers and gateway nodes will be approximately the same. From the comparison of test data in Tables 20.2 and 20.3, the packet loss rate of multi-point communication is higher than that of single-point communication.

In a word, the transmission distance of the controller can reach 2.5 km when the packet loss rate is less than 10%, which can fully meet the requirements of large-scale farmland automatic irrigation.

20.4 Conclusions

Aiming at the disadvantages of short communication distance and high power consumption of wireless controller in agricultural automatic irrigation, this paper adopts LoRa wireless communication technology to develop a kind of solenoid valve controller. The controller is composed of MSP430 microprocessor, charge and discharge control module, LoRa wireless communication module, electromagnetic valve control module, data acquisition module, and other peripheral structure circuits. Through repeated tests in the field, the controller can achieve a transmission distance of 2.5 km on the premise of less than 10% packet loss rate, which can fully meet the requirements of large-scale automatic irrigation of farmland and realize long distance and stable irrigation control operation.

References

1. Lin, C., Yu, B., Gao, P., Xie, J., Wang, W.: Design of intelligent water saving irrigation controller based on single chip microcomputer. Electron. Technol. Softw. Eng. **10**, 233–235 (2019)

2. Zhang, Y., Wen, Z., Li, L., Dong, X., Wang, Z.: Design of intelligent irrigation controller based on STM32. South Agric. Mach. **49**(20), 114 (2018)
3. Wang, M., Zheng, W., Tian, H., Zhang, X., Li, J.: Design and implementation of low-power real-time awake irrigation controller. J. Agric. Mech. Res. **38**(01), 113–118 (2016)
4. Chen, Y., Liu, Z., Yang, K., Xu, H.: Design and investigation on a novel intelligent irrigation controller. J. Anhui Agric. Sci. **43**(20), 359–361+382 (2015)
5. Ping, Y., Guo, L.: Design of low power automatic irrigation controller. Mod. Electron. Techn. **37**(10), 104–106 (2014)
6. Zhang, R.: Design and Implementation of Irrigation Controller Based on AT89S52. Kunming University of Science and Technology (2014)
7. Li, H.: Design and Implementation of Greenhouse Irrigation Controller Based on ZigBee. Northwest A & F University (2014)
8. Zhang, J.: Design of intelligent irrigation controller based on MC9S12NE64. Occupation **05**, 83–84 (2013)
9. Wang, X., Li, B., Shen, C., Xing, Z., Yan, H., Bao, F.: Design and implementation of automatic irrigation controller based on Wi-Fi. China Rural Water Hydropower **12**, 46–49+54 (2011)

Chapter 21
Research on Security Technology Requirement for Zero Trust Security Product

Haohao Song

Abstract Like "machine learning" and "AI," Zero Trust has become one of cybersecurity's latest buzzwords. Zero Trust security product becomes a new kind of network security products. Due to the lack of professional testing and evaluation system for Zero Trust security product, the basic security of Zero Trust security product cannot be guaranteed. Research on security technology requirement of Zero Trust security product is provided, and the security technology requirement would be used in the actual product testing, to further promote the development of network border security.

21.1 Introduction

Zero Trust is a strategic initiative that helps prevent successful data breaches by eliminating the concept of trust from an organization's network architecture. Rooted in the principle of "*never* trust, always verify," Zero Trust is designed to protect modern digital environments by leveraging network segmentation, preventing lateral movement, providing Layer 7 threat prevention, and simplifying granular user-access control [1–3].

Zero Trust was created by John Kindervag, based on the realization that traditional security models operate on the outdated assumption that everything inside an organization's network should be trusted. Under this broken trust model, it is assumed that a user's identity is not compromised and that all users act responsibly and can be trusted. The Zero Trust model recognizes that trust is a vulnerability. Once on the network, users—including threat actors and malicious insiders—are free to move laterally and access or exfiltrate whatever data they are not limited to. Remember, the point of infiltration of an attack is often not the target location [4].

Zero Trust security product is based on the continuous dynamic security access control technology to reduce the security risk of access process.

This paper provided the research on security technology requirement of Zero Trust security product.

H. Song (✉)
The Third Research Institute of Ministry of Public Security, Shanghai, China
e-mail: songhh@mctc.org.cn

L. C. Jain et al. (eds.), *Wireless Technology, Intelligent Network Technologies, Smart Services and Applications*, Smart Innovation, Systems and Technologies 258,
https://doi.org/10.1007/978-981-16-5168-7_21

21.2 Zero Trust Security Product

Zero Trust security product is based on the Zero Trust security architecture, which reduces the boundary of network defense to a single or smaller resource group. Its central idea is that enterprises should not automatically trust any person/matter/thing inside or outside, and should not grant full trust authority to the system according to the physical or network location, before authorization, any person/thing trying to access the enterprise system shall be verified, and access to data resources shall be granted only when resources need to be granted. The security policy of Zero Trust security product is not to trust any user. No one can access the network unless Zero Trust security product clearly knows the identity of the access subject. The existing traditional access control product only need to know the IP address or user/host information, but in the Zero Trust security product, more explicit information is needed, and requests without knowing the user identity or the authorization path are rejected.

The access rights of users will no longer be affected by the geographical location, but different users will have different resources to access due to their different permission levels, and the VPN used to log in from the external network to the internal network will also be abandoned. Zero Trust security product subvert the access control paradigm and guide the security architecture from "network centric" to "identity centric". Its essential demand is identity centered access control. This is similar to the concept of unbounded network. In the borderless network concept, not all end users are behind the office and firewall, but work remotely, using their iPad or other mobile devices.

Networks need to know more about their roles and identify which users are allowed to connect to the network to work. Zero Trust security product is a response to the development trend of enterprise network. Enterprise network begins to include remote users and cloud-based assets located at the boundary of enterprise network. Zero Trust security product focuses on protecting resources, not network segments, because network location is no longer considered a major component of resource security.

21.3 Research on Security Technology Requirement of Zero Trust Security Product

In order to test and evaluate Zero Trust security product completely and accurately, we need to study the security functions that Zero Trust security product should have, including the security functions provided for the network security and the security functions for own security. On this basis, the security technology requirement of Zero Trust security product is formed. According to this security technology requirement, we can test and evaluate Zero Trust security product.

21.3.1 Security Functions for the Network Security

There are six basic principles to be followed in the design and implementation of Zero Trust security product's security functions for network security: (1) All data sources and computing services are considered as resources. (2) All communications are secure, and the security is independent of network location. (3) Authorization for access to a single enterprise resource is authorized for each connection. (4) Access to resources is determined by policy, including the status of the user's identity and the required system, and possibly other behavior attributes. (5) The enterprise should make sure that all the affiliated and related systems are in the securest state as possible, and monitor the system to ensure that the system in the securest state. (6) User authentication is dynamic and strictly enforced before access is allowed.

Based on the six basic principles, the security functions of Zero Trust security product for the network security is researched.

21.3.1.1 Comprehensive Identification

To some extent, the traditional border-based network security architecture assumes or acquiesces that the intranet is secure, and considers that security is the digital wall of the enterprise. Through the border security products such as firewall, web application firewall (WAF), intrusion prevention system (IPS), etc., the enterprise network exports are heavily protected, while the enterprise intranet security is ignored.

Under the background of "the enterprise boundary is disintegrating, and the border based security protection system is failing", Zero Trust security product overthrows the architecture of the traditional border security and gives new suggestions to the security architecture. The core idea is by default, no one/device/system inside and outside the network should be trusted, and it needs to be based on the authentication and authorization reconstruct the trust foundation of access control. Zero Trust security product subverts the paradigm of access control and leads the security architecture from network centered to identity centered. Its essential demand is identity centered access control.

From the perspective of security technology requirement, Zero Trust security product should realize comprehensive, dynamic, and intelligent access control to people/devices/systems with the help of modern identity management platform.

Based on the above security requirement, Zero Trust security product needs its identity governance modular to achieve a comprehensive identity of devices, users, applications and other entities, using two key technologies, device authentication and user authentication, in order to build an identity-based trust system, and establish a brand-new identity boundary for enterprises from Zero.

21.3.1.2 Dynamic Access Control

Based on the comprehensive identification, the other security function of Zero Trust security product is dynamic access control, or conditional access control. In the beginning, all users should be granted the minimum access rights, whether the user is the account used by the leaders or the account used by ordinary employees. The Zero Trust product extends the boundary of the enterprise network to each user's terminal, not just the boundary firewall of the enterprise intranet network. All users should be accessed through the distributed application proxy gateway of Zero Trust product, and all applications should be hidden in the enterprise intranet network to effectively protect the assets of the enterprise.

On this basis, the Zero Trust security product controls the access of users who request access to the intranet resources based on conditions for risk assessment, including but not limited to the address, the security attributes of the user terminal, the role of the user, the time when the user requests, etc. Based on the satisfied and unsatisfied conditions of the user, the Zero Trust security product dynamically determines the type and quantity of system resources that the user can access. The whole Zero Trust security system is realized fully.

21.3.2 Security Functions for Own Security

In addition to providing comprehensive identification, dynamic access control and other security functions to ensure the security of new network architecture, the self-security of Zero Trust security product also needs to be developed and implemented.

In order to ensure that all security policies of Zero Trust security product are not accessed and changed unauthorizedly, the Zero Trust security product needs to have basic security functions such as identity authentication, access control, and security auditing.

Before any security policy of Zero Trust security product is configured, strict identity authentication is required to confirm that only authorized user can access the Zero Trust security product. After identity authentication, the access control function is needed to confirm the authorized user's limit of authority, that is, which security function the user can access and which security function he can't access. At the same time, all authorized operations are required to audit records for traceability. Only with these basic security functions, the Zero Trust security product can effectively and continuously provide comprehensive identification, dynamic access, and other security functions.

Figure 21.1 shows security technology requirement for Zero Trust security product. The researched security technology requirement for Zero Trust security product contributes to develop and implement Zero Trust security product as a whole product.

Fig. 21.1 Security technology requirement for Zero Trust security product

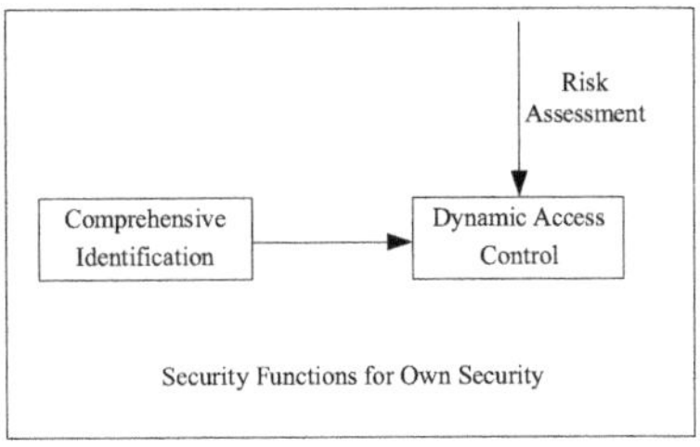

21.4 Conclusion

The security assessment of Zero Trust security product belongs to a new field, and only a few testing institutions in China have carried out relevant testing and evaluation work. In this paper, the research on security technology requirement of Zero Trust security product is provided. The security functions that Zero Trust security product should have, including the security functions provided for the network security and the security functions for own security, is studied firstly. On this basis, the security technology requirement of Zero Trust security product is formed. Based on the provided security technology requirement, we can test and evaluate Zero Trust security product completely and accurately.

References

1. Gilman, E., Barth, D.: Zero Trust Networks Building Secure Systems in Untrusted Networks, 1st edn. O'Reilly Media, State of California, USA (2017)
2. Rose, S. et al.: Zero Trust Architecture. National Institute of Standards and Technology (NIST) Draft, 2nd edn. Special Publication 800–207, Gaithersburg, Md., February 2020. Available: https://nvlpubs.nist.gov/nistpubs/SpecialPublications/NIST.SP.800-207-draft2.pdf. Last accessed 05/21 (2020)
3. Cunningham, C.: The Zero Trust eXtended (ZTX) Ecosystem: Extending Zero Trust Security Across Your Digital Business. Cambridge, MA (2018)
4. Kindervag, B.J., Shey, H., June, Mak, K.: The Future of Data Security: A Zero Trust Approach. FORRESRER (2014)

Chapter 22
Research and Application of Automatic Recognition and Extraction of Water Body in Remote Sensing Images Based on Deep Learning

Bo Wu, Yi Gao, Xiaolei Xie, and Ye Yang

Abstract In this paper, we proposed a deep learning-based method for automatic recognition and extraction of water body in remote sensing images. Taking an area Hunan in China as an example, a water body sample database is first established, and then, the currently popular OCNet algorithm is applied to conduct experiments on the water body elements in the images. Recall ratio, precision ratio, and Mean Intersection over Union (MIoU) are used as accuracy measurement indicators to carry out identification and extraction of water body elements. The experimental results show that the proposed algorithm achieves satisfactory recognition results on both the training and the test sets, and the recall ratio, precision ratio, and MIoU are all more than 95%. Therefore, the proposed method can be applied for the automatic identification and extraction of large-scale, high-resolution water bodies in remote sensing images in real time and also provides a feasible way for the automatic identification and extraction of other ground feature elements in remote sensing images.

22.1 Introduction

Water body, as one of the most important type of basic geographic information, is the most common type of ground feature in remote sensing images. According to the shape of the water body, it can be divided into two categories: banded water bodies, such as rivers, canals, and ditches; and regular or irregular polygonal water bodies, such as reservoirs, lakes, and cisterns. Quickly and accurately extracting water information is of great significance to the study on ecological issues such as surface water changes, protection of aquatic plants and animals, river regulation, and pollution control.

B. Wu (✉)
PLA Army Engineering University, Nanjing, China
e-mail: mr_wubo@163.com

Y. Gao · X. Xie · Y. Yang
PLA Troops of 61175, Nanjing, China

L. C. Jain et al. (eds.), *Wireless Technology, Intelligent Network Technologies, Smart Services and Applications*, Smart Innovation, Systems and Technologies 258,
https://doi.org/10.1007/978-981-16-5168-7_22

The traditional remote sensing water body extraction method involves drawing the outline along the water body boundary manually, which has the advantage of high accuracy, but also the disadvantages of manpower and time consumption. With the development of remote sensing technology, researchers have found that the water bodies can absorb most of the incident energy in the visible and the near-infrared bands, and have a characteristic of weaker reflectivity on the images. Based on this characteristic [1], Mcfeeters [2] proposed the classic Normalized Difference Water Index (NDWI) that could clearly distinguish the water bodies from other types of ground features. However, the water body extracted according to this index still contains more background. Xu [3] found that water body's spectrum in the mid-infrared band is more prominent. Therefore, based on NDWI, Xu Hanqiu replaced the near-infrared band with the mid-infrared band and proposed a Modified Normalized Difference Water Index (MNDWI), which improved the accuracy of water body extraction.

In addition, based on the spectrum similarity, some scholars proposed water extraction methods based on supervised and unsupervised classifications, such as maximum likelihood classification [4], minimum distance, *K*-MEANS [5], and iterative self-organized data analysis (ISODATA) [6]. Supervised classification methods rely on artificially distinguishing the spectral characteristics of water pixels from that of non-water. Unsupervised classification methods do not require human experience, but they are inferior to supervised classification methods in classification accuracy. The existing unsupervised classification methods set the classification rules through the program or the software itself, and thus exclude the human experience from the algorithm, with a further step on the road to intelligence. However, in the face of remote sensing data showing "exponential growth", all the above-mentioned methods have certain limitations.

In recent years, with the rapid development of big data and computer hardware, deep learning technology based on big data and high-performance computing has shown significant application value in the field of remote sensing. In this paper, a water sample database for remote sensing images is established (total of 8225 sample images of 512 × 512 pixels), and a deep learning convolutional neural network suitable for remote sensing images is designed and constructed. The validity of the model is verified by taking a certain area in southern China as an example. The implementation process is shown in Fig. 22.1.

Fig. 22.1 Extraction and application of remote sensing images

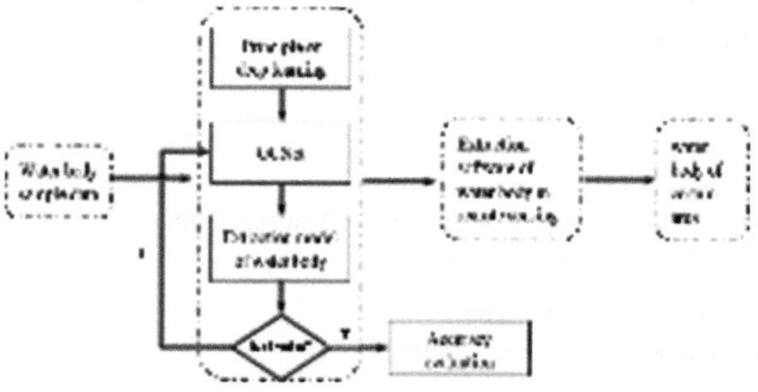

22.2 Construction of Remote Sensing Water Sample Database

Typical types of water bodies in remote sensing images include polygon-shaped lakes and reservoirs, long and small strip-shaped double-line and single-line rivers with obvious trunks, more concentrated and regular block-shaped paddy fields. In order to meet the generalization requirements of deep learning models, the samples must be as diverse as possible to build a relatively complete water body sample database. The specific implementation process is shown in Fig. 22.2.

The quality of the sample will directly affect the accuracy of the training model. Therefore, the image files and their corresponding vector files must meet the following conditions before being input to the model: (1) The remote sensing image is full and clear, with 5% for permitted cloudy areas or no cloud; (2) the band information and the spatial reference information are complete and correct; (3) the coordinate systems of the vector and the image files must be consistent; and (4) the vector file must be a polygon file. Generally, the image and the vector files need to be pre-processed to meet the above conditions. The purpose is to ensure that when using the remote sensing processing module gdal in Python to read the remote sensing images and the vector files, they can be trimmed at the same geographical location in order to generate labeled binary maps, which will be finally saved as samples.

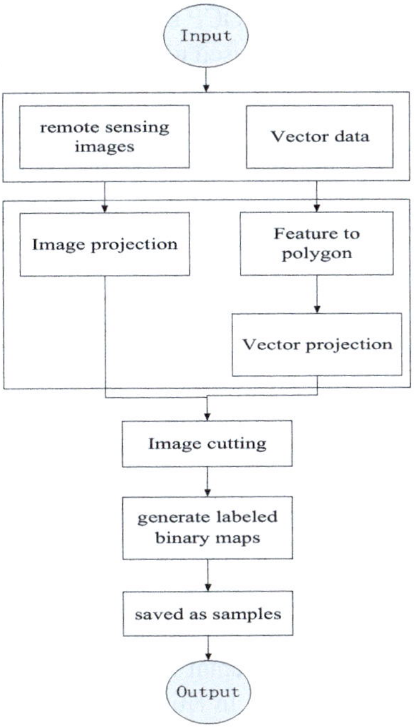

Fig. 22.2 Process of collecting remote sensing image samples

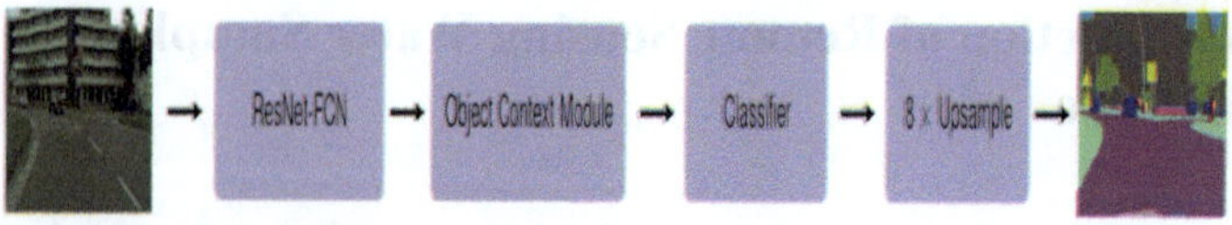

Fig. 22.3 Principle of OCNet

22.3 Principle of OCNet

Convolutional neural network (CNN) is a network structure designed specifically for image data in deep learning. In each layer of the CNN, the image features are obtained through convolution operation, and then enter the pooling layer, which reduces the amount of calculation, features, and parameters. However, data pooling will lead to a certain degree of information loss, affecting the accuracy of the results. In order to compensate for feature losses, some scholars have proposed the pyramid scene parsing network (PSPNet) [7], which explores the ability of global context information through the pyramid pooling module. The PSPNet divides the feature map into multiple regions, and all pixels in each region are considered as the context of this region. The object context network (OCNet) [8] proposes object context pooling (OCP), which collects information according to the object context. The schematic diagram of OCNet is shown in Fig. 22.3.

The feature maps are obtained from the input images through the ResNET-FCN network. In the OCP layer, the feature maps are transformed into new feature maps, which are then used to predict the label of each pixel. Finally, the output results are obtained by eight times of cyclic sampling. In OCNet, pixel-by-pixel prediction is not performed; instead, similar pixels are aggregated, and then, context segmentation is performed. By using the information of the pixel set belonging to the same object, a pixel label contained in the object is obtained, where the pixel set is called the object context.

22.4 Experiment and Analysis

22.4.1 Experimental Platform and Sample Data

The experiment uses Python and is implemented on an Ubuntu 16:04 operating system. The experimental hardware environment includes an Intel Xeon (R) Gold 6134 CPU@3.2 GHz, a 128 GB memory chip, and a NVIDIA Quadro P6000 graphics card.

The experimental sample data is high-score two remote sensing images of 75 different scenes in Hunan. A total of 8225 sample images of 512 $\times$ 512 pixels are generated, of which 1666 are used as the test set for testing the trained model, and

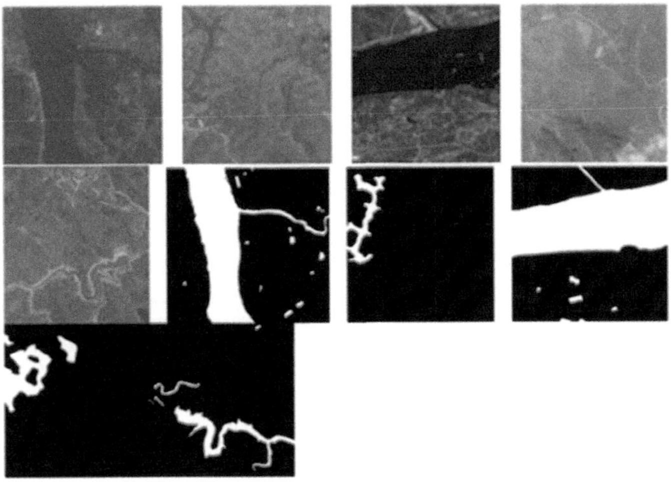

Fig. 22.4 An example of the training sample

the rest are used as the training set for training the model. An example of the training sample is shown in Fig. 22.4.

22.4.2 Accuracy Analysis

In deep learning, the standard metrics such as recall ratio, precision ratio, and Mean Intersection over Union (MIoU) [9] are often used as evaluation indicators. This paper also uses these three indicators to quantitatively evaluate the accuracy of the results.

The precision ratio refers to the number of positive predictions divided by the total number of predictions. The recall ratio refers to the proportion of predictions that are correctly predicted to be positive to the total number of real positive values.

The MIoU mainly describes the intersection average between the real and the predicted values. This indicator reflects the degree of capture of the target and the accuracy of the model. The formula for MIoU can be expressed as [10]:

$$\mathrm{MIoU} = \frac{1}{k+1}\sum_{i=0}^{k}\frac{P_{ii}}{\sum_{j=0}^{k}P_{ij}+\sum_{j=0}^{k}P_{ji}-P_{ii}}$$

where k is the number of extracted categories, P_{ii} is the number of matching pixels, P_{ij} is the number of i misclassified as j, and P_{ji} is the number of j classified as i.

Figure 22.5 shows the results of OCNet algorithm. Figure 22.5a shows the change of overall precision and recall ratios with the number of training epochs, on the test

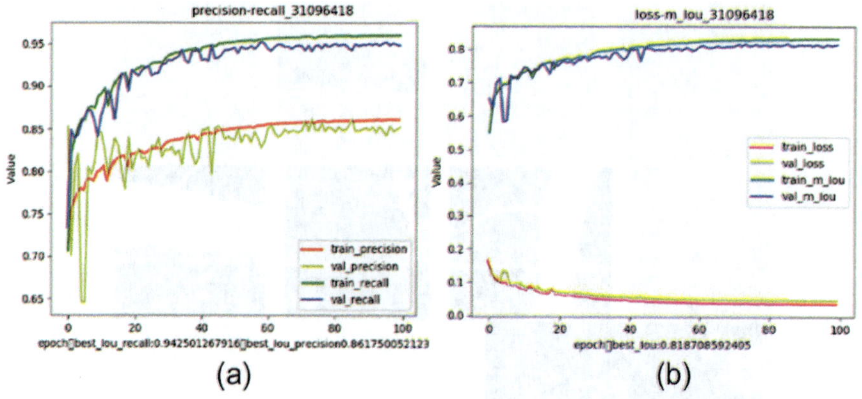

Fig. 22.5 Results of OCNet

and the training sets, respectively. Figure 22.5b shows the change of MIoU with the number of training epochs, on the test and training sets, respectively. The algorithm is trained 50 epochs, the batch size is 8, and the initial learning rate is 0.001.

The horizontal axis indicates the number of epochs, the vertical axis on the left indicates the precision and the recall ratios, while the vertical axis on the right indicates the MIoU. It can be seen from the figure that with the continuous increase of epochs, the algorithm converges at epoch 40, with the precision to 94%, the recall ratios to 86%, and the MIoU to 82%.

22.4.3 Water Extraction Results on the Test Set

The trained model is applied on the test data. Two images of the urban and the mountainous areas in southern China are selected as the test images. The test results are shown in Fig. 22.6. The identification of the polygon-shaped reservoir (a) in the urban area, the strip-shaped double-line river (b) in the mountainous area, and the case where there is part of land (c) is relatively accurate, and the result of the extracted boundary is good.

22.5 Conclusion

In this paper, through the training of OCNet model, a water body recognition system for remote sensing images is obtained. The experimental results show that the recognition and extraction results are good, but there is still a certain gap between the obtained and the ideal results. Further improvement can be done in the following two aspects:

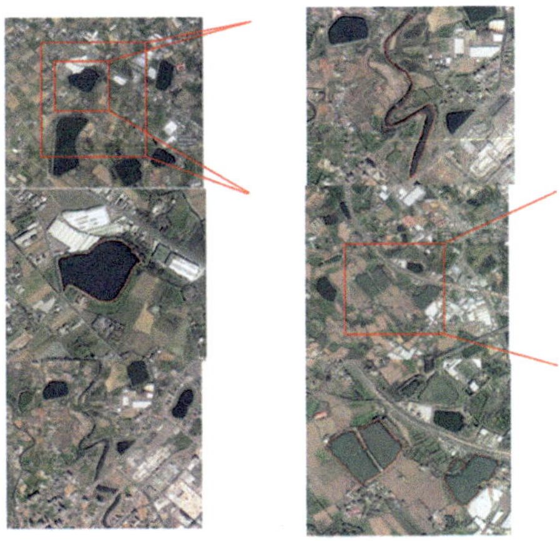

Fig. 22.6 Water extraction results on the test set

Firstly, diversity of the training sample should be increased. Currently, only water samples from a certain area in southern China have been established, which limits the application scope of the proposed model. Later, more samples in inland area should be added for training to increase the application scope of the proposed model.

Secondly, transfer learning (TL) can be introduced. By transferring the already trained weight parameters to new training tasks, TL avoids the random initialization of weights, accelerates the model training speed, and improves the training accuracy.

References

1. Chang, Q.R., Jiang, P.G., Zhou, Y., et al.: Introduction of Remote Sensing Technology. Science Press, Beijing (2004)
2. Mcfeeters, S.K.: The use of normalized difference water index (NDWI) in the delineation of open water features. Int. J. Remote Sens. **17**(7), 1425–1432 (1996)
3. Xu, H.Q.: A study on information extraction of water body with the modified normalized difference water index. J. Remote Sens. (05), 79–85 (2005)
4. He, Y.X., Zhou, J.: Comparison of the accuracy of some supervised classification methods in ENVI. Technol. Innov. Appl. **233**(13), 28–30 (2018)
5. Santos, A.C.S., Pedrini, H.: A combination of *K*-means clustering and entropy filtering for band selection and classification in hyperspectral images. Int. J. Remote Sens. **37**(13), 3005–3020 (2016)
6. Ma, C., Dai, Q., Liu, S.: A new method of remote sensing image segmentation based on PSO and isodata. Geomat. Inf. Sci. Wuhan Univ. **37**(1), 35–38 (2012)
7. Zhao, H., Shi, J., Qi, X., Wang, X., Jia, J.: Pyramid Scene Parsing Network. IEEE Computer Society (2016)
8. https://github.com/hszhao/PSPNET.2018.3.22
9. Yuan, Y., Wang, J.: OCNet: Object Context Network for Scene Parsing (2018)
10. Wang, J., Li, J., Ding, B., et al.: Performance analysis of semantic segmentation method based on deep learning in remote sensing image segmentation. Comput. Measure. Control **27**(7), 231–235 (2019)

Chapter 23
Khmer Named Entity Recognition Based on LSTM-CRF Model

Lei Teng, Xin Yan, Jun Xie, Feng Zhou, Guangyi Xu, and Yuanyuan Mo

Abstract Named Entity Recognition(NER), as a significant research content of information extraction, is broadly used in Natural Language Processing fields such as information retrieval, machine translation, question answering system, and so on. Named Entity Recognition in Khmer plays a supportive part in the study of Chinese–Khmer bilingual understanding. However, the differences between languages make it hard to transfer the common Chinese and English Named Entity Recognition into Khmer. Aimed at the problem that the output of long short-term memory neural network model does not consider the sequence of tags, the output of long short-term memory neural network model and Khmer entity feature is used to input features of conditional random field model to extract Khmer corpus entity. The result of experiment shows that the precision of the conditional random field model is increased by 1.32%, recall rate by 2.55% and the $F1$ value which is used to measure the precision and recall rate by 1.92% after using long short-term memory neural network as the inputted feature of conditional random field model, which unites the output of neural network model with the entities in Khmer. The outcome of experiment proves the effectiveness of this method.

L. Teng · X. Yan (✉) · J. Xie · F. Zhou
Faculty of Information Engineering and Automation, Kunming University of Science and Technology, Kunming 650500, China
e-mail: Kg_yanxin@sina.com

Yunnan Key Laboratory of Artificial Intelligence, Kunming University of Science and Technology, Kunming 650500, China

G. Xu
Yunnan Nantian Electronic Information Industry Co., Ltd., Kunming 650040, China

Y. Mo
School of Southeast and South Asia Languages and Culture, Yunnan Minzu University, Kunming 650500, China

L. C. Jain et al. (eds.), *Wireless Technology, Intelligent Network Technologies, Smart Services and Applications*, Smart Innovation, Systems and Technologies 258,
https://doi.org/10.1007/978-981-16-5168-7_23

23.1 Introduction

As a significant research content of information extraction, Named Entity Recognition is broadly used in information retrieval, machine translation, question answering system, and so forth. The study of Named Entity Recognition in Khmer plays an essential part in Chinese–Khmer bilingual understanding. In task of Khmer Named Entity Recognition, the lack of corpus resources and experts in Khmer linguistics makes the language processing research technology for the language lag behind. From the historical factors of Khmer language, we can see that Khmer language also has a variety of word formation problems, which makes its entity composition no single. Therefore, Named Entity Recognition in Khmer task is a more complex research work.

In research of Named Entity Recognition task, the main research methods are divided into rule-based, statistical model-based, and neural network-based methods.

Rule-based method: The main feature is to analyze corpus in a field, then ask experts in the language field to learn the corpus of the language, then summarize some characteristics of the language learned and some unique language rules of the language, and use rule matching and other methods to realize the Named Entity Recognition of the language at last. Example 1: Alfred et al. [1] presented a method on the basis of rules for Named Entity Recognition of Malay language. In this essay, they take part-of-speech tagging features method to select candidate entities from the corpus, and then identify entities according to the entity tagging in the manually tagged dictionary. Yan and Bi [2] also proposed a method on the basis of rule, which is mainly used for Vietnamese Named Entity Recognition. Through analysis of some characteristics of Vietnamese entities, they add artificial rules, and then use these rules to named entity recognition of Vietnamese. This kind of method needs to be written by linguists. This kind of work is not only time-consuming and labor-consuming, but also prone to conflict between rules due to rule inconsistency. Therefore, this method of Named Entity Recognition on the basis of rules is not the better method in practical work.

Research method based on Statistics: The main characteristics of this method is to integrate some features of artificial learning into models of statistical learning to realize entity recognition. The most important advantage of this method compared with the manual rule-making named entity recognition method lies in the use of easier model and less artificial annotation corpus for named entity recognition. Moreover, this method of machine learning entity features can not only be applied to a single language, but also to other languages that we have not been exposed to, we only need to linguistic this language Xi, we could have the model of Named Entity Recognition of languages, without any needs for experts to write rules for the language, saving resources. In this kind of statistical methods, the models which used for Named Entity Recognition contain Hidden Markov Model, maximum entropy, and conditional random fields. For instance, Yu et al. [3] used two-layer hidden Markov model to realize Chinese entity recognition. Zhang et al. [4] presented a multi feature cross entropy Chinese Named Entity Recognition model. By quoting a variety of external constraint features, the semantic information of the model prototype was added to obtain heuristic knowledge, and then this heuristic knowledge was combined with the maximum entropy model to obtain Chinese named entities. Pan et al. [5] presented a Named Entity Recognition method for Khmer language that integrates entity characteristics. This method firstly makes use of word segmentation model and part-of-speech tagging model to extract morphological and speech characteristics of Khmer language and then combines the unique language characteristics of Khmer language manually tagged. Then, the entity recognition of Khmer language is realized through conditional random field algorithm. This method has gotten good results in task of named entity recognition in Khmer. Pan et al. [6] presented a method for Vietnamese Named Entity Recognition on the basis of conditional random field. The method is to extract terms and lines of speech as feature templates according to the language characteristics of Vietnamese itself and then use conditional random field model to realize Vietnamese named entity recognition. However, this kind of learning method needs to understand the semantics to develop the characteristics of exercising model. The quality of feature setting directly affects the quality of Named Entity Recognition task results.

Method in the basis of neural network: The method solves the problem of using artificial feature engineering to train model and increase human labor. The method based on neural network can learn the semantic features of languages from the original corpus and then extract the information from the semantic features through neural network model to achieve Named Entity Recognition. The common neural network models for Named Entity Recognition include recurrent neural network (RNN) [7], long short-term memory (LSTM) [8, 9], and convolutional neural network (CNN) [10, 11]. Many scholars have used these methods in Named Entity Recognition research. Gregoire et al. [12] used cyclic neural network model to solve sequence annotation, and this model is more useful than the original statistical model in sequence annotation. Jin [13] proposed a method of biomedical Named Entity Recognition in the basis of recurrent neural network, which integrates sentence vector and word vector containing sentence information into bi-directional long short-term memory (Bi-LSTM) neural network. The result of experiment shows that the accuracy rate of the method is 1.40% higher than that of the traditional RNN model. Maimaitiayifu [14] proposed a method of Uyghur Named Entity Recognition on the basis of Bi-LSTM plus convolutional neural network and conditional random field (Bi-LSTM-CNN-CRF) model, which uses whirly neural network to train character features in Uyghur words, not only word vector and word vector are used, but also the part-of-speech features manually marked are input into neural network in the form of vector. Furthermore, it has gotten good outcome in Uighur Named Entity Recognition. Lample et al. [15] proposed a single-layer bi-directional cyclic neural network plus conditional random field to deal with the task of Named Entity Recognition and then achieved good results in datasets of multiple languages. Huang et al. [16] used the Bi-LSTM and conditional random field (Bi-LSTM-CRF) model to label sequences, connect the input context features and the spelling features of words directly with the output of the Bi-LSTM neural network and obtained good results through experiments.

To sum up, we can see that the deep learning method can abandon the traditional artificial feature engineering, thus saving a lot of human costs. After the practice of the research methods in Named Entity Recognition, the traditional Bi-LSTM model never consider order between the output tags when it obtains the complete context information, resulting in the poor effect of entity recognition. It is necessary to add conditional random field (CRF) model which can obtain the constraint rule from exercising data to ensure the correctness of the prediction tags. This essay presents a method of Named Entity Recognition on the basis of Bi-LSTM-CRF neural network [17]. The Bi-LSTM neural network model output and the unique entity features of Khmer are taken as the input features in CRF model; then, entity recognition of Khmer is realized by using CRF model in order to strengthen the influence of entity recognition of Khmer.

23.2 Khmer Named Entity Recognition on the Basis of Bi-LSTM-CRF

23.2.1 Khmer Entity Features

Khmer originated in the Cambodia countries of Southeast Asia. Due to historical reasons, Khmer has been influenced by many languages [18] in the process of development, resulting in the unique language characteristics of Khmer, which have the following impacts on the work of named entity recognition: First of all, Khmer vocabulary is different from English with clear boundary features, such as spaces, which leads to a high error rate in the work of word segmentation in Khmer, thus affecting the named entity recognition rate. Secondly, Khmer lacks the part-of-speech transformation features that play a great part in the recognition of naming entities, such as the feature that naming entities in English start with capital letters, which brings higher demands on mining deeper semantic information of Khmer. For the above reasons, it is very difficult for Khmer language to identify the named entity. However, linguists have found that the named entity of Khmer language still has some unique characteristics as follows.

Different from the position of Chinese entity deixis, the position of Khmer entity deixis in entity is mostly in the position of prefix. For example, the location of "ខេត្ត(province)" in the entity "Yunnan Province" is in front of the provincial name "Yunnan", that is to say, its composition is "ខេត្ត(province) យូណាន(Yunnan)";

The writing method of place names is similar to that in English. The place names and place names appear in the order of small to large, and the place names can appear next to each other. For example, the Khmer corresponding to "Pu'er, Yunnan Province" is "ទីក្រុង(city) Pu'er ខេត្ត(province) យូណាន(Yunnan)";

Different from the writing order of organization names in Chinese, organization names in Khmer are generally in the middle structure, and the attribute is in the post position, and the category of organization is in the position of prefix. For example, the Khmer language corresponding to the organization name "Kunming University of technology" is " សាកលវិទ្យាល័យ(University) វិទ្យាសាស្ត្រនិងបច្ចេកវិទ្យា(technology) គុនមីង(Kunming)";

In the Khmer name representation, the appellation words should be placed in front of the name, for example, "Auntie", "Li". The title indicating the position should be placed before the name and the space should be added before the position and the name. For example, " សាស្ត្រាចារ្យលី" means "Professor Li". People's names cannot appear next to each other. They have to be divided due to spaces, punctuations, and no notional words, such as: in the sentence of " ហ៊ុនសែនព្រះមហាក្សត្រអ្នកដឹកនាំសំខាន់នៅក្នុងប្រទេសកម្ពុជា(Hun Sen and Sihamoni are important leaders in Cambodia)", the names of " ហ៊ុនសែន" and " ព្រះមហាក្សត្រ(Sihamoni)" are separated by spaces;

The place names usually need to be separated by commas before and after, but if the place names are embedded in the organization structure names, they can appear next to each other. For example, the organization name "Kunming charity" corresponds to Khmer language " គុនមីងសប្បុរសធម៌".

23.2.2 Bi-LSTM-CRF Model

The model in this paper for Named Entity Recognition is Bi-LSTM-CRF model (see Fig. 23.1). It adds a layer of conditional random field model after the original hidden layer of Bi-LSTM so that model can not only obtain context information of a single word, but still take full advantages of the annotation information before or after the sentence. Because complex entity structure of Khmer language is to enhance the influence from entity recognition of Khmer language, this essay uses the output of Bi-LSTM neural network and entity features of Khmer language as the input features of CRF model, which is similar to the use of maximum entropy features. Finally, the named entity recognition of Khmer language is realized by learning CRF.

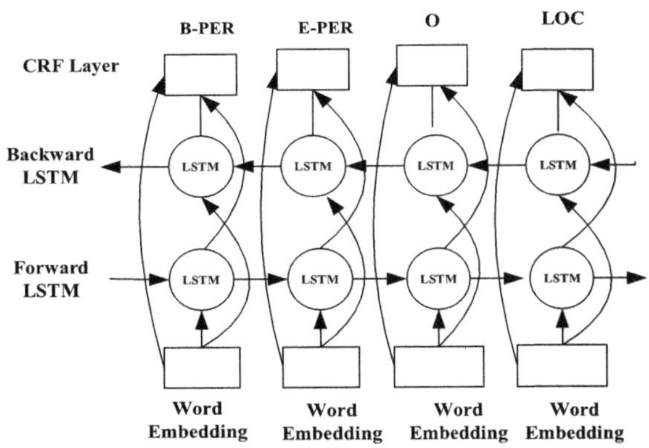

Fig. 23.1 Bi-LSTM-CRF mode

The process of Named Entity Recognition in the basis of Bi-LSTM-CRF network model is as follows:

23.2.3 Bi-LSTM Layer

In this essay, input layer of Bi-LSTM layer in Bi-LSTM-CRF network model is word vector, which can be expressed as $e = (e_1, e_2, \ldots, e_n)$. The hidden state of sequence in hidden layer of Bi-LSTM layer can be expressed as $H = \{h_1, h_2, \ldots, h_n\}$. According to the Bi-LSTM network structure chart, the hidden state sequence here is the forward hidden state sequence $\left(\overrightarrow{h_1}, \overrightarrow{h_2}, \ldots, \overrightarrow{h_n}\right)$ and backward hidden state sequence obtained by reverse encoding $\left(\overleftarrow{h_n}, \overleftarrow{h_{n-1}}, \ldots, \overleftarrow{h_1}\right)$. In the forward representation sequence $\left(\overrightarrow{h_1}, \overrightarrow{h_2}, \ldots, \overrightarrow{h_n}\right)$, $\overrightarrow{h_t}$ represents the forward notation of the t time of vector. The calculation is as follows:

$$\overrightarrow{h_t} = \text{LSTM}\big[(e_n, e_{n-1}, \ldots, e_t)\big] \tag{23.1}$$

where LSTM(·) represents the calculation of the LSTM model, namely e $=$ $(e_n, e_{n-1}, .., e_1)$, obtaining the backward notation sequence by backward coding $\left(\overleftarrow{h_n}, \overleftarrow{h_{n-1}}, \ldots, \overleftarrow{h_1}\right)$, $\overleftarrow{h_t}$ represents the forward of the t time of vector. The calculation is as follows:

$$\overleftarrow{h_t} = LSTM\big[(e_n, e_{n-1}, \ldots, e_t)\big] \tag{23.2}$$

The backward representation sequence receives the reverse sequence of the input vector as the input, where LSTM(·) represents the calculation of the LSTM model.

After the above sequence input, the hidden state representation sequence h_t of the hidden layer of Bi-LSTM network structure is composed of forward representation and backward representation: $h_t = \left[\overrightarrow{h_t}; \overleftarrow{h_t}\right]$, at this time, the hidden state sequence of the hidden layer can be obtained as $\{h_1, h_2, \ldots, h_n\} \in \mathbb{R}^{n \times m}$, where [;] is the connection.

23.2.4 CRF Layer

In Bi-LSTM-CRF network model mentioned in this essay, the CRF is mainly to assign markers to each word; then calculate the score of the whole marker sequence of the input sequence; obviously the score of the marker sequence is composed of two parts—one is the score of the word marker, the other is the score of the transfer between the marker sequences; finally, select the final marker sequence with high score as the final part of this paper tag results.

Firstly, from the description of the above Bi-LSTM layer, we can see that the hidden state sequence output by the hidden layer of the Bi-LSTM network model can be shown as $\sqrt{(h_1, h_2, \ldots, h_n)}$. Then a linear layer needs to be connected to the hidden layer. The main function of linear layer maps the hidden state representation sequence from m dimension to k dimension, where k represents the number of labels in the annotation set used in this paper. In this paper, the annotation method of BIOES is used. The specific content is shown in Table 23.1.

Khmer corpus has been pre-processed by word segmentation, part-of-speech tagging, and so on. Taking the basic feature of word form and part-of-speech as one of the input features of conditional random field can enrich the semantic information of input features and improve the effect of Khmer entity recognition. The basic feature template designed in this paper is shown in Table 23.2.

The basic feature template simply describes the part-of-speech or morphology of the core word and its up and down words, and its semantic information is very limited.

Table 23.1 BIOES tagging set

Entity tag	Start tag	Middle	End	Single entity tag
Person name	B-PER	I-PER	E-PER	S-PER
Organization	B-ORG	I-ORG	E-ORG	S-ORG
Place	B-LOC	I-LOC	E-LOC	S-LOC
Non-entity mark	O	O	O	O

Table 23.2 Basic feature template

Index	Template form	Template meaning
1	Current word(0)	Current word
2	Current word(−1)	The first word on the left of the current word
3	Current word(−2)	The second word on the left of the current word
4	Current word(1)	First word on the right of current word
5	Current word(2)	The second word on the right of the current word
6	Current POS(0)	The part-of-speech of current words
7	Current POS(−1)	The part-of-speech of the first word on the left of the current word
8	Current POS(−2)	The part-of-speech of the second word on the left of the current word
9	Current POS(1)	The part-of-speech of the first word on the right of the current word
10	Current POS(2)	The part-of-speech of the second word on the right of the current word

According to the entity characteristics of Khmer entity mentioned above, the entity indicators are collected manually and the word list of them is constructed, as shown in Table 23.3. These entity indicators can provide rich entity information for Khmer named entity recognition, so as to enhance the impact of Khmer Named Entity Recognition.

Based on the rich entity features in Khmer provided above and the entity indicators listed in Table 23.3, the feature templates of entity information are established.

Then, the $n \times k$ fraction matrix output by the Bi-LSTM network structure through the linear layer is defined as P, where $P_{i,j}$ in the matrix P represents the fraction marked as the jth label in the ith word of a sentence.

After that, the output matrix P in Bi-LSTM neural network and Khmer entity features is taken as the input in CRF model. And annotation information of input sentence sequence is obtained through CRF model learning. In this case, for the input sequence: $(x_1, x_2, \ldots, x_n)$, the possibility of prediction tag: $(y_1, y_2, \ldots, y_n)$. Conditional probability is expressed as:

$$p(y|x) = \frac{1}{Z(x)} \exp\left(\sum_k \lambda_k f_k(y, x)\right) \quad (23.3)$$

Among them,

$$Z(x) = \sum_y \exp\left(\sum_k \lambda_k f_k(y, x)\right) \quad (23.4)$$

Table 23.3 Khmer entity demonstrative words list

Entity classification	Pointer word
Person name	ព្រឹទ្ធបុរស (dean), សាស្ត្រាចារ្យ(professor), វេជ្ជបណ្ឌិត(doctor), គ្រូស្រី(female teacher), គ្រូប្រុសប្រុស(male teacher), មិត្តរួមថ្នាក់(classmate), ពូ(uncle) មីង(aunt) ជីតា(grandfather), យាយ(grandmother), ប្រធានាធិបតី(president), នាយករដ្ឋមន្ត្រី(prime Minister)
Place name	ខេត្ត(province), ប្រទេស(country), ទីក្រុង(city), ខោនធី(country), ស្រុក(district), ភូមិ(village), ផ្លូវហាយវេ(highway), ភ្នំ(mountain), ទន្លេ(river), បឹង(lake), សមុទ្រ(sea)
Organization name	សាកលវិទ្យាល័យ(university), វិទ្យាល័យ(high school), វិទ្យាល័យ(junior high school), សាលាបឋមសិក្សា(primary school), អង្គការ(organization), ក្រុមហ៊ុន(company), មន្ទីរពេទ្យ(hospital), គ្លីនិក(clinic), ស្ថាប័ន(organization)

Table 23.4 Khmer entity information feature template

Index	Template form	Meaning of templates
11	CUR_PER_SUF	Is current word a person name demonstrator
12	NEXT_PER_SUF	Does current word contain a person name indicator
13	CUR_LOC_SUF	Is current word a place name indicator
14	NEXT_LOC_SUF	Does two words on the left of the current word contain the place name indicator
15	CUR_ORG_SUF	Is current word an organization name indicator
16	NEXT_ORG_SUF	Does two words on the left of the current word contain the organization name indicator
17	CUR_PER_NAME	Is current word a common person name
18	CUR_LOC_NAME	Is current word a common place name
19	CUR_ORG_NAME	Is the current word a common organization name

$Z(x)$ expressed as normalization factor. λ_k get through training data. $f_k(y, x)$ represents the sum of transfer features and state features of each location.

Finally, the maximum likelihood estimation is used to calculate the parameter λ, and the Viterbi algorithm is to generate the best output sequence.

23.3 Experiments

23.3.1 *Datasets*

Khmer corpus needed for the experiment includes 138,396 Khmer words after segmentation and manual tagging. A Khmer corpus published by PAN localization Khmer (PCL), which contains 73,127 Khmer words and has been segmented and part-of-speech tagging, but the corpus does not have entity tagging. That is to say, the entities in the corpus need to be labeled manually. The remaining 66,269 Khmer words are divided tools. Finally, the Khmer language experts are invited to label all the corpus named entities to get corpus of Khmer named entity recognition. The annotation set for manual annotation mentioned in this essay is as same as shown in Table 23.4. 80% of the corpus is used as training set and rest of 20% as testing set. Proportion of datasets is shown in Table 23.5.

23.3.2 *Evaluation Metric and Parameter Setting*

The evaluation metric used in this paper is accuracy P, recall R and F_1 severally. This is the specific formula:

Table 23.5 Corpus of Khmer named entity recognition

Corpus classification	Total no. of words	No. of person names	No. of place names	No. of organization names
Total corpus	138,396	2672	3485	3027
Training set	110,717	2137	2788	2421
Test set	27,679	535	697	606

Table 23.6 Influence on different dropout parameters

Model	Accuracy (%)	Recall (%)	F_1 value (%)
Bi-LSTM-0.3	73.05	73.13	73.08
Bi-LSTM-0.5	73.03	73.16	73.09
Bi-LSTM-0.6	73.01	73.19	74.10

$$P = \frac{\text{Number of named entities correctly recognized}}{\text{Total number of named entities identified}} \times 100\% \tag{23.5}$$

$$R = \frac{\text{Number of named entities correctly recognized}}{\text{Total number of named entities in Corpus}} \times 100\% \tag{23.6}$$

$$F_1 = \frac{2 \times P \times R}{P + R} \tag{23.7}$$

The super parameter setting in the neural network model has a great impact on the implementation in named entity recognition task. Then, the text sets the super parameters as follows:

In this paper, word embedding is used to train fixed dimension word vectors. This method can capture semantic and syntactic information between word and can maintain semantic relationship between words. The dimension of word vector is set to 100. The word vector is trained by the word2vec tool published by Google. The selected optimization algorithm is the most widely used stochastic gradient descent (SGD) [19] on Bi-LSTM neural network, with the learning rate set to 0.01, and dropout parameter set to 0.3, 0.5, and 0.6 for comparison. The specific experimental data is shown in Table 23.6. In this paper, the dropout parameter is set to 0.6.

23.3.3 *Experimental Analysis*

To check the rationality of proposed model and the necessity of each module in model effectively, this paper conducts four groups of comparative experiments on the Khmer named entity recognition task, and these four experiments are performed in the corpus composed of above. The evaluation method is still the above-mentioned accuracy, recall rate, and F_1 value.

Experiment 1: The main objective of this experiment is to use named entity recognition result of CRF model as a standard, then to verify the performance difference of recurrent neural network model contrasted with the traditional statistical model. In experiment, CRF ++ is used as the tool to realize CRF model, and the basic features composed of morphology and part-of-speech are used as input features. The model accuracy is 70.85%, recall rate is 71.98%, and value of F_1 is 71.41%.

Experiment 2: This experiment is to prove the influence of CRF model on the recognition of Bi-LSTM neural network model. Main function of CRF is to optimize output result label of neural network according to the relationship between adjacent labels. The specific operation of the experiment is to add a CRF layer after Bi-LSTM neural network model to output optimal tag sequence. The specific experiment data is shown in Table 23.7.

From Table 23.7, the recognition effect of neural network model is significantly improved after adding CRF. For the Bi-LSTM neural network model, after adding CRF model, its F_1 value is increased by 1.28%. Compared with the basic experiment 1, its F_1 value increased by 6.48%. Therefore, it can be concluded that adding the linear CRF module is helpful for improving the recognition rate of the model.

Experiment 3: The objective of experiment is to verify the effect of Khmer entity features on Khmer entity recognition of Bi-LSTM-CRF neural network model. The specific method is to take output of Bi-LSTM neural network model and Khmer entity features as the input features of CRF model for Khmer named entity recognition. The outcome of this experiment is shown in Table 23.8.

From the above table, the CRF model with entity features significantly improves the recognition influence of the neural network model. For the Bi-LSTM-CRF neural network model used in this essay, after using Bi-LSTM neural network, the output of the neural network model and the entity features in Khmer language are taken as the input features of the CRF model; accuracy rate, recall rate, and F_1 value are increased by 1.32%, 2.55%, and 1%, respectively, 1.92%. Compared with the basic experiment 1, accuracy, recall, and F_1 values of this method are increased by 7.83%, 8.99%, and 8.4%, respectively. Therefore, it can be concluded that adding Khmer entity features to CRF model can enhance the recognition rate of the model.

Table 23.7 Effect of adding CRF model on model performance

Model	Accuracy (%)	Recall (%)	F_1 value (%)
Bi-LSTM	75.67	77.58	76.61
Bi-LSTM-CRF	77.36	78.42	77.89

Table 23.8 Effect of adding Khmer entity features to CRF model on model performance

Model	Accuracy (%)	Recall (%)	F_1 value (%)
Bi-LSTM-CRF	76.89	78.03	77.46
Bi-LSTM + CRF	78.68	80.97	79.81

23.4 Conclusion

In the chapter, the Bi-LSTM-CRF model is proposed for Named Entity Recognition of Khmer. Among them, on the basis of Bi-LSTM-CRF model, the output of Bi-LSTM neural network and entity features of Khmer are taken as the inputted features of CRF model, which enriches the input of CRF model and improves effectiveness of model in the recognition of Khmer named entities. However, the training time is too long due to the effect of internal parameters of the model. How to optimize the model and shorten the training time is the next problem to be studied. The research results of this paper are not only applied to the study of Named Entity Recognition in Khmer, but also have an important reference for other languages.

Acknowledgements This work is supported by National Nature Science Foundation under Grant: No. 61562049 and No. 61462055.

References

1. Alferd, R., Leong, L.C., On, C.K., et al.: Malay named entity recognition based on rule-based approach. Int. J. Mach. Learn. Comput. **4**(3), 300–306 (2014)
2. Yan, D.H., Bi, Y.D.: Rule-based recognition of vietnamese named entities. J. Chin. Inf. Process. **28**(5), 198–205 (2014)
3. Yu, H.K., Zhang, H.P., Qun, L., Lv, X.Q., Shi, S.C.: Chinese named entity identification using cascaded hidden Markov model. J. Commun. **27**(2), 87–94 (2006)
4. Zhang, Y.J., Xu, Z.T., Xue, X.Y.: Fusion of multiple features for Chinese named entity recognition based on maximum entropy model. J. Comput. Res. Dev **45**(6), 1004–1010 (2008)
5. Pan, H.S., Xin, Y., Yu, Z.T., Guo, J.Y.: A Khmer named entity recognition method by fusing language characteristics. J. Chin. Inf. Process (1), 4003–4007 (2014)
6. Pan, Q.Q., Zhou, F., Yu, Z.T., Guo, J.Y., Xian, Y.T.: Recognition method of Vietnamese named entity based on conditional random fields. J. Shandong Univ. (Nat. Sci.) **49**(1), 76–79 (2014)
7. Rafal, J., Wojciech, Z.: An empirical exploration of recurrent network architectures. In: Proceedings of the 32nd International Conference on Machine Learning (ICML-15), pp. 2342–2350 (2015)
8. Hu, Z.T., Ma, X.Z., Liu, Z.Z., Eduard, H., Eric, P.: Harnessing deep neural networks with logic rules. In: Proceedings of ACL-2016, pp. 21–30 (2016)
9. Wu, Y.G., Jiang, M., Lei, J.B., et al.: Named entity recognition in Chinese text using deep neural network. Studies in Health Technology and Information, pp. 624–628 (2015)
10. Santos, C.D., Guimaraes, V., Niteroi, R.J., Janeiro, R.D.: Boosting named entity recognition with neural character Embedding's. In: Proceedings of NEWS 2015 The Fifth Named Entities Workshop, pp. 25–30 (2015)
11. Zhang, J., Song, Y., Zhang, C., Liu, S.: Evolutionary hierarchical Dirichlet processes for multiple correlated time varying corpora. In: Proceeding of the 16th ACM SIGKDD International Conference on Knowledge Discovery and Data Mining, Washington, USA, pp. 1079–1088 (2010)
12. Gregoire, M., Yann, D.H., et al.: Using recurrent neural networks for slot filling in spoken language understanding EEE/ACM. Trans. Audio, Speech, Lang. Process 530–540 (2015)
13. Jin, L.K.: Named entities recognition based on recurrent neural network in biomedical literature. Dalian University of Technology, pp. 1–76 (2016)

14. Maimaitiayifu, Silamu, W., Muhetaer, P., Yang, W.Z.H.: Uyghur named entity recognition based on Bi-LSTM-CNN-CRF model. Comput. Eng. (44), 230–236 (2018)
15. Lample, G., Ballesteros, M., Subramanian, S., et al.: Neural architectures for named entity recognition. ArXiv preprint arXiv:1603.01360 (2016)
16. Huang, Z., Xu, W., Yu, K.: Bidirectional LSTM-CRF models for sequence tagging. ArXiv preprint arXiv:1508.01991 (2015)
17. Chen, G.Q.: On the evolution of Wa prepositions in Khmer. Minority Lang. Chin. (4), 32–37 (1999)
18. Mikolov, T., Sutskever, I., Chen, K., et al.: Distributed representations of words and phrases and their compositionality. ArXiv preprint arXiv:1310.4546 (2013)
19. Bottou, L.: Large-scale machine learning with stochastic gradient descent. In: Proceedings of COMPSTAT'2010, Physica-Verlag HD, pp. 177–186 (2010)

Chapter 24
Identification of Transfer Function of Position Servo System

Deng Ming Zhang, Da Wei Gu, Fu Men Cai, Shu Heng Chen, and Lun Kuan Ye

Abstract In the three closed-loop servo system, the position loop is the key consideration object, and the transfer function directly reflects the relationship between the position instruction signal and the feedback signal. Because the servo system contains controller, motor, mechanism, control algorithm, and so on, the whole system is more complex, and all kinds of coefficient constant cannot be obtained accurately, so it is difficult to obtained transfer function accurately by modeling and simulation method. In order to improve the identification accuracy of the system transfer function, the amplitude-frequency and phase-frequency data are obtained by experimental method, and the algorithm program is written by MATLAB. The algorithm can identify the system transfer function quickly and accurately, and can achieve the identification accuracy that the absolute error of amplitude-frequency characteristic and phase-frequency characteristic is very small. It has great practical value for the research of servo control system.

24.1 Introduction

Servo system is also called follow-up system. However, the control strategy used in most follow-up systems, designed in practical application is based on the mathematical model of the controlled object, so it is usually required to measure the transfer function of the servo system. Therefore, in the research of servo control system, in order to better control the controlled object and understand the performance of the system, it is usually necessary to measure the transfer function [1] of the servo control system. Because the process of establishing the transfer function of the system according to the method of theoretical analysis is more complicated, the determination of each parameter of the motor also needs some estimates and assumptions, and the influence of friction torque cannot be ignored, and the calculation is inaccurate.

D. M. Zhang (✉) · D. W. Gu · F. M. Cai · S. H. Chen · L. K. Ye
Shanghai Aerospace Control Technology Research Institute, Shanghai 201109, China

Shanghai Servo System Engineering Technology Research Center, Shanghai 201109, China

L. C. Jain et al. (eds.), *Wireless Technology, Intelligent Network Technologies, Smart Services and Applications*, Smart Innovation, Systems and Technologies 258,
https://doi.org/10.1007/978-981-16-5168-7_24

Finally, the accuracy of transfer function is not good. Therefore, the test method is often used to measure the transfer function of servo control system.

The dynamic characteristics of the system are mainly analyzed in time domain and frequency domain. Therefore, there are two methods to measure the transfer function, one is to use the unit step response measurement method in time domain, the other is to use the sweep feedback signal tracking method. The unit step response measurement method is to add a known transient disturbance to the identification system, test the step response curve of the system, and analyze the step response curve to obtain the transfer function of the system. The tracking method of sweep feedback signal is to analyze the data of feedback signal by inputting sinusoidal signal of the same amplitude and different frequency to the system, in which each frequency point inputs not less than 3 cycles of sinusoidal signal. Data analysis of feedback signals, the amplitude-frequency characteristic curve, and phase-frequency characteristic curve are obtained, and the transfer function of the system is obtained.

In view of the identification transfer function of amplitude-frequency characteristic curve and phase-frequency characteristic curve, many scholars have put forward various identification methods. Among them, the least square method is used to fit the amplitude-frequency and phase-frequency curves to determine the undetermined parameters [2]. The method is only effective for simple and smooth curves, and the fitting accuracy for complex and tortuous curves is low or even invalid. The method of derivation and construction of Simulink control system model is used to obtain the of transfer function [3]. The method is complex, theoretical, and difficult to be used in engineering practice. For this reason, the MATLAB algorithm program is written. The algorithm can quickly and accurately identify the transfer function of the system, and it is simple and reliable. The optimal solution of the transfer function can be obtained by repeated iteration.

24.2 Test Principles

The transfer function is based on the Laplace transform, the relationship between input and output of a linear constant system described by its own parameters, which expresses the inherent characteristics of the system. The transfer function of servo system takes servo controller and servo mechanism as a whole and obtains the relationship between position instruction input and load position feedback output. The working principle of servo system is shown in Fig. 24.1. The main principle of transfer function is obtained by measuring the frequency characteristics of the system. A set of 5 sine wave position signals with the same amplitude are transmitted at different frequencies using the IPC, and the controller solves the position signal and controls the servo mechanism to drive the load motion and records the load position feedback curve at each frequency; according to the relationship between position instruction and position feedback curve, the amplitude-frequency and phase-frequency data of the system are calculated, and the Bode diagram is drawn, identification of transfer

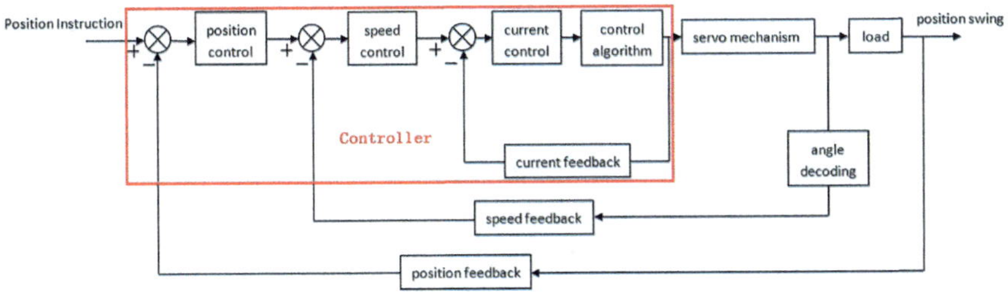

Fig. 24.1 Schematic diagram of servo system

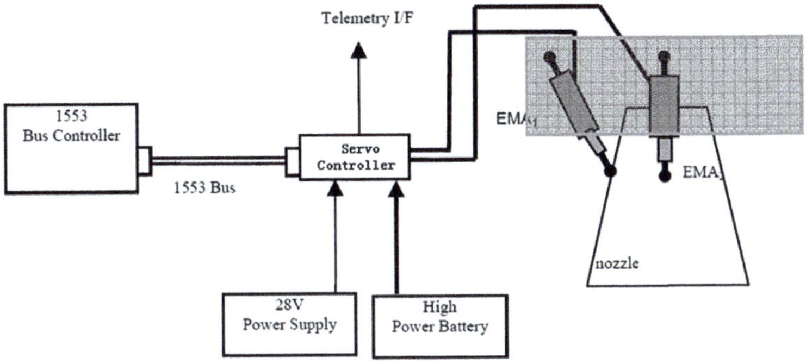

Fig. 24.2 Composition of servo system

function for frequency domain data, and the transfer function of the system can be identified by iterative program.

The test hardware which includes power, computer, industrial control computer, controller, servo mechanism, load table, and servo system composition is shown in Fig. 24.2. The position instruction is sent to the controller by 1553B bus. The controller calculates and outputs six PWM waves according to the algorithm in the virtual wire frame of Fig. 24.1. The PWM wave drives the servo mechanism to move back and forth under the action of the power amplifier circuit, a load nozzle swings according to the instruction under the action of the servo mechanism; during the working process of the system, the servo controller continuously collects signals such as voltage, current, position feedback, speed feedback, and so on, and these signal data flow through 1553B bus input to the computer.

24.3 Design of Transfer Function Identification Algorithm

According to the position instruction, position feedback, amplitude frequency, and phase-frequency data, with the frequency characteristic data of the controlled object, the transfer function can be obtained by using the frequency domain method. Considering that there are complex resonant links in the actual system, because of the high parameter order, in order to reduce the computational complexity, the higher order system is decomposed into multiple subsystems with smaller dimensions and fewer parameters. According to the measured frequency characteristic data, each frequency link is identified in turn, so that the complete transfer function can be obtained.

The principle of transfer function identification is as follows [4–9]:

$$G(s) = \frac{U_o}{U_i} = \frac{b_0 + b_1 s + b_2 s^2 + \cdots + b_m s^m}{1 + a_1 s + a_2 s^2 + \cdots a_n s^n} n > m \tag{24.1}$$

$$\begin{aligned} G(jw) &= \frac{\left(b_0 - b_2 w^2 + b_4 w^4 - \cdots\right) + j\left(b_1 w - b_3 w^3 + b_5 w^5 - \cdots\right)}{\left(1 - a_2 w^2 + a_4 w^4 - \cdots\right) + j\left(a_1 w - a_3 w^3 + a_5 w^5 - \cdots\right)} \\ &\triangleq \mathrm{Re}(w) + j\mathrm{Im}(w) \end{aligned} \tag{24.2}$$

Formula, U_o as feedback output, U_i as Input instruction, b_m as molecular coefficient, a_n as denominator coefficient, s as a factor, m and n as order, w as frequency, $\mathrm{Re}(w)$ as reality, $\mathrm{Im}(w)$ as imaginary.

Formula (24.2) both sides multiplied $\left(1 - a_2 w^2 + a_4 w^4 - \cdots\right) + j\left(a_1 w - a_3 w^3 + a_5 w^5 - \cdots\right)$ achieved

$$\begin{aligned} &\left(b_0 - b_2 w^2 + b_4 w^4 - \cdots\right) + j\left(b_1 w - b_3 w^3 + b_5 w^5 - \cdots\right) \\ &= [\mathrm{Re}(w) + j\mathrm{Im}(w)] \\ &\left[\left(1 - a_2 w^2 + a_4 w^4 - \cdots\right) + j\left(a_1 w - a_3 w^3 + a_5 w^5 - \cdots\right)\right] \\ &= \mathrm{Re}(w) + j\mathrm{Im}(w) + \mathrm{Re}(w)\left(-a_2 w^2 + a_4 w^4 - \cdots\right) \\ &\quad - \mathrm{Im}(w)\left(a_1 w - a_3 w^3 + a_5 w^5 - \cdots\right) \\ &\quad + j\left[\mathrm{Re}(w)\left(a_1 w - a_3 w^3 + a_5 w^5 - \cdots\right) + \mathrm{Im}(w)\left(-a_2 w^2 + a_4 w^4 - \cdots\right)\right] \end{aligned} \tag{24.3}$$

The real and imaginary parts of the equation are:

$$\begin{aligned} \mathrm{Re}(w) = &\left[\, 1\ 0\ -w^2\ 0\ w^4 \ldots \mathrm{Im}(w)w\ \mathrm{Re}(w)w^2\ -\mathrm{Im}(w)w^3\ -\mathrm{Re}(w)w^4 \ldots \right] \\ &\left[\, b_0 \ldots b_m\ a_1 \ldots a_n \,\right]^T \end{aligned} \tag{24.4}$$

$$\begin{aligned} \mathrm{Im}(w) = &\left[\, 0\ w\ 0\ -w^3\ 0 \ldots -\mathrm{Re}(w)w\ \mathrm{Im}(w)w^2\ \mathrm{Re}(w)w^3\ -\mathrm{Im}(w)w^4 \ldots \right] \\ &\left[\, b_0 \ldots b_m\ a_1 \ldots a_n \,\right]^T \end{aligned} \tag{24.5}$$

Assumption $G(s)$ measured amplitude-frequency characteristics are $G(\omega)$, and phase-frequency characteristics are $\varphi(\omega)$. Then the corresponding real and imaginary frequency characteristics can be calculated from the following formula:

$$\mathrm{Re}(w_i) = 10^{[G(w_i)/20]} \cos[\varphi(w_i)] \; i = 0, 1, 2 \tag{24.6}$$

$$\mathrm{Im}(w_i) = 10^{[G(w_i)/20]} \sin[\varphi(w_i)] i = 0, 1, 2 \tag{24.7}$$

Set the number of measured frequencies is *L*, use $\widehat{\mathrm{Re}}(w)$ replacement $\mathrm{Re}(w)$ in the formula (24.4), use $\widehat{\mathrm{Im}}(w)$ replacement $\mathrm{Im}(w)$ in the formula (24.5). Considering the test error and calculation error, the matrix Eq. (24.8) is obtained by arranging (24.4) and (24.5):

$$\begin{bmatrix} \hat{\mathrm{Re}}(w_1) \\ \cdots \\ \hat{\mathrm{Re}}(w_L) \\ \hat{\mathrm{Im}}(w_1) \\ \cdots \\ \hat{\mathrm{Im}}(w_L) \end{bmatrix} = \begin{bmatrix} 1 & 0 & -w_1^2 & 0 & w_1^4 & \cdots & \hat{\mathrm{Im}}(w_1)w_1 & \hat{\mathrm{Re}}(w_1)w_1^2 & -\hat{\mathrm{Im}}(w_1)w_1^3 & \cdots \\ \cdots & \cdots & \cdots & \cdots & \cdots & \cdots & \cdots & \cdots & \cdots & \cdots \\ 1 & 0 & -w_L^2 & 0 & w_L^4 & \cdots & \hat{\mathrm{Im}}(w_L)w_L & \hat{\mathrm{Re}}(w_L)w_L^2 & -\hat{\mathrm{Im}}(w_L)w_L^3 & \cdots \\ 0 & w_1 & 0 & -w_1^3 & 0 & \cdots & -\hat{\mathrm{Re}}(w_1)w_1 & \hat{\mathrm{Im}}(w_1)w_1^2 & \hat{\mathrm{Re}}(w_1)w_1^3 & \cdots \\ \cdots & \cdots & \cdots & \cdots & \cdots & \cdots & \cdots & \cdots & \cdots & \cdots \\ 0 & w_L & 0 & -w_L^3 & 0 & \cdots & -\hat{\mathrm{Re}}(w_L)w_L & \hat{\mathrm{Im}}(w_L)w_L^2 & \hat{\mathrm{Re}}(w_L)w_L^3 & \cdots \end{bmatrix} \begin{bmatrix} b_0 \\ \cdots \\ b_m \\ a_1 \\ \cdots \\ an \end{bmatrix} + \begin{bmatrix} \varepsilon_R(W_1) \\ \cdots \\ \varepsilon_R(W_L) \\ \varepsilon_I(W_1) \\ \cdots \\ \varepsilon_I(W_L) \end{bmatrix} \tag{24.8}$$

Simplify the above formula to:

$$Y = HX + \varepsilon$$

The *Y* is the observation vector, the *H* is the coefficient matrix, the *X* is the unknown parameter vector, and the ε is the measurement error vector. Assume $E(\varepsilon) = 0$, According to the matrix transformation, the *X* are as follows:

$$\hat{X} = \left[\hat{b}_0 \ldots \hat{b}_m \; \hat{a}_1 \; \ldots \hat{a}_n \right]^T = \left(H^T H\right)^{-1} H^T Y \tag{24.9}$$

The coefficient of the molecular denominator of the transfer function is calculated according to formula (24.9), and the transfer function *G*(*s*) is obtained.

24.4 Example Analysis of Transfer Function Identification

A sinusoidal position signal of 0.5° and different frequencies is sent to the servo mechanism by the controller, so that the servo mechanism works at different frequencies

[10, 11]. Locator software automatically saves position instruction and position feedback data, by formula (24.1) calculate $G(s)$, and according to formula (24.10) and formula (24.11), the amplitude-frequency data and phase-frequency data of servo mechanism at different frequencies are calculated.

$$L(w) = 20\lg|G(jw)| \tag{24.10}$$

$$\emptyset(w) = \angle G(jw) \tag{24.11}$$

Matlab write transfer function identification program, program iteration steps are as follows:

Suppose the transfer function molecule and denominator are m order and n order functions, the m and n parameters are brought into the program for calculation and get the transfer function $G(s)$;
Draw the Bode graph using $G(s)$, use test data to map Bode graph;
By comparing the identified transfer function Bode diagram with the experimental data Bode diagram, the identification error is calculated;
Compare the identification error with the set error threshold. If the error meets the requirements, $G(s)$ is the system transfer function;

If the error between the identification curve and the test curve is large and does not meet the requirements, execute step (1)–step (4), and finally get the transfer function $G(s)$.

The flowchart of the program algorithm is shown in Fig. 24.3.

The servo mechanism driven by the controller is tested in the frequency range of 1–50 Hz. The amplitude and phase attenuation of different frequencies are measured, and the amplitude-frequency and phase-frequency data are obtained. Using MATLAB program to iteratively, finally get the servo system transfer function molecular order $m = 12$, denominator order $nm = 15$, the system transfer function as follows:

$$\frac{\begin{array}{c}-2.9508e07\left(s^2+2.746s+790.1\right)\left(s^2+13.97s+3733\right)\\ \left(s^2-4.322s+8997\right)\left(s^2+152.3s+2.829e04\right)\\ \left(s^2-12.1s+2.371e04\right)\left(s^2-83.09s+6.605e04\right)\end{array}}{\begin{array}{c}(s-2307)(s+324.5)(s+31.12)\left(s^2+2.973s+793\right)\left(s^2+14.39s+3761\right)\\ \left(s^2-4.675s+9019\right)\left(s^2-12.35s+2.379e04\right)\left(s^2+162.5s+3.626e04\right)\\ \left(s^2-88.46s+6.446e04\right)\end{array}}.$$

The amplitude-frequency and phase-frequency data at different frequencies are obtained from the transfer function. The comparison between the frequency domain characteristic test data and the transfer function identification data is shown in Table 24.1, and the test curve and the identification curve are shown in Fig. 24.4. Figure 24.4 shows that in the range of 90–200 rad/s angular frequency, there are many frequency turning points, which leads to the difficulty of identification of transfer function and the high order. After repeated iterations, the program identifies the transfer function

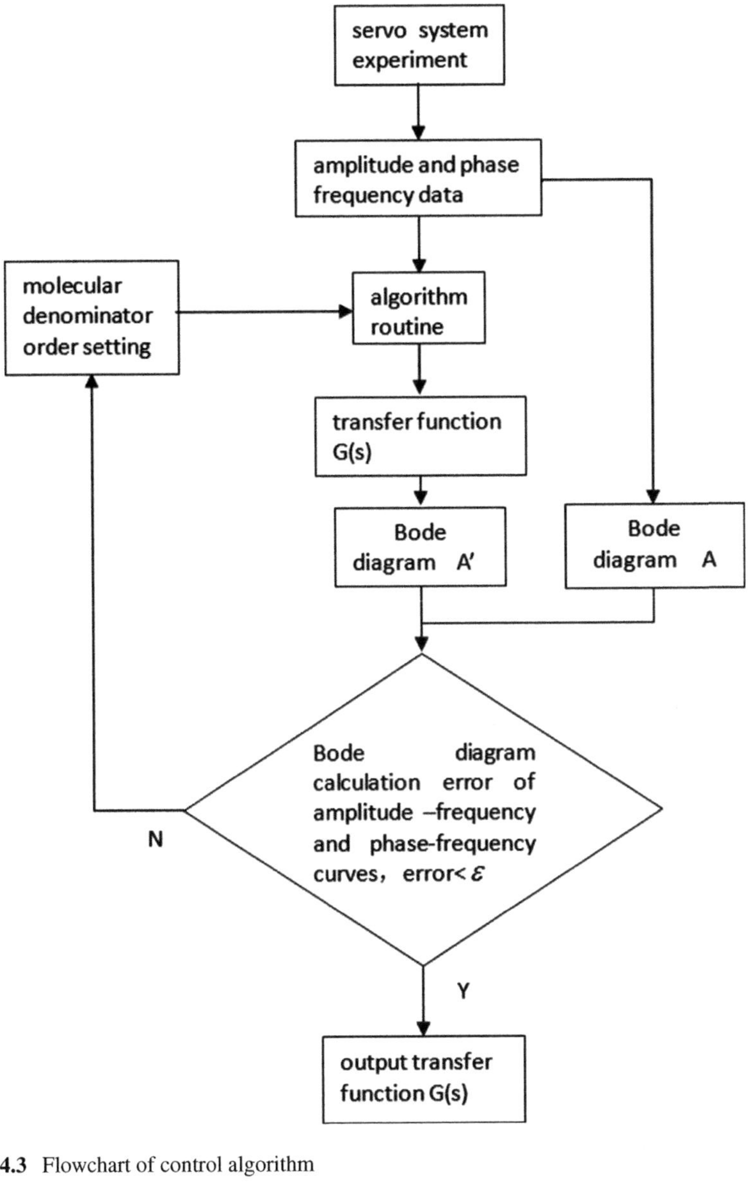

Fig. 24.3 Flowchart of control algorithm

well. Table 24.1 shows that the amplitude error and phase error are very small (other model identification errors about 30%). Figure 24.4 shows that the identification accuracy of transfer function is high, and the program is reliable (Fig. 24.5).

Table 24.1 Comparison of test and identification frequency domain data

Frequency (Hz)	Angle frequency (rad/s)	Test amplitude (dB)	Identification amplitude (dB)	Amplitude error (dB)	Test phase (°)	Identification phase (°)	Phase error (°)
1	6.28	−0.2	−0.222	0.022	−12	−12.1	0.100
1.6	10	−0.5	−0.489	−0.011	−19	−19	0.000
2	12.6	−0.8	−0.723	−0.077	−24	−23.4	−0.600
3	18.8	−1.5	−1.47	−0.030	−34	−33.5	−0.500
4	25.1	−2.4	−2.5	0.100	−42	−42.6	0.600
5	31.4	−3.1	−3.17	0.070	−46	−46.1	0.100
6	37.7	−4.3	−4.06	−0.240	−54	−53.1	−0.900
7	44	−5	−4.98	−0.020	−58	−58.1	0.100
8	50.2	−6	−5.9	−0.100	−62	−62.1	0.100
9	56.5	−6.6	−6.81	0.210	−65	−64.7	−0.300
10	62.8	−7.5	−7.45	−0.050	−67	−66	−1.000
11	69.1	−8	−7.99	−0.010	−69	−68.9	−0.100
12	75.4	−8.6	−8.66	0.060	−71	−71.4	0.400
13	81.6	−9.4	−9.34	−0.060	−73	−73.3	0.300
14	87.9	−10	−10.1	0.100	−75	−74.5	−0.500
15	94.2	−11.2	−11.2	0.000	−78	−78.2	0.200
20	126	−12.8	−12.8	0.000	−81	−81	0.000
24	151	−14.2	−14.2	0.000	−84	−84	0.000
30	188	−15	−15	0.000	−88	−88	0.000
35	220	−16.3	−16.3	0.000	−94	−94	0.000
40	251	−18.2	−18.2	0.000	−101	−101	0.000
45	283	−19.8	−19.8	0.000	−110	−110	0.000
50	314	−21	−21	0.000	−116	−116	0.000

24.5 Conclusion

For the electric servo system, the minimum error method is iteratively used to identify the transfer function. To solve the problem that the identification error is large and the function is not suitable in the high-frequency segment, the m program is written by the MATLAB, and the high order iteration is carried out by the computer to improve the identification accuracy, and the error inflection point of the amplitude phase is automatically corrected in the high frequency band. The method is used to identify the transfer function of the existing test data, to predict the amplitude and phase attenuation at unknown frequency by the identified transfer function, and to test the unknown frequency. By comparing the prediction data of the transfer function with the experimental data many times, the correctness of the transfer function is fully

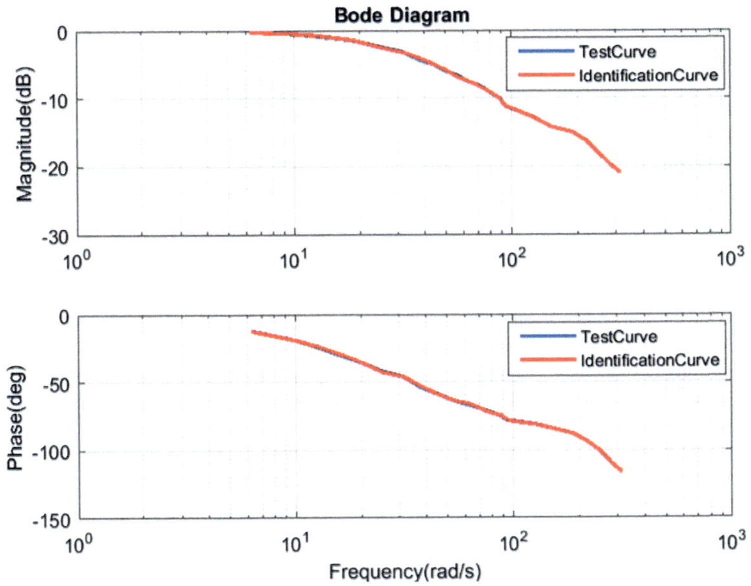

Fig. 24.4 Comparison of test and identification curves

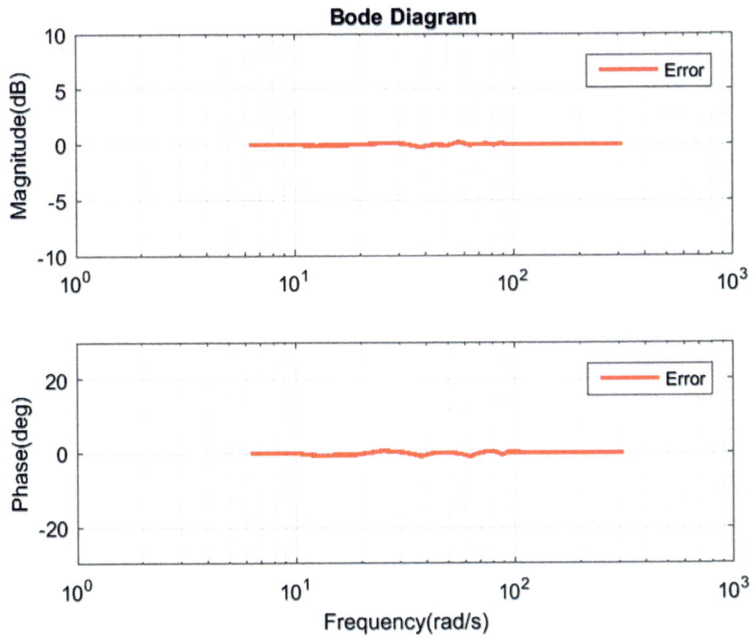

Fig. 24.5 Identification error curve

verified, and the method can accurately and effectively identify the transfer function of the servo system.

References

1. Wang, W.G., Chen, T., Shen, X.: Research on testing method of mechanical time constant of DC servo system. J. Instrum. **12** (2005)
2. Kaoru, I., Tsutomu, O.: Reliable control using equivalent transfer function for position servo system. In: 12th International Power Electronics and Motion Control Conference (IEEE Cat. No. 06EX1282), pp 1184–1189 (2006)
3. Imaizumi, T., Ohmae, T.: Reliable control using equivalent transfer function for control system with multiple minor-loop. In: The 29th Annual Conference of IEEE Industrial Electronics Society, pp. 611–616 (2003)
4. Lu, D.Y., Wang, X.J., Li, H.N., Liu, Y.: The design of the transfer function identification system of the servo system. Measurem. Control Technol. **4** (2015)
5. Vlasov, N.P.: A method for obtaining transfer functions of automatic control systems on alternating current. In: Automatic and Remote Control (1960)
6. Fang, C.Z., Xiao, D.: Process Identification. Tsinghua University Press (1988)
7. Hughes, F.M., Aldred A.S.: Transient characteristics and simulation of induction motors. Proceedings of IEEE, vol 111, pp 2041–2050 (1964)
8. Ling, X.: The damping iterative parameter identification method for dynamical systems based on the sine signal measurement. Signal Process. 660–667 (2016)
9. Chen, J., Wang, X.P.: Identification of Hammerstein systems with continuous nonlinearity. Inf. Process. Lett. **115**(11), 822–827 (2015)
10. Liu, H., Zhang, X.H., Duan, X.S., Xiao, Z.Z., Zhou, W., Huo, X.J.: Study on space motion decoupling of axisymmetric thrust vector control servo mechanism. Flight Control Detect. **2**(04), 83–88 (2019)
11. Wang, G.L., Chen, X., Zeng, F.Q., Xu, Y.B, Yu, G.: PMSM space vector control system and MATLAB simulation. Flight Control Detect. **1**(03), 59–62 (2018)

Chapter 25
Research on the Sound Insulation Measurement of Building Elements Based on Intelligent Laboratory Measurement System

Zhikai Zhang, Jian Wu, and Delin Wu

Abstract According to the traditional sound insulation measurement method of building elements, a sound insulation intelligent laboratory measurement system based on the Internet of Things was developed, by using of wireless measurement scheme, reducing the use of measurement equipment and developing intelligent mobile terminal measurement control software to meet the requirements of automatic and accurate measurement of sound insulation of building elements. It made the sound insulation measurement more intelligent and convenient. The experimental measurement results showed that the intelligent laboratory measurement system meets the requirements of ISO 10140-2-2010 Acoustics-Laboratory measurement of sound insulation of building elements—Part 2: Measurement of airborne sound insulation, which can realize accurate and rapid measurement of sound insulation of building elements, and the measurement efficiency was also improved with half time saving.

25.1 Introduction

When it was necessary to eliminate or reduce noise measures such as noise elimination, sound absorption and sound insulation can be used. Among them, sound insulation as an effective means of noise reduction was widely used, such as sound

Z. Zhang (✉) · J. Wu · D. Wu
Zhejiang Province Institute of Metrology, Hangzhou, China
e-mail: zzkhello520@163.com

Z. Zhang
Key Laboratory of Acoustics and Vibration Precision Measuring Technology of Zhejiang Province, Hangzhou, China

D. Wu
State Key Laboratory of Acoustics, Institute of Acoustics, Chinese Academy of Sciences, Beijing, China

L. C. Jain et al. (eds.), *Wireless Technology, Intelligent Network Technologies, Smart Services and Applications*, Smart Innovation, Systems and Technologies 258,
https://doi.org/10.1007/978-981-16-5168-7_25

insulation, sound insulation cover, and sound insulation screen. In order to evaluate the sound insulation performance of building element objectively and quantitatively, we must test the building element. At present, the sound insulation theory had become mature, and some new building elements were constantly appearing and being applied. Accurate evaluation of the sound insulation performance of elements has a very important significance in application [1–4].

The sound insulation test of building element could be generally divided into laboratory measurement and field measurement [5, 6]. The experimental condition of laboratory measurement is very demanding, special sound insulation room is needed, and the measurement accuracy is high. The field test equipment is relatively simple, which is greatly affected by the environment. The side perspective phenomenon of the elements in the field test cannot be avoided, and the measurement accuracy is poor.

The traditional sound insulation measurement equipment of building element is mainly composed of multi-channel signal analyzer, microphone, sound source, PC, and a series of other supporting equipment [7, 8]. The equipment between multiple sound insulation rooms is connected by cable, which has large volume, cumbersome wiring, high degree of human intervention, poor controllability, and repeatability, which could not fully meet the test requirements. With the development of the Internet of Things technology, the application of wireless communication technology and computer technology in acoustic testing has become more and more extensive, making it possible for the measurement data of various measuring instruments to be transmitted wirelessly, and the measurement results can be obtained intelligently through software. In the environment of the Internet of Things, intelligent and accurate measurement of the sound insulation performance of elements is a development trend of structural sound insulation measurement.

In this paper, aiming at the laboratory measurement method of sound insulation of building elements, a set of laboratory intelligent measurement system of sound insulation was developed, and the Internet of Things technology was used to communicate wireless data of various measuring instruments and equipment, so that the measurement room did not need to be connected to cables. Traditional sound signal analyzers and digital sound level meters were replaced by smart mobile terminals and smart sound level meters, which realize the intelligent measurement of sound insulation of building element and make the Internet of Things technology fully applied in building acoustics measurement. The limitation of laboratory measurement conditions was solved, and the interference of human factors in the measurement process was reduced. The intelligent control software could efficiently process and analyze the collected measurement data, realize the automation of the measurement process, make the sound insulation measurement more convenient and intelligent, and improve the accuracy and reliability of the sound insulation measurement results of building element.

25.2 System Design Scheme

25.2.1 System Composition

The measurement system is mainly composed of sound source room sound equipment, receiving room measurement equipment and smart mobile terminals (as shown in Fig. 25.1).

Sound source room sound equipment mainly includes intelligent signal generator and power amplifier-integrated machine, regular dodecahedron sound source, intelligent sound level meter, and router. Receiving room measurement equipment mainly includes intelligent sound level meter and router. The technical indicators of the measuring equipment are shown in Table 25.1.

In the sound source room, the signal generator and power amplifier-integrated machine could be remotely controlled through the smart mobile terminal, and the signal generation type (sine wave, white noise, pink noise, MLS, and other signals), frequency, amplitude, and other parameters could be adjusted. The power amplifier

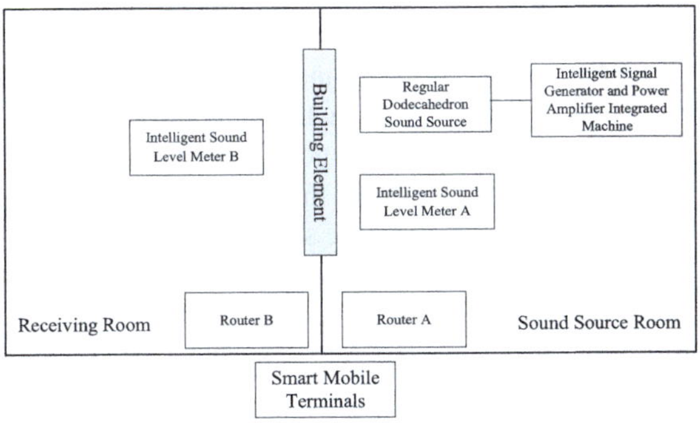

Fig. 25.1 Structure of system

Table 25.1 List of measuring equipment

Serial number	Device name	Technical requirements	Number	Function
1	Intelligent signal generator and power amplifier-integrated machine	MPE: ± 0.3 dB	1	Standard device
2	Omnidirectional sound source	Stability: ± 0.02 dB	1	Standard device
3	Intelligent sound level meter	Class 1	2	Standard device
4	Router	–	2	Signal enhancement
5	Smart mobile terminal	–	1	Control analysis

drives the regular dodecahedron sound source to produce sound. The intelligent sound level meter A measures the acoustic signal from the sound source in real time and transmits the acoustic signal to the intelligent mobile terminal through router A for various acoustic signal analysis.

In the receiving room, the smart sound level meter B could measure the received acoustic signals from the sound source room in real time and transmit them to the smart mobile terminal through router B for various acoustic signal analysis.

25.2.2 System Design

In order to realize the intelligent measurement and accurate measurement of the sound insulation in the laboratory measurement system of the sound insulation of building elements, the system design concept is mainly considered from the following aspects: (1) In order to improve the measurement efficiency, the wireless measurement scheme is introduced, and each measurement room does not need to be connected; (2) in order to avoid the change of sound pressure level caused by manual intervention, the traditional indoor sound pressure level measurement equipment is replaced with a smart sound level meter with wireless function and a smart mobile terminal, and the measurement network is connected through a router to measure. During the measurement, it can not only automatically control the sound source to produce sound, but also automatically receive measurement data; (3) by using smart mobile terminals and measuring instruments, the development of measurement software based on the Android system can achieve sound pressure levels. The measurement is more intelligent and convenient. Accurate measurement, remote reading of measurement data from wireless network, and analysis and calculation of building component sound insulation loss are supported. The flowchart of measurement process is shown in Fig. 25.2.

25.3 Measurement Steps and Methods

25.3.1 Test Schedule

25.3.1.1 Measurement Room

The laboratory consists of two adjacent sound source rooms and receiving rooms. There is a test hole between the two rooms (as shown in Fig. 25.1) for the installation of building element. The volume and size of the two rooms should not be exactly the same. The appropriate room size ratio should be selected to make the normal frequency of the low-frequency measurement as uniform as possible. In order to obtain a better diffusion sound field, the diffuser can be installed in the receiving room.

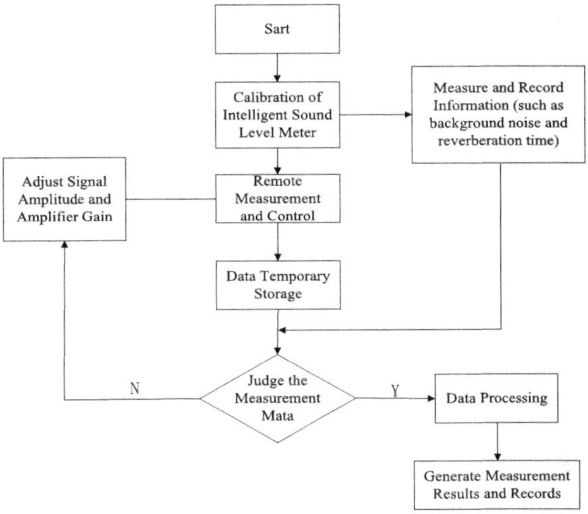

Fig. 25.2 Flowchart of the measurement process

At the same time, it can also avoid the influence of sound insulation measurement due to large changes in the sound pressure level in the room. According to the requirements of ISO 10140-2-2010, the volume of the receiving room should not be less than 50 m^3, and the volume V of the receiving room should be measured and recorded.

Before measuring the background noise of the room, the intelligent sound level meter should be placed in the geometric center of the room or in the normal working position (the measuring points should be far away from the edge interface, and the minimum distance from the edge interface is 0.7 m), or it can be arranged according to actual needs. During the measurement, the airtight soundproof doors of the sound source room and the receiving room should be closed, and other places with sound leakage should be sealed. The sound calibrator is used to calibrate the intelligent sound level meter, and then the 1/3 octave band sound pressure level and A-weighted sound pressure level at the measuring point are measured and recorded remotely through the intelligent mobile terminal as the background noise sound pressure level of the sound source room and the receiving room.

When measuring the reverberation time of the room, the airtight soundproof doors of the sound source room and the receiving room should be closed, and other places with sound leakage should be sealed. The intelligent mobile terminal remotely controls the signal generator to output a pink noise signal, the signal passes through a 1/3 octave band-pass filter to become narrow-band noise, and then it is output to a regular dodecahedron sound source through a power amplifier, and finally, it makes sound in receiving room. The intelligent mobile terminal controls the signal generator to measure the reverberation time with the interrupted sound source method or MLS method [9]. The smart mobile terminal records the decay curve of the sound pressure level over time through the smart sound level meter, which is approximately a straight line within the range of (5–25) dB below the steady-state sound level. The reverberation time should be the average slope of the line segment. The corresponding sound pressure level at the bottom of the selected line segment should be at least 15 dB higher than the background noise, and the reverberation time T of the receiving room should be recorded.

Under normal test conditions, the reverberation time T of the receiving room generally meets the requirements of formula (25.1) in the frequency range:

$$1s \leq T \leq 2\left(\frac{V}{50}\right)^{\frac{2}{3}} s \tag{25.1}$$

where V is the volume of the receiving room, in cubic meters (m^3); T is the reverberation time, in seconds (s).

In laboratory measurement facilities for measuring sound insulation, any indirect sound transmission should be negligible compared to the sound transmission through the sound insulation test piece.

25.3.1.2 Sound Insulation Test Piece

For the selection and installation of different building elements, the representative size of the appropriate building elements will be selected and installed at the opening of the test piece in strict accordance with the requirements of ISO 10140-2-2010, and the area S of the sound insulation test piece will be recorded.

25.3.2 Sound Field Generation in the Source Room

In the sound source room, the dodecahedron sound source should be placed at the position that provides the diffusion sound field as much as possible, and it is required to keep a certain distance from the sound insulation specimen. The intelligent mobile terminal control software sends instructions to remotely control the intelligent signal generator to use pink noise to generate sound. The sound generated is required to be continuous and stable and have a continuous spectrum within the measurement frequency range. The sound pressure level difference between the adjacent 1/3 octave bands should not be greater than 6 dB. The spectrum acoustic signal measured by the receiving room smart sound level meter is fed back to the smart mobile terminal, and the signal generator amplitude and power amplifier gain are adjusted so that the sound pressure level of all frequency bands measured in the receiving room is higher than the background noise by more than 15 dB.

25.3.3 Average Sound Pressure Level Measurement

Before measuring the average sound pressure level of the sound source room and the receiving room, the smart sound level meter should be reasonably placed indoors, to ensure that most of the indoor space that can be measured are covered, and the distance between the smart sound level meter and the room boundary should be more than 0.7 m, and the distance between the smart sound level meter and the building specimen should be more than 1.0 m.

When measuring the average sound pressure level, the measurement time should not be less than 30 s, which is calculated according to formula (25.2),

$$L = 10\lg \frac{\frac{1}{T_{\mathrm{m}}}\int_0^{T_{\mathrm{m}}} p^2(t)\mathrm{d}t}{p_0^2}\,\mathrm{dB} \tag{25.2}$$

where p is the sound pressure, and the unit is (Pa); p_0 is the reference sound pressure, and the value is 20 μPa; T_{m} is the integration time, and the unit is (s).

25.3.4 Measurement of Sound Insulation of Building Element

When the system is used to measure the sound insulation of building element, the sound source shall be far away from the test piece, and the distance between the sound source and the test piece shall not be less than 1.0 m, as shown in Fig. 25.1. The sound source room and the receiving room should be closed. The soundproof door should be closed, and other places with sound leakage should be sealed. According to the requirements of 3.3, the smart mobile terminal receives the sound signal sent by the intelligent sound level meter and calculates the average sound pressure level L_1 and L_2 of the sound source room and the receiving room.

According to formula (25.3), the sound insulation R of building components in each frequency band is obtained, and the expression [10, 11] is:

$$R = L_1 - L_2 + 10\lg \frac{\mathrm{ST}}{0.16V} \tag{25.3}$$

where R is the sound insulation of building components, in (dB); L_1 is the average sound pressure level in the sound source room, in (dB); L_2 is the average sound pressure level in the receiving room, in (dB); S is the area between the sound source room and the receiving room, in (m^2); T is the reverberation time in the receiving room, in (s); V is the volume of the receiving room, in (m^3).

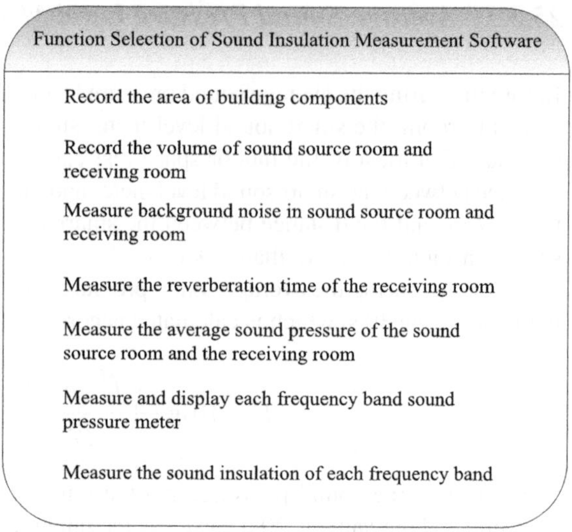

Fig. 25.3 Function selection of the measurement software

25.4 Sound Insulation Measurement Software and Measurement Results

25.4.1 Sound Insulation Measurement Software

In the measurement process of sound insulation of building element, the measurement software of intelligent mobile terminal plays a key role, which can realize remote one-key control of sound source room, average sound pressure level, and sound insulation loss measurement, and real-time monitor the operation status, and measurement process of sound insulation loss measurement system. The function selection of measurement software is shown in Fig. 25.3.

25.4.2 Measurement Result

The steel sound insulation door is selected as the object to be tested. The measurement process is carried out in accordance with the steps in Sect. 25.3. The experimental measurement data are shown in Tables 25.2 and 25.3. Different software functions are selected to test various parameters through the intelligent mobile terminal control software, and the sound insulation factor measurement results are shown in Fig. 25.4.

From the measurement process and data, it is found that compared with the traditional wiring measurement, the sound insulation loss of the measured object measured by the intelligent measurement system of the laboratory can solve the influence of the laboratory measurement conditions, shorten the measurement time by about 1/2, and the system has better repeatability and stability.

Table 25.2 Measurement data of the average sound pressure level of the source room

Frequency/Hz	Background noise/dB	Measurement noise/dB
100	35.0	70.6
125	34.4	76.5
160	33.5	76.5
200	35.9	83.4
250	32.2	77.6
315	31.0	80.7
400	31.7	80.0
500	31.4	78.8
630	30.3	77.7
800	33.1	77.3
1000	32.8	75.4
1250	33.4	74.8
1600	30.5	78.6
2000	32.0	76.4
2500	29.9	75.0
3150	26.8	73.8
4000	25.2	72.6
5000	22.2	75.7

Table 25.3 Measurement data of the average sound pressure level of the receive room

Frequency/Hz	Background noise/dB	Measurement noise/dB
100	36.9	47.2
125	37.6	51.6
160	31.0	50.4
200	36.9	52.9
250	31.5	51.5
315	32.9	52.7
400	32.7	49.0
500	33.2	48.0
630	31.5	47.0
800	30.3	45.4
1000	28.8	41.9
1250	27.7	39.4
1600	27.5	45.0
2000	27.2	44.8
2500	27.7	44.9
3150	26.4	44.4
4000	21.2	39.6
5000	20.3	37.8

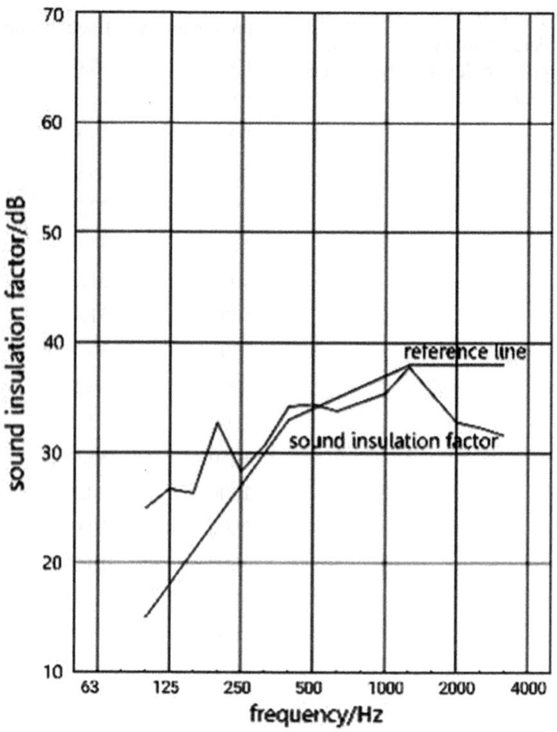

Fig. 25.4 Measurement results of the sound insulation factor

25.5 Conclusion

Aiming at the measurement of sound insulation loss of building elements, an intelligent laboratory measurement system based on intelligent mobile terminal is proposed. It meets the requirements of ISO 10140-2-2010 laboratory measurement of sound insulation of building elements-Part 2. It has the advantages of high automation control, simple operation, convenient measurement, and intelligence and improves the accuracy and reliability of sound insulation measurement results. The system realizes the application of the Internet of Things technology in the measurement of acoustic sound insulation. The measurement data of each acoustic measuring instrument is transmitted wirelessly, and the measurement results can be obtained through remote intelligence, which helps to simplify the measurement method and improve the measurement efficiency. In the environment of Internet of Things, the intelligent measurement system can be extended to the measurement of acoustic physical quantities of other building elements. Intelligent measurement will be a development trend of architectural acoustic measurement in the future.

Acknowledgements This work is supported by the State Key Laboratory of Acoustics, Chinese Academy of Sciences (Grant No. SKLA202009), and Zhejiang Provincial Quality and Technical Supervision System Research Project (Grant No. 20190106).

References

1. Cai, L.G., Zhu, J.: Analysis on selection of sound insulation property for external windows in existing buildings. Constr. Technol. **48**(15), 33–36 (2019)
2. Chen, Q.S., Ji, Z.J., Chen, J.H., et al.: Analysis on the approaches of improving sound insulation performance of lightweight wall. New Build. Mater. **38**(5), 31–34 (2011)
3. Cai, L.G., Zhu, J., Chen, Y.: Experimental study on sound insulation performance of single layer windows with different structural types and parameters. Noise Vibr. Control **38**(1), 171–176 (2018)
4. Li, M.Z., Liu, A.B., Zhang, J.G., et al.: Sound insulation performance analysis of different enclosure structures in residential buildings. Build. Energy Effi. **47**(11), 127–130 (2019)
5. General Administration of Quality Supervision, Inspection and Quarantine of the People's Republic of China: GB/T 19889.3-2005 Acoustics Measurement of Sound Insulation in Buildings and of Building Elements Part 3: Laboratory Measurements of Airborne Sound Insulation of Building Elements (2005)
6. General Administration of Quality Supervision, Inspection and Quarantine of the People's Republic of China: GB/T 19889.4-2005 Acoustics Measurement of Sound Insulation in Buildings and of Building Elements Part 4: Field Measurement of Sound Insulation Between Rooms (2005)
7. Dong, M.L., Chen, D.Y.: Sound transmission measuring system based on four-microphone method. Comput. Meas. Control **16**(7), 909–913 (2008)
8. Liu, K.L.: On measurement method for sound insulation weak parts of ledge wall, door and window. Res. Explor. Lab. **37**(8), 29–33 (2018)
9. Xiong, W.B., Yuan, F., Zhou, L.B., et al.: Principle and application of impulse response measured by maximum length sequence signal. China Environ. Protect. Ind. **10**, 139–141 (2019)
10. Ministry of Construction of the People's Republic of China: GB/T 50121-2005 Rating Standard of Sound Insulation in Buildings (2005)
11. International Standard Organization: ISO 10140-2-2010 Acoustics Laboratory Measurement of Sound Insulation of Building Elements Part 2: Measurement of Airborne Sound Insulation (2010)

Chapter 26
Research on Application of Somatosensory Interactive Technology in Teaching

Xiaoqiang Hu, Zhimei Ma, Xiaoting Fan, and Yuanyuan Wang

Abstract As a new human–computer interaction technology, somatosensory interaction technology enables people to reach a new height in the way of controlling multimedia. People do not need to directly touch the equipment, but operate multimedia equipment through natural language such as body movements. At present, somatosensory technology has some frontier explorations and experiments in the field of teaching applications. It is used in the classroom as a teaching tool to make up for the shortcomings of the traditional teaching model. It can not only help teachers create a more realistic teaching environment, but also optimize the sense of experience of students. This article focuses on the teaching application cases of LeapMotion equipment development and design and analyzes the form of somatosensory interaction technology used in teaching, and comparisons are made in terms of teaching environment, teaching methods, teaching subjects, and teaching content. It also provides ideas for the subsequent research on somatosensory technology in teaching.

26.1 Introduction

With the continuous development of science and technology, the informatization and digitization of modern educational technology are rapidly spreading, and the form of education is changing in the era of intelligence [1]. The traditional teaching mode can no longer meet the necessary conditions for teaching, and more efficient human–computer interaction technology is needed to innovate the teaching mode. By analyzing the existing somatosensory teaching cases, this article discusses the application form and effect of somatosensory interaction technology in teaching and provides ideas for subsequent research.

X. Hu (✉) · Z. Ma · X. Fan · Y. Wang
Jiangxi Science and Technology Normal University, Nanchang, China
e-mail: chinavrmm@163.com

L. C. Jain et al. (eds.), *Wireless Technology, Intelligent Network Technologies, Smart Services and Applications*, Smart Innovation, Systems and Technologies 258,
https://doi.org/10.1007/978-981-16-5168-7_26

26.2 Overview of Somatosensory Interactive Technology

26.2.1 Concept of Somatosensory Interaction Technology

Somatosensory interaction technology is a technology that uses computers to capture user actions to obtain commands and then uses algorithms to give users feedback [2]. It can realize three-dimensional space depth measurement, human bone point detection, movement, gestures, face and voice recognition, and other functions, allowing operators to use human posture and movement to interact with computer software. The most commonly used devices for somatosensory interaction technology include data gloves, Kinect, and LeapMotion. LeapMotion has many experienced users due to the advantages of cheap price, small size, high accuracy, and low latency.

26.2.2 The Way of Somatosensory Interaction

The use of somatosensory interaction technology depends on the appropriate interface performance and interaction mode. According to the situation, it can be divided into the following three scenarios: (1) Virtual scene—the software interface uses a two-dimensional or three-dimensional virtual interface to form the scene required for learning content and operation the person uses body movements to interact with the props in the virtual interface. (2) Augmented reality scene—the software interface uses real-time video collected by a somatosensory camera to superimpose virtual props to form a real and virtual scene. (3) Mixed reality scenes—the software interface uses superimpose the operator's portrait matting in the real video captured by the motion-sensing camera in real time onto the virtual background and props scene to reconstruct the mixed reality scene of the real portrait and the virtual scene.

26.3 Case Analysis of Somatosensory Interactive Technology in Teaching

In recent years, our team has been devoted to studying the application of somatosensory interaction technology based on LeapMotion equipment in teaching and has developed many teaching application cases. The following mainly through three representative cases analyze how somatosensory interaction technology is used in teaching.

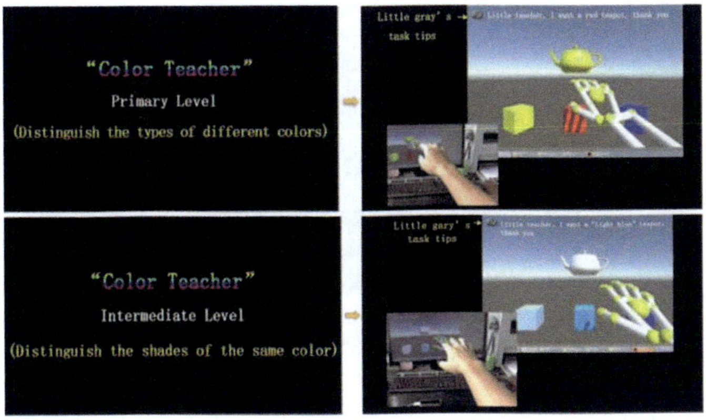

Fig. 26.1 Effect of color discrimination application

26.3.1 Preschool Color Discrimination Teaching

This case is a somatosensory application designed to assist young children learn color discrimination. The application is divided into two parts. The first part is the "primary level," learn to distinguish the types of different colors; the second part is the "intermediate level," learn to distinguish the shades of similar colors. In this application, students do not need to operate the mouse and keyboard to select colors, but instead grab the corresponding colors through gesture interaction and add them to the kettle, simplifying hand movements, and allowing students to focus on the task itself [3], as shown in Fig. 26.1. In this case, the somatosensory application is used as a teaching tool for "ability training," which reduces the learning burden of students at lower ages. Students do not need to learn new operation methods [4] and only need to use gestures to operate the learning software flexibly and reduced increased frustration caused by lack of computer operation experience; at the same time, it can also cultivate concentration. Students are immersed in realistic application scenarios, enjoying the freshness brought by somatosensory learning methods, which can effectively improve learning efficiency.

26.3.2 Primary School Science Curriculum Teaching

This case is based on the elementary school "science" course "understand the weather" content as an entry point, designed a meteorological popular science application based on somatosensory interactive technology [5]. The application has three parts. The first part belongs to the operation experience type. Students enter in the role of the first person and change the weather through the switching of gestures, which can enable students to experience different virtual weather effects. The second part belongs to the animation narrative type, taking the adventures of the bear as the

Fig. 26.2 Scene effect diagram of meteorological science application

clue of the whole story. Students can watch animated short films interspersed with meteorological science knowledge. The third part belongs to answering questions. The scene is a farm paradise, where there are many small animals, and each animal carries a customs clearance book. After finding the small animal, get the problem title through gesture interaction, only answer all questions correctly to pass the customs successfully. In the entire application, students can click the buttons on the menu bar through gesture interaction to switch between different weathers. Similarly, jumping between levels and receiving rewards after each level can also be achieved using gesture interaction, as shown in Fig. 26.2. In this case, the somatosensory application is used as a "knowledge teaching" teaching tool. On the one hand, assisting teachers in teaching new courses, changing the monotonous teaching methods in traditional classrooms, and turning the classroom into an immersive game scene; On the other hand, students are immersed in the virtual weather scene through gesture interaction, which deepens their understanding of weather.

26.3.3 Middle School Innovation Curriculum Teaching

This case is a somatosensory interactive application designed to solve the problems existing in the teaching of ceramic relic restoration [6]. In the current ceramic relic restoration courses, when students are training for restoration of ceramic cultural relics, they need to conduct a detailed and comprehensive observation of the ceramic cultural relics, know the current status and extent of damage to the cultural relics, and then judge the location of the fragments. If there is a splicing error during bonding, the ceramic cultural relics will suffer irreversible artificial damage. This ceramic cultural relic repair application based on somatosensory interaction technology can solve these problems well. In the virtual scene, students use gesture interaction to grab and assemble the fragments of ceramic cultural relics, so there is no need to worry about the problem of damage to it caused by misalignment and can be practiced repeatedly for many times, as shown in Fig. 26.3. In this case, the somatosensory application is used as an "operational practice" teaching tool; on the one hand, it

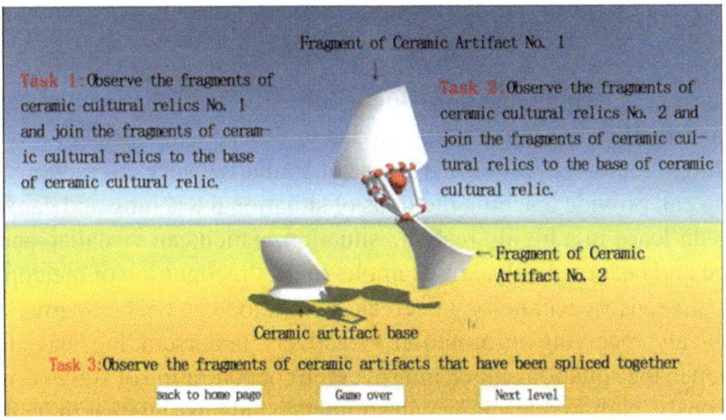

Fig. 26.3 Effect picture of ceramic relic restoration application

reduces the teacher's teaching tasks, and the teacher does not need too much actual operation demonstration and instruct students only when they practice; On the other hand, it solves the problem of insufficient teaching equipment, and every student can practice freely. There is no need to worry about that improper operation will damage the cultural relics.

26.4 Discussion on the Effect of Somatosensory Interactive Technology in Teaching

26.4.1 Teaching Environment

From the traditional "blackboard + chalk" to "multimedia integrated machine + push–pull blackboard" to the current "electronic whiteboard" [7], although the traditional mouse and keyboard control has been changed to touch control, but still teachers are restricted to the podium. There is a sense of distance between teachers and students, and the entire teaching atmosphere becomes rigorous and rigid due to distance. If you use an electronic whiteboard combined with somatosensory interaction technology, this problem can be solved effectively. Teachers can achieve page turning, circle, erasing, and other actions of PPT with simple gestures. Teachers can fully integrate into the students, with good interaction with students, and the entire teaching atmosphere will become relaxed and active. At the same time, the use of teaching resources developed by somatosensory interactive technology can expand the classroom into a variety of experience environments, only need to change the software to switch learning content.

26.4.2 Teaching Methods

In the traditional teaching, the most commonly used teaching method is the lecture method, and the teacher imparts knowledge to the students through oral language. This teaching method is very dull. The use of somatosensory interaction technology in teaching can provide teaching methods of situational teaching and game teaching. Students can learn in a highly realistic situation, which can visualize and visualize knowledge and become lively. For example, in the "recognition of meteorology," the use of somatosensory technology to create a situation can enable learners to change from passively receiving information to actively understanding and constructing information in the situation, which improves the concentration of students in learning; simultaneously, gamified teaching can stimulate students' interest in learning.

26.4.3 Teaching Subject

In traditional teaching, the teacher acts as the leader and controls the rhythm of the entire lesson, and students receive information more passively. Under the application of somatosensory interaction technology, the focus of teaching is shifted. Students are the main body of learning, and teachers are more of the designers and promoters of learning. For example: In the ceramic relic restoration course, after the teacher imparts the theoretical knowledge to the students, he uses this ceramic relic restoration somatosensory application to help students practice exercises. The teacher does not need to explain too much, but allows the students to immerse them and practice repeatedly. Classrooms using somatosensory interaction technology can effectively promote behavioral interaction between teachers and students.

26.4.4 Teaching Content

Traditional teaching content refers to objective materials such as textbooks, videos, and text materials. Students need teachers to explain before they can absorb and understand knowledge. Under the application of somatosensory interaction technology, somatosensory application has become the teaching content. It is an application developed with teaching as the goal. For example: "Forest Adventure" somatosensory game [8], using the geographical knowledge of junior high school as the educational content of the entire game, students answer the geographically relevant questions in the game problems come to pass, but also learn related knowledge points. Under the application of somatosensory technology, the content learnt by students is expanded from the text in the textbook to the learning of activities and skills. Students do not need to rely entirely on the teacher's explanation to understand knowledge, but can independently explore and learn.

26.5 Summary

With the advancement of technology and the development of the times, modern tools based on artificial intelligence, big data, VR, and other technologies have made people's lives more and more convenient and efficient. During the epidemic period, the concept of "stopping classes and not stopping school" promoted the development of online education, and at the same time, it also attracted attention to somatosensory education. For example: The epidemic urged the whole country to "home fitness." Somatosensory technology's accurate recognition of limb movements and digital courses developed by 3D technology enable students to exercise at home and learn knowledge without leaving home. It can be seen from this that future smart applications focus on experience. Therefore, the combination of somatosensory interaction technology and artificial intelligence technology in teaching will have good development prospects.

Acknowledgements Special funds for visiting Scholars in the Young and Middle-aged Teacher Development Program of ordinary Universities in Jiangxi Province (2016).

The key project of Jiangxi Teaching and Research Project, "the research on cultivating VR talents in Jiangxi Universities" (JXJG-18-10-2).

Humanities and Social Sciences projects of Colleges and Universities in Jiangxi Province, "Research and Development of Restoration Teaching system of Haihun Hou Cultural relics based on VR" (JD17051).

References

1. Liu, L.: The effect of somatosensory interaction technology applied to basic education. China Inform. Technol. Educ. **11** (2019)
2. Wang, S.: The application of somatosensory interaction technology in science popularization demonstration-Take the somatosensory interaction device of "Everything is Soul" as an example. Design **33**(7), 32–34(2020)
3. Li, Q., Wang, Q.: Review of the application status of somatosensory interactive technology in education. J. Distance Educ. **33**(1), 48–56 (2015)
4. Sang, Y.R., Lu, J.L.: Application of somatosensory interaction technology in middle school physics teaching. Friends Phys. **33**, 6 (2017)
5. Zhu, Z.Y.: Development of VR Teaching Resources for Elementary Science Based on Somatosensory Interaction. Jiangxi Science and Technology Normal University (2019)
6. Su, R.: Design and Development of Educational Education of Ceramic Cultural Relic Restoration Based on Somatosensory Interaction Technology. Jiangxi Science and Technology Normal University (2017)
7. Li, Q.L., Yang, X.D.: Exploring the application of somatosensory interaction technology in the field of education. Comput. Program. Skills Maintenance **11**, 82+90 (2016)
8. Jiang, Y.H.: Design and Development of Educational Games Based on Somatosensory Interaction. Jiangxi Science and Technology Normal University (2015)

Chapter 27
The Design and Implementation of the Identification System for Primary Protected Birds in China

Xinhua You, Chunxiu Xiong, Lu Chen, Haiyan Wu, and Yu Dai

Abstract This paper briefly introduces the composition of identification system for primary protected birds in China and the functions of each module, and puts forward the realization of the design ideas for the bird's identification system, and gives its working process, that is, how the inference engine module obtains the best identification decision. Combined with computer recognition technology, the system can use forward and reverse inference to obtain the identification decision, solve the problem that forest public security organization is difficult to quickly and accurately identify rare birds in illegal events of protecting rare birds, and fill the blank of computer identification technology in forest police handling cases. Finally, the importance of the recognition system and its application prospect are summarized.

27.1 Introduction

With the more and more attention of human beings for protecting the ecological environment, the forest public security organs encountered more and more illegal events in the process of protecting wild animal [1]. However, the forest public security organ has not attached great importance to the application of criminal technology in forest cases for a long time, and has little research, and has no relatively complete technical support system. How to apply the criminal technology of local public security to the actual case handling of forest public security and better serve the forest public security is still blank.

Rare birds are one of the important protection objects of forest public security organs. In recent years, illegal cases of destroying wild bird resources have been reported in newspapers or media from time to time. The phenomenon of illegal hunting, selling, transporting and smuggling birds occurs frequently. In some provinces and regions of China, there are even special gangs who live by hunting birds.

X. You (✉) · C. Xiong · L. Chen · H. Wu · Y. Dai
Computer Science and Information Engineering, Zhixing College of Hubei University, Wuhan, P. R. China
e-mail: yxh687273@163.com

L. C. Jain et al. (eds.), *Wireless Technology, Intelligent Network Technologies, Smart Services and Applications*, Smart Innovation, Systems and Technologies 258,
https://doi.org/10.1007/978-981-16-5168-7_27

There are many kinds of wild animals in China, which is one of the countries with the most abundant species of wild animals in the world, including 1319 species (or subspecies) of the birds, and nearly 300 species of rare birds are protected by the state [2]. These birds live in different environments, their ecological habits are different, and their external morphology also has its own characteristics, even if experts without the help of reference books, it is difficult to identify them all. Although there are local records or maps about birds all over the country to help identify them, it is very difficult for forest public security organs with weak technology to identify wild birds. In the process of sending them to other specialized agencies for identification, there are problems of sample preservation, which costs manpower, material and financial resources, and sometimes delays in handling cases [3]. The study of the computer identification system is particularly important for law enforcement officers to quickly identify the captured birds on the spot and determine whether they belong to the rare birds are protected by the state so as to carry out the next step of case handling procedures [4].

27.2 Composition of the Bird Identification System

In the bird identification system, these functions are completed by inference engine module, knowledge base module, working memory module and user interface module. The structure is shown in Fig. 27.1.

In order to make the bird recognition system have real interpretation function, we use more knowledge base modules than rules, annotate each rule, and code more knowledge. We use inference module engine and interpretation engine to use this knowledge base module at the same time [5].

In the structure of the bird identification system, knowledge base module, inference engine module and working memory module are the core of the recognition system. The main parts of the system are knowledge base module and inference engine module. According to the reasoning system discussed so far, the knowledge

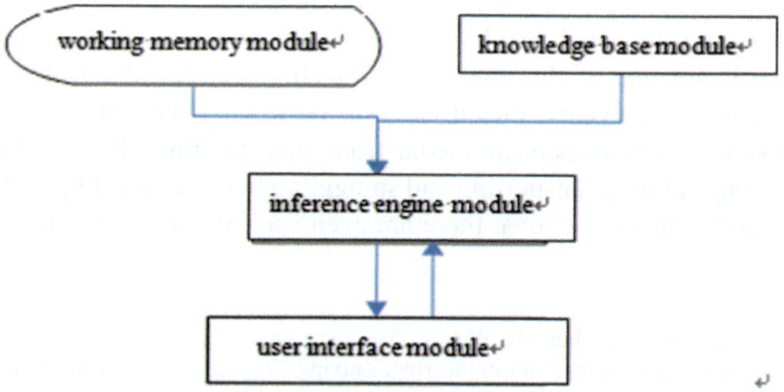

Fig. 27.1 Structure of the bird identification system

base module is composed of predicate calculus facts and rules related to the characteristics of China's primary protected birds. The inference engine module consists of all operating knowledge base modules to deduce the information of protected birds required by users, such as search, forward chain or reverse chain. User interface may include some kind of natural language processing system. It allows users to interact with the system in a limited natural language form. It is also possible to use a graphical interface with a menu to explain the subsystem, analyze the reasoning structure executed by the system, and interpret its recognition results to the user.

27.3 Realization of the Bird Identification System

In this system, the inference engine module is connected with the user interface module, and the recognition decision algorithm is described. Based on the characteristics of retrieval, forward reasoning I and reverse reasoning level II [3] can be used to obtain the recognition decision.

Realize forward reasoning I, first of all, extracted the feature parameters of the primary protected birds in China, and then compared with the knowledge base module, found out the best recognition method: the knowledge base module is SDB1, the bird feature parameter is EDB2, and the working memory is module TDB3. The algorithm can be described as follows:

```
Procedure Date_Drive (TDB3, SDB1, EDB2)
BEGIN
S ← ScanFirst (TDB3, SDB1, EDB2)
//the function ScanFirst() is to search bird feature database
//If a bird characteristic library that meets the requirements is found, then sets
Solving_flag 1;
While(NOT(S = Φ)) AND Solving_flag = 0 Do
// If more than one bird feature libraries are found, then loop
R: = Conflict_Resolution(S)
// Clear up the conflict, and then select the identification decision with the highest
priority
Excute(R)
// Select the operational part of the identification decision
S ← ScanFirst(TDB3, SDB1, EDB2);
// call the function ScanFirst(), find the next identification decision
EndWhile
END
```

Using this algorithm, the recognition can be divided into different priorities, and then organizing the content of teaching decision in the order of priority from high to low, to ensure that the bird feature library selected each time has the highest priority. This algorithm can implement a variety of recognition decisions for a certain bird feature, so as to better help users to identify the national primary protected birds.

Realize reverse reasoning II, It is suitable for the user to ask for the bird recognition system actively. In this retrieval process, the recognition decision takes a certain bird feature hypothesis target (M) proposed by the user, finds the set (A) of all the antecedent knowledge points in the working memory module (TDB3), and then finds out all the bird features in the knowledge base module (SDB1), and takes the knowledge point as the target to carry out reverse reasoning. The algorithm can be described as follows:

```
Procedure Goal_Drive (M, TDB3)
BEGIN
Scan (TDB3, M)
//Search for the above bird feature set: A = {A1, A2... An}
If(A = Φ)OR(all(A) ≥ passed)then
// If the above bird characteristics set is empty
Date_ Drive (SDB1, TDB3, M)
//then study the target knowledge points M
Else
// the above bird characteristics set is not empty
While i
//check the characteristics of each birds
Goal_ Drive (A, TDB3)
//then recursive invocationGoal_ Driv(),study the knowledge points
Endif
Endwhile
Endif
END
```

The algorithm recursively searches the bird features to find out the best recognition decision. The retrieval process can be realized step by step, and the individual characteristics of birds provided by users will be identified. As shown in Table 27.1, after collecting enough birds' characteristic data, different bird photos are analyzed to show the comparison with common identification systems. It can be seen from the figure that the system has a certain improvement in the speed and accuracy of bird identification, and through further optimization and design, it can reduce the cost of equipment in practical application [6].

Table 27.1 Comparison of common systems

	Recognition speed	Accuracy
opencv	0.7	0.8
TensorFlow	0.6	0.8
Forward reasoning I/reverse reasoning II	0.7	0.83

27.4 Conclusions

The system collects a large number of national primary protected birds' characteristic database, especially the detailed photos of different characteristics. Many detailed photos of birds do not exist in the current data, so it need to take photos by the research team. However, wild animals are active, and it is difficult to get close to them under normal conditions, so it is hard to capture the detailed photos that meet the requirements. Only when the birds' feature database with enough data support can it have enough accuracy in extracting bird features and making the best recognition decision.

The system is composed of inference engine module, knowledge base module, working memory module and user interface module, which can solve the problem that it is difficult for forest public security organs to identify rare birds quickly and accurately. Including the design of computer recognition algorithm, the compilation of retrieval map is also the difficulty of this system. The diversity of bird species and characteristics make it more difficult to draw up a retrieval map by grasping the combination of features which are well-circumscribed, intuitive and easy to operate.

The example given by this bird identification system is only an embryonic form of the bird identification system model. The emphasis is on how to design and identify the system. The design of the bird identification system generally adopts the prototype method, and many functions or knowledge need to be enriched and improved in the future. Through the development of the system, and constantly supplement and improve, especially combined with the existing system, it can quickly and conveniently identify the primary protected birds involved in the case.

Acknowledgements This work was financially supported by Prject of Scientific Research Program of Hubei Provincial Department of Education (No. B2020329), "Research on GPS signal acquisition and ad-hoc network transmission method based on multi-target continuous motion position monitoring" and Prject of Scientific Research Program of Hubei Provincial Department of Education (No. B2020330), "Computer Search for Non-Isomorphic Extremal Trees of Harmonic index with Given Degree Sequence".

References

1. Sun, P.: Computer Assisted Instruction. Inner Mongolia University Press, Hohhot (2010)
2. Hao, J., Xu, D.: Moving object detection based on video surveillance, science & technology information. **2** (2008)
3. Wang, L. et al.: A face location method in complex background based on optical flow field. Comput. Eng. Appl. **8**, 68–71 (2017)
4. Lin, H.W. et al.: Research on moving target detection method based on background subtraction technology. J. Ntl. Univ. Defense Technol. **v01.25**(3), 66–69 (2003)
5. Wang, Y.Q.: Artificial Intelligence Application of Principles and Methods. Xi'an Jiaotong University Press, Xi'an (2012)
6. Yi, W., et al.: Morphological description matrix and its application in color image retrieval and recognition. Scientia Sinica Technologica **34**(3), 337–344 (2016)

Chapter 28
The Construction of Online and Offline Golden Course—Network Cabling and Testing Course as an Example

Lei Wang

Abstract The evaluation standard of golden course is high-level, innovative and challenging, i.e., one degree for both sexes. Combined with the network cabling and testing course, this paper deeply discusses how to construct the online and offline mixed golden course from the initial design to the teaching implementation, to the complete evaluation of the course, so as to realize the cultivation of applied undergraduate talents who are ready to take up their posts and get employed after graduation.

28.1 Introduction

In June 2018, the Ministry of Education held a national conference on undergraduate education in Colleges and universities in the new era. At the meeting, the concept of "golden course" was put forward for the first time, pointing out that it is necessary to increase the difficulty of undergraduate courses, expand the depth of courses and turn "water course" into an abstruse and challenging course. Subsequently, "wastewater innovation fund" was officially written into the Ministry of education document for the first time [1]. For the first time, the Ministry of education put forward the evaluation standard of "high level, innovation and challenge" for "golden class" [2].

Network cabling and testing course is an independent experimental course. The courses include network engineering specialty and Internet of things engineering specialty. It belongs to the professional basic course, with a total of 32 class hours. The teaching content of the course comes from the actual talent demand of the enterprise. By analyzing the talent demand, designing the course content and benchmarking the corresponding post recruitment requirements, the talent training mode is established. First, the content of pre-job training will be moved to colleges and universities so as to achieve the goal of starting work and graduation employment. In order to achieve this goal, it is necessary to actively cooperate with enterprises and deepen the cooperation mode between schools and enterprises. Through the form of inviting in and going

L. Wang (✉)
Shanghai Jianqiao University, 1111 Hucheng Ring Road, Shanghai, China
e-mail: 03010@gench.edu.cn

L. C. Jain et al. (eds.), *Wireless Technology, Intelligent Network Technologies, Smart Services and Applications*, Smart Innovation, Systems and Technologies 258,
https://doi.org/10.1007/978-981-16-5168-7_28

out, enterprises are encouraged to participate in the formulation of course teaching content, and the employment practice posts of enterprises are reserved in advance to provide convenience for students' subsequent employment practice.

In the construction of golden course, the description of gender is high-level, innovative and challenging. The so-called high-level refers to the organic integration of knowledge, ability and quality and is to cultivate students' comprehensive ability and advanced thinking in solving complex problems. The so-called innovation refers that the contents of the curriculum should reflect the frontier and the times, teaching form presents advanced and interactive, and the learning results are exploratory and personalized. The so-called challenge refers to the difficulty of the course, which can only be achieved by skipping. Teachers and students have higher requirements in preparing lessons and after class. In this paper, network cabling and testing course as an example, respectively, from the teaching content, teaching objectives to reflect the high-level of the course, from the teaching methods and teaching mode to reflect the innovation, from the evaluation system, reflect the challenge of the course, introduce the relevant content in detail.

28.2 Clear Teaching Objectives, Benchmarking Post Talent Needs

Teaching objectives are very important for curriculum teaching. Only with clear teaching objectives can the effective training of talents be realized. According to Bloom's classification theory of educational objectives (revised edition), the cognitive process is divided into six levels from low to high: memory, understanding, application, analysis, evaluation and creation [3]. In order to reflect the high-level requirements, in the course design of the network cabling and testing course, the curriculum training goal is the employment goal. Through the analysis of the talent demand of the industry and enterprise, and the corresponding knowledge point analysis of the post demand, the knowledge points are divided into different cognitive levels and treated differently in the course design. Network generic cabling and testing course take generic cabling engineering as the overall course design idea. The course content is divided into engineering design project, engineering construction project and engineering acceptance project. Four subprojects are designed in the engineering design project, including generic cabling design scheme formulation, design drawing etc., and eight subprojects are designed in the engineering construction project, including twisted pair connection technology, module pressing technology, trunking bridge connection technology, optical fiber connection technology, etc. In the project acceptance project, seven subsystems are designed, including basic use of certification tester, certification test of channel permanent link, certification test of optical fiber link, network troubleshooting, basic use of OPV XG network analyzer, etc.

In the specific implementation process, combined with the superstar course platform, through teaching videos, online homework, online examination, some basic, simple knowledge points can be learned and mastered by students in the early stage through online teaching, and through the online and offline mixed teaching mode, students are mainly guided to cultivate the ability to analyze and solve problems in class The practical ability of high school students, for example, when teaching to the completion acceptance of generic cabling project, will analyze various fault phenomena in detail, requires students to use the relevant knowledge they have learned to solve all kinds of engineering problems encountered so as to help students successfully solve relevant problems in future work.

28.3 Optimizing Teaching Content and Training Applied Undergraduate Talents

In terms of teaching content, the specific goal is to cultivate applied undergraduate talents, integrate the latest technology of industry and enterprise into the classroom, introduce the experts of industry and enterprise into the classroom and offer special lectures for students so as to improve the students' acceptance of new technology and the cutting edge of the course content. The following methods are mainly used to improve the high-level requirements of the course content.

28.3.1 Introduction of National Standard Requirements to Improve Students' Awareness of Standards

In the course of teaching, the national standards will be introduced in detail. Relying on the contents of the national standards, the relevant contents involved in the generic cabling project will be introduced in depth and detail. At present, the national standards followed in China's generic cabling project are GB 50,311-2016 "code for design of generic cabling system engineering" and GB. In the course of 50,312-2016 code for acceptance of generic cabling system engineering, students will be required to complete a course analysis report through network survey, and compare the latest version of national standard with the previous version of GB 50,311-2007 and GB. The differences between 50,312-2007 and the contents added or revised in the new standard are summarized so as to improve the students' ability of data access and document collection.

28.3.2 *Hold Regular Course Discussion Meeting, from Enterprise to Enterprise*

In order to better optimize the curriculum resources, the person in charge of the course will actively visit the industry and enterprises, listen to the opinions of all parties, introduce the latest technology and the most classic cases into the course and understand the talent demand of the corresponding positions of the enterprise through visiting the enterprises, integrate the post requirements into the course teaching in real time, and arrange students to participate in post-practice and employment practice so as to achieve the goal. Now from the enterprise talent demand, to the school personnel training, and finally realize the closed-loop operation of students' internship and employment.

28.3.3 *Introduce High-End Lectures, Expand Students' Horizons and Realize Ideological and Political Education*

Through the in-depth school enterprise cooperation mode, experts from industry and enterprises are introduced to provide relevant lectures for students. The lecture contents include case sharing of generic cabling engineering, recruitment requirements of enterprises, professional ethics of enterprises and the latest advanced technologies of generic cabling. Through such lectures, on the one hand, students' vision is expanded, and on the other hand, students' professional ethics are improved and students are optimized. To train students to become an applied talents who are proficient in all kinds of technical ability, comply with professional ethics and be responsible for their own life.

28.4 Using a Variety of Teaching Methods to Promote Students' Learning Enthusiasm

Network cabling and testing course is an independent experimental course, which includes a large number of experimental content. Therefore, it is very important to design various teaching methods to enhance students' learning enthusiasm and bring their innovation into full play. The following is a brief summary of several commonly used methods.

28.4.1 *Mix Online and Offline, Expand Course Hours*

The number of class hours of a course has always been a pain point that restricts the difficulty and depth of the course. Through reasonable curriculum design, teachers can achieve the established teaching objectives in a reasonable number of class hours. However, if they want to be able to speak more, teach more, improve some teaching difficulties, let students play a creative role and make the course more challenging. They need to spend a lot of extra time in the explanation of relevant content and sharing of relevant experience. This state was difficult to achieve before, but now based on the emergence of various online platforms, teachers can publish a large number of basic teaching videos, basic assignments and basic examinations directly through the online platform and require students to complete the learning of relevant content outside the classroom, which will virtually expand the total course learning. In class, the teacher only needs to summarize the online content and answer questions, instead of explaining it from the beginning. By saving these class hours, teachers can explain more cutting-edge technologies, more practical cases, arrange more discussion and sharing so that more students can further participate in the learning of the course and improve students' innovation [4].

28.4.2 *Writing Course Analysis Report to Promote Students' Independent Thinking Ability*

Course analysis report refers to the specific analysis of a certain problem and an event through the study of the course and the investigation of students and summarizes the relevant suggestions, opinions and results. Through this method, students' learning situation can be evaluated, and the teacher's lecture content and methods can be adjusted to achieve the optimal learning effect. In the course of network cabling and testing, students will be required to have five course analysis reports, including national standard analysis report, electrical parameter analysis report, transmission medium analysis report, data center construction analysis report and wireless network construction analysis report, etc. Through these reports, students can summarize the relevant content they have learned and can also make personal analysis report based on the course content. Data collection, summary and improvement of relevant knowledge reserves, such as the contents of the last two reports, data center construction analysis report and wireless network construction analysis report are not involved in the actual course, and the teacher will only explain through a simple case in the course of teaching, but students need to collect and investigate the data themselves to complete the specific points analyze the content of the report.

28.4.3 The Operation of Group Mode Can Improve Students' Team Cooperation Ability

The generic cabling project is a team cooperation project, through which the project can be successfully completed. Therefore, how to instill this concept into students is very important. In the course design process, students will be required to divide into groups, with each group in groups of seven, and each group will be required to complete a group cooperation after half a semester of cooperation and communication. Take a building as an example to complete the integration. Each student will have a division of labor in the formulation of wiring engineering design scheme. For example, with 7 people as an example, member 1 completes the investigation of buildings, member 2 completes the design of work area subsystem, member 3 completes the design of horizontal trunk subsystem, member 4 completes the design of other subsystems, member 5 completes the drawing of relevant charts in the design scheme, and member 6 completes the design scheme phase For document writing, team member 7 is the project manager, who needs to complete the report and display of group design scheme in class. Through this kind of team cooperation, students' enthusiasm for learning and participation can be greatly improved, and their creativity can be brought into play.

28.4.4 Classic Case Analysis, Help Students Quickly Accumulate Experience

Project experience needs time to accumulate. The relevant knowledge points learned by students in the school and in the classroom are not enough for the situation that they need to deal with when they go to work in the future. How to help students quickly accumulate experience is also a problem that needs to be considered in the construction process of golden course, especially for the course of network generic cabling, which attaches great importance to engineering experience. Therefore, the course principal mainly solves this problem through two methods: one is to visit the industry and enterprise regularly to understand the latest practical cases and bring the cases back into the classroom teaching; the other is to require industry and enterprise experts to set up relevant lectures so as to help students accumulate project experience and help students integrate the knowledge they have learned.

28.5 Using 1 + X Assessment Mode to Optimize the Evaluation System

Formative evaluation, also known as process evaluation, is often combined with summative evaluation to evaluate the teaching effect and quality. In the past, the way of evaluation is to make a lifelong test, and the final examination will determine the source of all the results. However, this evaluation method is difficult to really judge a student's learning situation and teaching effect. Therefore, it is necessary to use a variety of methods to comprehensively evaluate the students' learning situation and the teacher's lecture situation.

28.5.1 Using 1 + X Mode to Evaluate Students' Learning Situation

At the beginning of the design of the network cabling and testing course, the overall evaluation mode is considered. Through the 1 + X mode, the learning effect of students is evaluated in a process. Among them, 1 accounts for 40% of the total, which refers to the final test. The main assessment content is various knowledge points that must be mastered in the course, with a total of 66 knowledge points. In the course of teaching, the teacher will integrate these knowledge points into the course teaching. It refers to the process score, accounting for 60%, including course analysis report score, operation test score and experimental report score, each accounting for 20%. Through the accumulation of these process scores, students' learning situation can be judged from daily learning, and students' participation and enthusiasm can be improved, so that students can truly meet the requirements of post skills required by the course.

28.5.2 Students' Mutual Evaluation is Adopted to Truly Reflect Students' Learning Situation

In the learning process of network cabling and testing course, students' mutual evaluation is also introduced. Taking the previous team work as an example, after the students complete the preparation of generic cabling engineering design scheme in the form of group, the project team will report and display, and other groups will give a basic score to the exhibition group, and the teacher will also give a professional score so as to achieve double evaluation subsystem, more fair evaluation of students' learning effectiveness.

28.5.3 Using Mycos Teaching Quality Platform to Comprehensively Evaluate the Teaching Situation of Teachers

At the end of the course, students will be required to use the Mycos teaching quality platform to comprehensively evaluate the teaching level of teachers and comprehensively evaluate the high-level, innovative and challenging degree in the course process so as to urge teachers to continuously improve their teaching level and further help students master relevant curriculum knowledge. In addition, there will be a large number of additional evaluation methods to comprehensively evaluate the teaching level of teachers during the epidemic period, such as supervising and listening to classes and learning between teachers.

28.6 Conclusion

Gold curriculum construction is the product of the development of the times. However, how to build a good golden course and how to evaluate the construction effect of the golden course is a problem that needs to be discussed together. Reasonable use of online and offline resources, effective use of all kinds of high-tech means, under the premise of achieving the goal of basic courses, to achieve gender equality, is a long time in the future need to continue to explore and correct the problem. The key point of the construction of golden course lies in the teachers and students, and the teachers must have a sense of responsibility. The preparation of the golden course construction needs to spend a lot of energy on the design, production, research and development. In the process of operation, it needs constant adjustment and improvement. Therefore, a high-quality golden course needs a lot of time and experience. The construction of golden course also depends on students because the goal of all courses is to serve students. Therefore, only if students are willing to devote their energy to study can the construction of golden course be successful. Continuous improvement, constant deliberation and constant discussion are the process that the golden course construction must go through.

References

1. Chen, B.S.: Speech at the National Conference on Undergraduate Education in the New Era, pp. 4–10. China Higher Education (2018)
2. Wu, Y.: Construction of China's "Golden Class", pp. 4–9. Chinese University Teaching (2018)
3. Lin, L.W.A. et al.: Bloom's Taxonomy of Educational Objectives: Learning and Teaching and Its Evaluation from the Perspective of Taxonomy (Revised Edition). Translated by Xiaoping Jiang, Meiqin Zhang, Jingjing Luo. Foreign Language Teaching and Research Press, Beijing (2009)

4. Lai, Z.X., Wang, X., Xiao, Y.P., Yang, X.P.: Current Situation and Trend of New Hybrid Teaching Based on Intelligent Teaching Platform: Visual Analysis of Literature in 2016–2017. Software Guide, pp. 219–221 (2018)

Chapter 29
Simulation and Experimental Research on Vibration Magnetic Effect of High-Speed Rail Wheel Set Material

Zhenfa Bi and Zongkai Wang

Abstract In this paper, the magnetic memory signal of high-speed iron wheel set material is studied by combining simulation and experiment with vibration characteristics. In the simulation process, the vibration frequency is 50 Hz, the vibration time increases from 4000 to 8000 h, the mean tangential component of magnetic memory signal decreases from −343.33 to 810.4 A/m, the difference is increased from 769.72 to 1544.37 A/m, and the average value of normal component decreases from 191.5 to 785.8 A/m, and the difference increases from 450.32 to 2074.04 A/m. The variation of the above values show that the different vibration time of the same frequency affects the tangential and normal components of the signal. The results indicate that the change of magnetic signal is consistent with the simulation results. This paper has practical reference significance for the application of high-speed railway wheel set fault detection by establishing the corresponding relationship with magnetic memory signal.

29.1 Introduction

The wheel set is an important component of the entire vehicle system, and the safety of train operation depends on the normal operation of the wheel set. When the train passes through a series of special road conditions such as uneven track, the wheel set will vibrate violently [1], which intensifies the dynamic force of the track on the wheel set. Long-term vibration will cause the wheel hub, rim and spokes. The macroscopic performance gradually deteriorates, and fatigue accumulation damage is inevitable, which seriously affects the safety of the train [2]. Therefore, researching the vibration characteristics of the wheel set system is an important method to ensure the safe operation of the high-speed railways [3].

Based on the metal magnetic memory detection technology, with the help of finite element software numerical simulation and laboratory sampling experiment research method, the vibration magnetic effect of wheel set material is studied, and

Z. Bi (✉) · Z. Wang
School of Railway Transportation, Shanghai Institute of Technology, Shanghai, China
e-mail: bizhenfa@sit.edu.cn

L. C. Jain et al. (eds.), *Wireless Technology, Intelligent Network Technologies, Smart Services and Applications*, Smart Innovation, Systems and Technologies 258,
https://doi.org/10.1007/978-981-16-5168-7_29

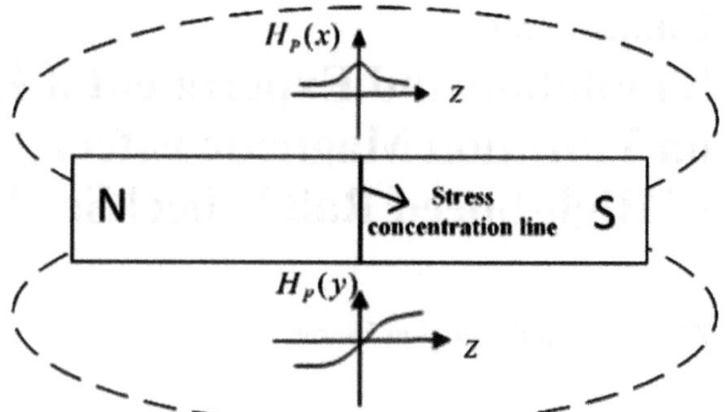

Fig. 29.1 Metal magnetic memory detecting schematic

the practicability of magnetic memory technology in the field of wheel set safety detecting is verified.

29.2 The Principle of Metal Magnetic Memory Detecting

The metal magnetic memory detecting technology is based on the residual magnetic fields (RMF) of the component or equipment local defect (such as crack) or stress concentration area [4]. The ferromagnetic material forms an effective magnetic field under the working load, the intensity of the magnetic field generated by the stress is:

$$H_\sigma = \frac{3\sigma}{2\mu_0}\left(\frac{\partial\lambda}{\partial M}\right) \tag{29.1}$$

In the formula, μ_0 is the space permeability, M is the magnetization; λ is the expansion coefficient [5] and σ is the stress.

The principle of the metal magnetic memory detection technology is essentially to pick up the leakage magnetic fields information from the superficies of the ferromagnetic member under the function of the geomagnetic field. In the stress concentration region, the inherent form of leakage magnetic field appears, that is, tangential H_p (x) has maximum value, normal H_p (y) has null point and large gradient value ($K = d$ $H_P(y)/dy$), as illustrated in Fig. 29.1.

29.3 ANSYS Model Simulation Analysis

"ANSYS" is an engineering simulation and 3D design software tool. For the sake of research the effect of different vibrance characteristics on the magnetic signal of the specimen, combined with the actual operation of the train, the finite element analysis software "ANSYS" is used. The sequential coupling method is used to carry out the finite element modeling. The model parameters: length 195 mm, width 50 mm, and height 1.5 mm.

In the transient dynamics simulation with time history, firstly, a 50 Hz sine function is applied to the model surface and the vibrance time history is set as 4000, 6000 and 8000 h, respectively. A magnetic field intensity of 39.8 A/m is applied in the *Z*-axis direction of the reconstructed model. After calculation, the distribute of magnetic field intensity under the joint load of three vibration times and magnetic field is obtained, as illustrated in Fig. 29.2.

The number in Fig. 29.2 represents the value of magnetic field intensity. From the figure above, it can be found that there is obvious change of magnetic field intensity signal in the middle edge area of the simulation model. This change trend is in line with the change law that the stress in the middle of the vibration test plate increases first. In order to conveniently observe the change trend of magnetic memory signal, we extract the characteristic value of magnetic field intensity signal in the figure above, as shown in Table 29.1

As shown in Table 29.1, when the vibrance rate remains unchanged, the magnetic signal of the model changes regularly with the change of the vibrance time. From 4000 to 8000 h, the tangential mean value of the magnetic signal is reduced from −343.3.3 to −810.4 A/m, and the difference between the maximum and minimum values is increased from 769.72 to 1544.37 A/m; the normal tangential mean increases from 191.5 to 785.8 A/m, and the difference from 450.32 A/m Increase to 2074.04 A/m. The above data changes show that the same vibration frequency and different vibrance time have an impact on both the tangential value and the normal value of the magnetic signal, and the tangential and normal show opposite trends in the mean value. It can

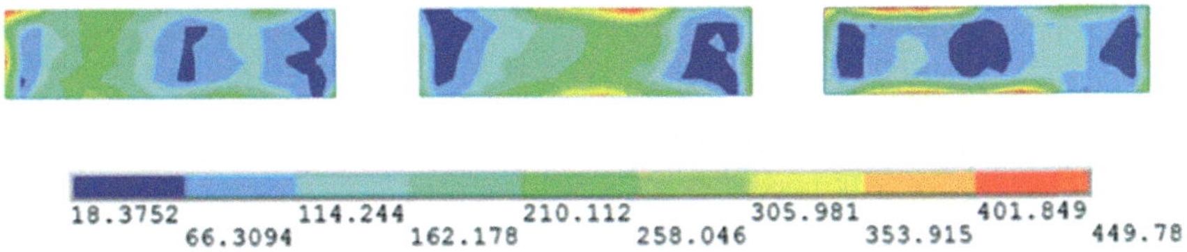

Fig. 29.2 Vibration simulation results

Table 29.1 50 Hz simulation signal value

Component direction	Time (h)	Maximum (A/m)	Minimum (A/m)	Difference (A/m)	Mean (A/m)	Variance (A/m)
Tangential	4000	20.22	−749.5	769.72	−343.3	8.03e5
	6000	95.30	−843.05	938.35	−346.3	8.92e5
	8000	−74.72	−1619.1	1544.37	−810.4	2.22e6
Normal	4000	−58.86	−509.19	450.32	191.5	6.33e5
	6000	467.21	−627.76	1094.97	228.07	5.49e5
	8000	1735	−339.04	2074.04	785.8	3.85e6

Fig. 29.3 25CrMo4 alloy structural steel specimen size

be seen from the change of variance that as time increases, the signal convergence degree of the tangential component of the magnetic signal of the test piece increases significantly, indicating that the influence of external interference signals gradually decreases along with the vibrance time increases. The normal component of magnetic signal also shows the same changing law.

The above phenomenon shows that while keeping other variables constant, changing the vibration time has a certain effect on the metal magnetic memory signal. The variation law can be used to evaluate the vibration magnetic properties of specimens based on metal magnetic memory parameters.

29.4 Vibration Experiment

29.4.1 Experimental Materials

The metal material used in the test is 25CrMo4 alloy structural steel of the same model as the CRH3 wheel set material. The dimensions of the plate shown in Fig. 29.3 (unit: mm).

29.4.2 Experimental Data Collection

This experiment uses a vibration exciter to apply vibration to the demagnetized specimen. Magnetic memory detector was used to detect the magnetic signal of the vibrance specimen along the regular direction, and the basic data was read and saved after the test. Figure 29.4 shows the experimental equipment.

29.4.3 Data Processing and Analysis

According to the principle of accelerated vibration, the test time is reduced by magnify the vibrance acceleration and vibrance frequency. The magnetic signal of the specimen with vibrance frequency of 50 Hz and vibration time of 4000, 6000 and 8000 h is analyzed, and the change of tangential value and normal value of magnetic signal on a certain path is analyzed.

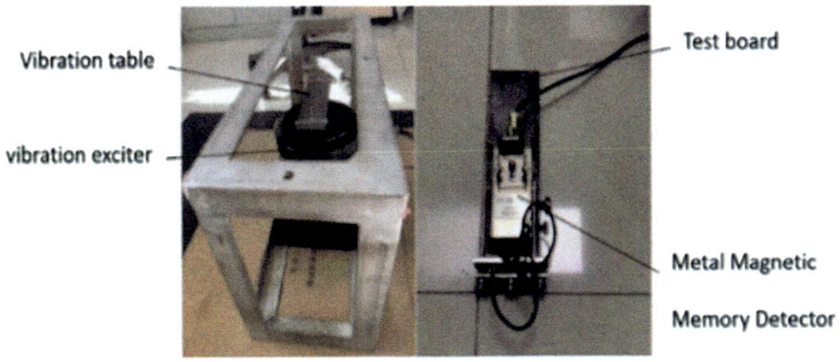

Fig. 29.4 Vibrance test equipment and collection instruments

Figures 29.5 and 29.6 show the magnetic field distribution in diverse vibrance periods in the frequency of 50 Hz.

The characteristic quantity of the signal is extracted, and its numerical characteristic under different vibrance characteristics is shown in Table 29.2.

According to Figs. 29.5, 29.6 and Table 3, during 4000–8000 h, the tangential mean value of the magnetic signal is −75.26 A/m changes to −57.69 A/m, the difference between the maximum and minimum is increased from 243 to 347 A/m; the average value changed from 21.04 to 20.03 A/m, the change was not obvious, and the difference increased from 52 to 104 A/m. Both tangential and normal variances gradually increase along with the raise of vibrance frequency, indicating that the influence of external interference signals gradually reduce along with the increase of vibrance time, which is uniformity with the above simulation results.

The above phenomenon shows that while keeping other variables constant, changing the vibration time has a certain effect on the metal magnetic memory signal.

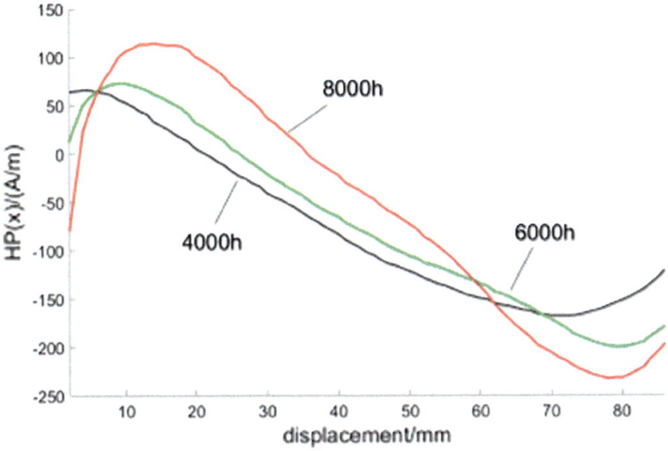

Fig. 29.5 Tangential component

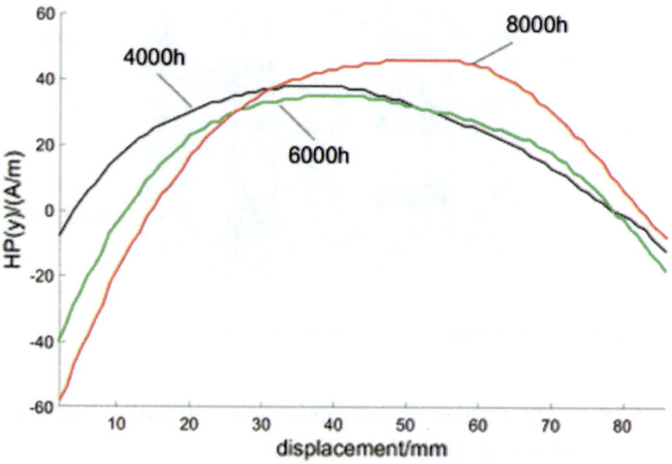

Fig. 29.6 Normal component

Table 29.2 50 Hz test signal value

Component direction	Time (h)	Maximum (A/m)	Minimum (A/m)	Difference (A/m)	Mean (A/m)	Variance (A/m)
Tangential	4000	66	−168	234	−75.26	6.47e3
	6000	73	−200	273	−71.9	8.4e3
	8000	114	−233	347	−57.69	1.44e4
Normal	4000	38	−14	52	21.04	2.35e2
	6000	35	−44	79	15.64	4.14e2
	8000	46	−58	104	20.03	7.87e2

29.5 Conclusion

Through the simulation and experimental analysis of high-speed rail wheel set material, the general variation law of magnetic signal with vibration characteristics is obtained.

(1) Through ANSYS simulation, between 4000 and 8000 h, the average value of the magnetic signal is reduced from −343.3.3 to −810.4 A/m, the difference between the maximum and minimum values is increased from 769.72 to 1544.37; the mean value of the normal component changes from 21.04 to 20.03 A/m. The phenomenon shows that while keeping other variables constant, changing the vibration time has a certain effect on the metal magnetic memory signal.

(2) Through the comparison of simulation and test, the simulation signal and test signal have the identical variety trend with the variety of vibration characteristics, which verifies the above rules. It has a certain scientific research value for

the application of magnetic memory detection technology in the reality wheel set vibrance state detection.

Acknowledgements The study work was supported by the National Natural Science Foundation of China under Grant No. 51405303, the Special Fund for the Selection and Training of Excellent Young Teachers in Shanghai Universities(ZZyy15110), and the Fund for the Development of Science and Technology Talents for Young and Middle- aged Teachers of Shanghai Institute of Technology (ZQ2019-21), and the Fund for Shanghai University of Applied Sciences Applied Research Achievements and Graduate Dissertations (1021GN203005002-B20).

References

1. Zhu, Q.L., Wang, H.S., Qian, X.L., et al.: Analysis of the influence of railway passenger wheels on failure modes. Railway Rolling Stock **37**(03), 100–103 (2017)
2. Wu, Y., Han, J., Liu, J., et al.: The influence of high-speed train wheel polygon wear on wheel-rail force and bogie vibration behavior. J. Mechan. Eng. **54**(04), 37–46 (2018)
3. Zhong, W.C.: Theoretical basis of metal magnetic memory diagnosis elastic plastic strain magnetization of ferromagnetic materials. Non Destruct. Test. **10**, 424–426 (2001)
4. Wan, S.Y.: Research on the Principle of Magnetic Memory Detection and Its Application Technology. Huazhong University of Science and Technology, Wuhan (2011)
5. Liu, Y.Z., Chi, Y.B., Liu, Y.: Research status and development of metal magnetic memory testing technology. Nondestruct. Test. **24**(6), 18–20 (2002)

Chapter 30
Establishment of Curriculum Evaluation System Based on OBE Concept and Employment Oriented

Lei Wang

Abstract At present, the education concept based on OBE mode and the teaching mode construction of promoting students' employment through curriculum learning are more and more recognized and implemented. How to achieve effective evaluation in the process of curriculum construction, so as to achieve the expected goal, is a problem that university teachers have been discussing. Taking the course of network cabling and testing as an example, this paper introduces how to integrate the concept of OBE and the demand of post talents into classroom teaching, puts forward four evaluation methods, and introduces a variety of evaluation methods so as to provide the basis for judging students' real learning results.

30.1 Introduction

The concept of outcomes-based education (OBE) is a kind of educational model based on learning outcomes [1]. The education of OBE concept mainly solves five key problems: what are the corresponding learning achievements expected of students, why they need to achieve these achievements, how to help students to achieve these achievements, how to understand that students have achieved achievements, and how to ensure that students can obtain effective results. These problems should consider the evaluation system, learning objectives, learning process, social needs, and improvement measures [2]. Combined with the characteristics of application-oriented curriculum, the OBE education concept is combined with the application-oriented curriculum evaluation, and a learning effect-oriented curriculum evaluation system is established to give full play to the development and construction functions of the evaluation system.

In addition, "the decision of the State Council on accelerating the development of modern vocational education" clearly pointed out: to accelerate the development of modern vocational education to "promote employment oriented." Curriculum is the carrier of education. Modern vocational education oriented by promoting

L. Wang (✉)
Shanghai Jianqiao University, 1111 Hucheng Ring Road, Shanghai, China
e-mail: 03010@gench.edu.cn

L. C. Jain et al. (eds.), *Wireless Technology, Intelligent Network Technologies, Smart Services and Applications*, Smart Innovation, Systems and Technologies 258,
https://doi.org/10.1007/978-981-16-5168-7_30

employment needs to be realized by promoting employment-oriented curriculum [3].

Network cabling and testing course is an independent experimental course. In the course design, it takes the post demand as the construction goal, introduces the OBE teaching mode to evaluate the students' learning results by teaching production, and evaluate the students' learning effect in the whole process through a variety of evaluation methods so as to achieve the expected effect of the course.

30.2 Establishment of Curriculum Objectives

Curriculum objectives are the core and key of the curriculum. Only when the reasonable and feasible curriculum objectives are defined can we achieve them. In the process of curriculum objectives construction, we should first consider the school running orientation of the school, understand the school-level talent training objectives, and need to integrate the ideological and political content into the curriculum construction. It is necessary to combine the knowledge goal, ability goal, quality goal, and education goal organically, and need to meet the talent demand of the standard industry enterprise, train the personnel for the enterprise, through the in-depth cooperation between the school and the enterprise, and jointly compile the curriculum objective with the experts of the enterprise. In addition, the curriculum objectives must also have a certain high-level, requiring students to highlight their own innovation and increase the challenge of the course through personal research on the basis of curriculum learning.

The network wiring and testing course combines the University's goal of cultivating application-oriented undergraduate talents, combined with enterprise experts, analyzes the post ability requirements according to the needs of industry post talents, integrates the knowledge points of the course, integrates the concept of professional ethics and professional ethics in the process of Ideological and political education into the course, enhances students' Ideological and political awareness, and provides final work support for students.

30.3 Description of Evaluation System

The purpose of evaluation is to test students' learning achievements. In the past, the evaluation methods were one time and summative activities at the end of teaching. The purpose of evaluation is to judge the learning results. However, the evaluation of courses under OBE teaching mode pays more attention to the learning effect of students. After learning this course, whether the students can meet the needs of future jobs. According to the basic requirements of the general secretary of internship on who, how, and for whom to train people, we should establish a reasonable evaluation system, introduce a variety of evaluation methods, and adopt the whole life cycle evaluation method. From the beginning of students' course learning, to the examination of learning effect in the course teaching process, to the comprehensive evaluation at the end of the course, and to the courses in the subsequent graduation internship knowledge tracking research. From the whole process of continuous evaluation of students' learning effectiveness, truly achieve the construction of process evaluation system.

Taking online and offline teaching as an example, the integration of online and offline assessment, taking online and offline teaching as an example, and taking online and offline assessment as an example, the integrated assessment system is formed. In the course, 1 + X are: 1 accounts for 40% of the total score, which is mainly the final written examination. The main assessment is the knowledge objectives of students. During the course teaching, 66 knowledge points involved in the course will be explained. Students are required to fully master these basic knowledge points. X1 is the operation assessment, accounting for 20% of the total score, that is, the assessment of students' ability objectives. The curriculum integrates the ability objectives students are required to complete four operation assessment contents. Through the assessment, students are required to master all kinds of operation skills. X2 is the course analysis report and online learning, accounting for 20% of the total score, that is, to assess the achievement degree of students' quality objectives. Under the premise of classroom learning, students are required to complete the research report on the cutting-edge technology of the course through self-study, extracurricular investigation and innovation. Combined with Superstar platform to complete the online learning, online homework, and online examination content. X3 is the course experiment report, accounting for 20% of the total score, to evaluate students' participation in the daily experiment process and provide the basis for the realization of process evaluation.

30.4 Description of Evaluation Methods

The establishment of evaluation system is based on various evaluation methods. Only by using diversified evaluation methods and evaluating students' learning effect from multi-dimensional can we reflect the teaching effect and students' learning effect most truly.

To establish a correct evaluation concept, what kind of students can be regarded as achieving the course learning objectives, and how to correctly evaluate the students' learning effectiveness will be the key to the curriculum evaluation system considered by teachers. Teachers should comprehensively evaluate students' learning effects from all aspects, and the evaluation methods should adopt various ways so as to achieve a fair and more efficient evaluation system. Taking the network cabling and testing course as an example, the evaluation methods included in the course include teacher evaluation, student mutual evaluation, off campus part-time teacher evaluation, employer internship evaluation, etc. Teacher evaluation refers to the teacher according to the curriculum objectives, combined with the actual situation of students, through various means of curriculum evaluation, including arranging online homework, online examination, offline operation experiment, Offline operation examination, experimental report submission, course analysis report submission, etc., through the way of quantitative evaluation to evaluate the overall course. Students' mutual evaluation refers to the self-evaluation of students by using the method of students' mutual evaluation in some curriculum arrangements. Through self-evaluation, students can supervise and encourage each other, find problems in the teaching process together and reflect problems from the perspective of students so as to further improve the teaching level. The third evaluation method is the evaluation of part-time teachers outside the school. The final source of students is employment, which is to move to enterprise posts. Therefore, the evaluation of enterprises is very important. In the course, part-time teachers from enterprises will be introduced to give lectures for students, and corresponding evaluation scores will be given according to the feedback of enterprises. Through the introduction of enterprise experts, students can learn from the perspective of enterprise talent demand evaluation of effectiveness. The fourth evaluation method is the employer's internship evaluation. After the course is completed, the students will still track the students' learning situation after they go to the enterprise to participate in the employment practice so as to further understand whether the course learning content can meet the demand in the actual work post. Although it is beyond the course performance evaluation, it is helpful for the course improvement.

The objective of curriculum evaluation is to evaluate students' learning achievements, but from another perspective, it can also improve the curriculum objectives and optimize an important environment of the subject system. From the formulation of curriculum objectives to the operation of curriculum teaching, the final evaluation of curriculum is carried out to realize a closed-loop operation. In this closed-loop system, teaching contents are continuously optimized, teaching methods are improved, and teaching experience is condensed and to achieve the goal of high-quality curriculum construction.

30.5 Conclusions

The construction of curriculum evaluation system is a gradual process, which is a huge system engineering. It is constantly self-correcting and improving method. According to the actual needs of talents in the industry and enterprises, the teaching content is constantly optimized and the evaluation method is constantly improved. According to the construction concept of OBE and the construction goal of promoting employment, the curriculum evaluation system can be really effective, guaranteeing the goal of talent training and promote the common progress of teachers and students.

References

1. Yang, J.: Research on Application-oriented curriculum evaluation system based on OBE mode. In: Wireless Internet Technology, pp. 82–83 (2019)
2. Wu, D., Zhang, W.: Curriculum reform of application-oriented undergraduate specialty based on results—taking modern logistics information management as an example. J. Higher Educ. 136–138 (2020)
3. Liu, S.: Exploration of project-based curriculum development oriented to promoting employment. In: China Vocational and Technical Education, pp. 64–68 (2014)

Chapter 31
Research on Evaluation Method of Transformer Operation State in Power Supply System

Jiajian Wang, Hu Liu, Zhiqun Pan, Shinmin Zhao, Weilong Wang, and Shen Geng

Abstract This paper studies the status evaluation of the power supply transformer in operation. The combination of improved group analytic hierarchy process (IGAHP) and index correlation method (CRITIC) is used to determine the weight of each evaluation factor. Based on the nature of the evaluation factor, an appropriate membership function is selected for matching then calculating the state of the transformer through fuzzy comprehensive evaluation. This method is verified by an example in a transformer in a power supply section of the Shanghai Metro. The test result is close to the actual situation. Has practical reference value.

31.1 Research Background

The research on the state assessment methods of subway traction power supply system mainly includes the following. Literature [1] proposes a health assessment method based on the principle of multi-layer immunity, which uses the characteristics of biological immune mechanism to monitor and predict the power supply system. This method uses the characteristics of biological immune mechanism to monitor and predict the power supply system. The literature [2] gives the evaluation index of the transformer and studies the weight of each index based on the CRITIC method. Literature [3] proposes a reasonable model for the optimization of transformer maintenance. The above method considers the status and indicators of power transformers in a reasonable and detailed manner, and the model is also more realistic. However, for traction power supply transformers in operation, in addition to internal factors, external environment and human factors cannot be ignored.

J. Wang · H. Liu (✉) · Z. Pan · S. Zhao · W. Wang · S. Geng
Shanghai Institute of Technology, Shanghai, China
e-mail: liuhu@sit.edu.cn

L. C. Jain et al. (eds.), *Wireless Technology, Intelligent Network Technologies, Smart Services and Applications*, Smart Innovation, Systems and Technologies 258,
https://doi.org/10.1007/978-981-16-5168-7_31

31.2 Evaluation System of Transformers in Operation

31.2.1 Evaluation Indicators of Transformers in Operation

Influencing factors selection of traction transformers in operation. This article analyzes the methods in various documents, combined with the opinions of experts from traction power supply companies, and proposes that the factors that have a greater impact on the operating status of traction transformers are mainly divided into the following four categories. (A) Maintenance ability: testing, inspection, and repair; (B) Equipment acquisition: cylinder, insulator, iron core, busbar, Buchholz relay, transformer oil, and central grounding knife; (C) Characteristic parameters: winding insulation resistance, winding DC resistance, dissolved gas content in the oil, absorption ratio, winding dielectric loss tangent value, and polarization coefficient; (D) Environment: humidity, temperature.

31.2.2 Weight Assignment

The weight assignment method of the classic analytic hierarchy process (AHP) is to assign a reasonable weight based on expert experience and then conduct a consistency test. This weight distribution method does not consider the differences between different experts. For experts with different experiences, ages, and knowledge levels, this difference can effectively reduce the weight distribution error and at the same time carry the expert's subjective knowledge. Therefore, this paper proposes an improved analytic hierarchy process (IGAHP) subjective assignment model and an objective assignment method based on the index correlation method (CRTTIC) to effectively optimize the above problems and make the results closer to the true value [4].

Improved analytic hierarchy process (IGAHP) subjective weight assignment.

Suppose the similarity between the decision of the pth expert and the qth expert is represented by $d_{pq}(p, q = 1, 2, \ldots, T)$ distance, then

$$d_{pq} = d(Y^{(p)}, Y^{(q)}) = \left[\sum_{i=1}^{n} (y_i^{(p)} - y_i^{(q)})^2 \right]^{1/2} \tag{31.1}$$

where $Y^{(p)}$ and $Y^{(q)}$ are the decision values of expert p and expert q, respectively, and $y_i^{(p)}$ and $y_i^{(q)}$ are the decision values of expert p and expert q for the ith index, respectively. The smaller the value of d_{pq}, the smaller the disagreement between the two experts.

Suppose the similarity between the tth expert and other experts' decision-making is denoted by d_t, then:

$$d_t = \sum_{j=1}^{T} d_{ij} \; j = 1, 2, ..., T \tag{31.2}$$

It can be seen that the final decision weight coefficient λ_i of the tth expert is calculated as formulas (31.4) and (31.5):

$$\lambda_1 = \lambda_2 = \cdots = \lambda_T = \frac{1}{T}, \text{ where } \; d_{pq} = 0 \tag{31.3}$$

$$\lambda_t = \frac{1/d_t}{\sum_{j=1}^{T} (1/d_j)}, \text{ where } \; d_{pq} \neq 0 \tag{31.4}$$

In the formula, T is the total number of experts, the smaller d_t means the smaller the difference between the expert and other experts, and vice versa.

Suppose the final index weight vector $\overline{Y} = (\overline{y_1}, \overline{y_2}, ..., \overline{y_n})$, then:

$$\overline{Y} = \sum_{i=1}^{T} \lambda_t Y^{(t)} \tag{31.5}$$

Objective weighting model is based on index correlation method (CRTTIC). The objective weight assignment method uses the standard deviation of the data to reflect the difference of factors, that is, the degree of variation, and uses the correlation coefficient to express the correlation of different indicators, and then use the conflict quantification formula to determine the degree of conflict between an indicator i and other indicators. To quantify the degree of conflict between index i and other indexes, the conflict quantification formula of index r_{ij} can be obtained as:

$$c_i = \sum_{j=1}^{n} (1 - r_{ij}) \tag{31.6}$$

where r_{ij} is the correlation coefficient between index i and index j, and its expression is:

$$r_{ij} = \frac{\sum_{k=1}^{m} (\lambda_{ik} - \overline{\lambda_i})(\lambda_{jk} - \overline{\lambda_j})}{\sqrt{\sum_{k=1}^{m} (\lambda_{ik} - \overline{\lambda_i})^2} \sqrt{\sum_{k=1}^{m} (\lambda_{jk} - \overline{\lambda_j})^2}} \tag{31.7}$$

where λ_{ik} and λ_{jk} are the normalized values of the Kth object under the ith index and the jth index, respectively, and $\overline{\lambda_i}$ and $\overline{\lambda_j}$ are the mean values of index i and index j, respectively.

Let C_i be the amount of information provided by index i, and its expression is:

$$C_i = \sigma_i c_i \tag{31.8}$$

where $\sigma_i = \sqrt{\sum_{j=1}^{m} (\lambda_{ik} - \overline{\lambda_i})^2/(m-1)}$. The larger C_i is, the more information provided by index i, and the greater the role in the evaluation process, so the weight ω_i of index i can be expressed as:

$$\omega_i = C_i / \sum_{i=1}^{n} C_i \tag{31.9}$$

IGAHP-CRTTIC comprehensive weight. The comprehensive weight is the coupling of the subjective weight and the objective weight, and the comprehensive weight ω_j of the evaluation index is calculated by the multiplicative normalization method:

$$\omega_j = (\alpha_j \cdot \beta_j) / \sum_{j=1}^{m} (\alpha_j \cdot \beta_j) \tag{31.10}$$

where α_j is the subjective weight of the jth index; β_j is the objective weight of the jth index;ω_j is the comprehensive weight of the j-th index; n is the number of indexes.

31.3 Fuzzy Comprehensive Evaluation of Power Supply System Substation

31.3.1 Fuzzy Evaluation Criteria

According to the overhaul regulations of the power supply company and the normative instructions of the equipment manufacturer, the evaluation of the traction transformer equipment status can be expressed in four levels, namely excellent, good, medium, and poor.

31.3.2 Determination of Membership

The degree of membership refers to the degree to which the evaluation indicators fit each evaluation level. After quantification processing, a certain value in [0,1] is used as the degree of fit of a single factor to a certain evaluation level, 0 means the lowest degree of fit, and 1 means the highest degree of fit.

Maintenance ability. The annual frequency of testing, inspection [5], and repair is used as a quantitative description of the maintenance capability. 450 times are the comprehensive daily inspections counted by the maintenance company. A value lower than this value indicates that the equipment is in good operating condition.

This value represents the decline in the operating state of the equipment, and its degree of membership is:

$$U_1(x) = \begin{cases} 1\,(0 < x < 450) \\ \dfrac{450}{x}\,(x \geq 450) \end{cases} \tag{31.11}$$

Equipment acquisition. The timely supplement of core spare parts is directly related to the long-term stable operation of the traction transformer. The storage capacity of the core components of the transformer, such as the cylinder, insulator, iron core, busbar, Buchholz relay, transformer oil, and grounding knife, is used as the reference index of the membership degree of equipment acquisition. According to a large amount of statistical analysis of data, when the storage of each core component reaches 30% of the total, the supply requirements for spare parts are basically met, and its membership function is:

$$U_2(x) = \begin{cases} \dfrac{x}{0.3}\,(0 \leq x < 0.3) \\ 1\,(x \geq 0.3) \end{cases} \tag{31.12}$$

Characteristic parameters. Characteristic parameters are mainly used to reflect the influence of the internal characteristic structure of the traction transformer on the operating state. Representative parameters include winding insulation resistance, winding DC resistance, winding dielectric loss tangent, dissolved gas content in oil, absorption ratio, and polarization coefficient. Among them, the winding DC resistance, winding insulation resistance, winding dielectric loss tangent value, absorption ratio, and polarization coefficient satisfy the conditions for stable operation as $\leq 4\%$, $\geq 1\ \text{M}\Omega/\text{kv}$, $\leq 2\%$, ≥ 1.3, and ≥ 1.5. The above-mentioned characteristic indexes are related to each other. The DC resistance of the winding and the insulation resistance of the winding directly affect the last three indexes. Therefore, the membership models of the above indexes are similar, and the membership functions are:

$$U_{31} = \begin{cases} 1(0 \leq x \leq c_1) \\ \dfrac{c_1}{x}\,(x > c_1) \end{cases} \text{ or } U_{32} = \begin{cases} \dfrac{x}{c_2}\,(0 \leq x < c_2) \\ 1\,(x \geq c_2) \end{cases} \tag{31.13}$$

In formula (31.13), when describing the winding DC resistance, $c_1 = 4$; when describing the winding dielectric loss tangent value, $c_1 = 2$; when describing the winding insulation resistance, $c_2 = 1$; when describing the winding dielectric loss tangent value, $c_2 = 1.3$; When describing the polarization coefficient, $c_2 = 1.5$.

31.3.3 Fuzzy Comprehensive Evaluation

According to the comprehensive weight set and the membership degree set of the evaluation index in Sect. 31.2, the fuzzy comprehensive evaluation [6] is obtained through vector operations:

$$B = \omega \cdot R \tag{31.14}$$

31.4 Engineering Case

Taking a traction transformer in operation at a station in the Shanghai Metro traction power supply system as an example, the affiliation of each layer index is shown in Table 31.1. Judge the condition of the transformer.

Table 31.1 Index weights and membership degrees of each layer

Target layer	Evaluation layer/weight	Index layer/weight	Parameter
Transformer status	Maintenance ability/0.343	Testing/0.355	275
		Inspection/0.305	98
		Repair/0.34	177
	Equipment acquisition/0.145	Cylinder/0.1	0.255
		Insulator/0.1	0.255
		Iron core/0.2	0.195
		Busbar/0.1	0.255
		Buchholz relay/0.1	0.24
		Transformer oil/0.2	0.255
		Center grounding knife/0.2	0.27
	Characteristic parameters/0.342	Winding insulation resistance/0.150	0.9
		Winding DC resistance/0.150	4.4
		Dissolved gas content in oil/0.201	0.075
		Absorption ratio/0.177	0.7
		Tangent of winding dielectric loss angle/0.103	0.97
		Polarization coefficient/0.219	1.2
	Environment/0.170	Temperature/0.449	36.75
		Humidity/0.551	84.5

Take the two-level index under the maintenance capability as an example. The index weight has been given by formulas (31.1)–(31.10), and the degree of membership is calculated by formula (31.11) as shown in Table 31.1.

Weight matrix

$$\omega_{21} = \begin{bmatrix} 0.355 & 0.305 & 0.34 \end{bmatrix}$$

Membership matrix

$$R_{21} = \begin{bmatrix} 0.15 & 0.7 & 0.15 & 0 \\ 0.75 & 0.25 & 0 & 0 \\ 0.15 & 0.45 & 0.3 & 0.1 \end{bmatrix}$$

According to formula (31.7), the state evaluation of the upper layer can be obtained

$$B_{21} = \begin{bmatrix} 0.333 & 0.478 & 0.155 & 0.034 \end{bmatrix}$$

In the same way, the state evaluation matrix of all indicators at the same level can be obtained. Through formula (31.11), the final evaluation state matrix can be obtained as $B = \begin{bmatrix} 0.459 & 0.463 & 0.021 & 0.057 \end{bmatrix}$. According to the principle of maximum membership, the status of the traction transformer in operation should be "good." The result is consistent with the actual evaluation result, which shows that the method is suitable for the evaluation of traction transformers in camp, and the results are scientific and reliable.

31.5 Summary and Outlook

Based on the characteristics of the traction transformer in the camp, this paper proposes an evaluation model with system characteristics, expands the multi-element evaluation index, and explores the membership function for some indexes at different levels, which provides a reference method for the research of other indexes. Combining specific engineering data, the calculation between levels is described in detail. The calculation results are similar to the actual situation, indicating that the method is practical, and the evaluation method and model have good adaptability.

References

1. Wang, Q., He, Z.-Y.: Health assessment method for metro traction power supply system based on multi-layer immune principle and fuzzy statistics. J. China Railway Soc. **37**(12), 31–39 (2015)

2. Du, J., Sun, M.-Y.: Research on multi-index evaluation method of transformer design scheme. Transformer, **56**(07), 19–23 (2019)
3. Zhang, Y., Liao, R.: Maintenance decision making of power transformer by integrated considering reliability and economy. J. Electrotech. Technol. **29**(11), 191–200 (2014)
4. Hua, Z.: Research on dissolved gas analysis and fault diagnosis in transformer oil. China Equip. Eng. **23**, 80–82 (2019)
5. Cheng, L. Application of oil chromatographic analysis in transformer fault identification. Electromech. Technol. Hydropower Station **40**(4), 63–65+75 (2017)
6. Zhong, Q., Guo, C.: Comprehensive evaluation method of helicopter power supply system based on multi-hierarchy fuzzy evaluation model. Foreign Electron. Measur. Technol. **34**(11), 27–31 (2015).

Chapter 32
Electrochemical Numerical Simulation of Atmospheric Corrosion Sensors Covered by Thin Liquid Film

Tingting Wang, Xiao Wei, Daiming Yang, and Yating Wang

Abstract To study the performance of atmospheric corrosion, a mathematical model of galvanic corrosion under equilibrium state was established, and the Tafel formula was used to describe the relationship between electrode overpotential and current density. Based on a galvanic corrosion sensor with a multilayer structure, the parameters of its simulation model were set and the electrochemical numerical simulation was carried out. The results showed that the corrosion current density was not uniformly distributed, the electrode potential was higher at the junction of anode and cathode, and the corrosion current was larger. Covered with thin liquid film, oxygen diffusion was the main factor limiting the corrosion rate. With the increase of the solution conductivity, the electrode potential and current density both increased and tended to be stable slowly. The simulation could be contributed to optimization of corrosion protection for smart wireless communication devices.

32.1 Introduction

Smart wireless communication devices are sensitive to components corrosion when exposed in the atmosphere. The atmospheric environment corrosivity is an extremely important issue in device life estimation and maintenance control. As factors affecting metal corrosion in atmospheric environment, temperature, humidity, rainfall, salt particles in the air, corrosive gases, etc., can be related to corrosion. Atmospheric corrosion sensor is a terminal to obtain metal corrosion rate under the action of multiple factors [1–5]. The sensor tracks the microcurrent signal which is transferring from the metal during the corrosion process and evaluates the corrosion rate. The integral of the corrosion current with respect to time leads to the amount of corrosion charge, which indicates the corrosion degree of the metal.

T. Wang · X. Wei
Electric Power Research Institute, China Southern Power Grid, Guangzhou, China

D. Yang (✉) · Y. Wang
School of Construction and Engineering, Shenzhen Polytechnic School, Shenzhen, China
e-mail: yangdaiming@szpt.edu.cn

L. C. Jain et al. (eds.), *Wireless Technology, Intelligent Network Technologies, Smart Services and Applications*, Smart Innovation, Systems and Technologies 258,
https://doi.org/10.1007/978-981-16-5168-7_32

Galvanic cell-based corrosion sensors were proposed in 1986 and applied in automobile industry, electric power industry and other fields. In the literature, many scholars have proposed corrosion sensors with different structures and parameters, while the characteristics did not compare for optimization. In fact, no one has compared any two sensors with different structures or parameters. The comparative test is costly and time-consuming. A better method is to obtain the performance parameters of different sensors through theoretical analysis. With the development of simulation technology, the effect of different parameters can be obtained by numerical calculation.

In this paper, an atmospheric corrosion sensor with an iron anode and a gold cathode is developed for electrochemical numerical simulation. The steady state model of galvanic corrosion is established. With the aid of COMSOL, the numerical simulation of the corrosion behavior of the sensor is conducted, combined with the prototype experiment for verification. The simulation process is meaningful for sensor selection and parameter optimization.

32.2 Principle of Galvanic Corrosion

The working principle of the sensor is galvanic corrosion on the anode surface in electrolyte, which can be used for electrochemical numerical simulation. Compared with the charged ions in the electrolyte, the corrosion current in equilibrium state is smaller, so it can be assumed that the conductivity of the electrolyte is constant. According to Ohm's law, the current density, conductivity and potential gradient form the following equation:

$$\mathbf{i}_\mathbf{l} = -\sigma_l \nabla \varphi_l \tag{32.1}$$

where $\boldsymbol{i}_l$ is the electrolyte current density, expressed as vector, and the unit is A/m^2; σ_l is the electrolyte conductivity, the unit is S/m; $\nabla \varphi_l$ to potential gradient, unit V/m.

When the electrochemical reaction keeps stable, that is, the current density at each point remains constant, and equation can be evaluated as follows:

$$\nabla \mathbf{i}_\mathbf{l} = 0 \tag{32.2}$$

The liquid film is thin, the vertical current of electrolyte is approximately equal to 0, the current density can be simplified as a two-dimensional distribution on the metal surface, and a thin shell without thickness is established on the electrode surface to simulate the current distribution. The current change in the vertical direction of the thin shell structure is ignored, and only the conduction of tangential ion current along the surface is considered. Equation (32.2) can be written as:

$$\sigma_l\left(\frac{\partial^2\varphi_l}{\partial x^2}+\frac{\partial^2\varphi_l}{\partial y^2}\right)=0 \tag{32.3}$$

On the electrode surface, the direction of electrolyte current is perpendicular to the electrode. The boundary conditions of current distribution are as follows:

$$\mathbf{n}\cdot\mathbf{i}_l=\sum i_m \tag{32.4}$$

In the formula, i_m is the small reactive current density on the local electrode surface.

The current distribution boundary conditions outside the electrode surface are as follows:

$$\mathbf{n}\cdot\mathbf{i}_l=0 \tag{32.5}$$

For the oxidation reaction on the surface of metal iron, the chemical reaction is iron corrosion from Fe to Fe^{2+}, where the standard potential of iron is $E^{\ominus}Fe/Fe^{2+} = -0.409$ V. The oxidation current density of iron was described according to the anodic reaction part of Tafel expression [6]:

$$i_{Fe}=I_{a0}\exp\left(\frac{\eta_a}{\beta_{aa}}\right) \tag{32.6}$$

where I_{a0} is the exchange current density of the anode phase, β_{aa} is the Tafel slope of the anodic reaction, and η_a is the anode overpotential.

For the reduction reaction of oxygen on the surface of the gold-plated layer, the chemical reaction is changing O^2 and H_2O to OH^-, where the standard potential of oxygen is $E^{\ominus}OH^-/O_2 = 0.401$ V. According to the cathode reaction part of Tafel expression, the current density of oxygen reduction reaction is described:

$$i_{O2}=-I_{c0}\exp\left(\frac{\eta_c}{\beta_{cc}}\right) \tag{32.7}$$

where I_{c0} is the exchange current density of the cathode phase, β_{cc} is the Tafel slope of the cathodic reaction of the cathode phase, and η_c is the cathode overpotential.

32.3 Simulation Model Parameters

The electrochemical simulation model is consistent with the actual sensor as shown in Fig. 32.1.

Due to the gold-plated layer of the sensor is only 35 μm, the size of the sensor differs by more than a thousand times from the length (50 mm) of the sensor. In order

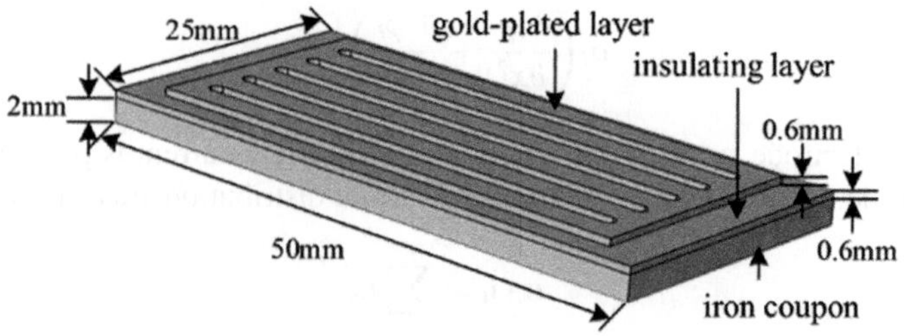

Fig. 32.1 Electrochemical simulation model of corrosion sensor

to facilitate grid division for simulation, the thickness of the gold-plating layer was increased to 0.6 mm. Simulation parameters of the model are set as follows (Table 32.1):

The electrochemical corrosion process of atmospheric corrosion sensor was simulated by using COMSOL Multiphysics.

Table 32.1 Parameters of electrochemical simulation model of corrosion sensor

Name	Symbol	Unit	Value
Equilibrium potential of gold	E_{Au}	V	1.498
Equilibrium potential of iron	E_{Fe}	V	−0.409
Equilibrium potential of oxygen	E_{O2}	V	0.401
The exchange current density of gold oxidation	I_{Au}	A/m^2	10^{-4}
The exchange current density of iron oxidation	I_{Fe}	A/m^2	10^{-4}
The exchange current density of oxygen reduction on the surface of gold	I_{O2_Au}	A/m^2	10^{-3}
The exchange current density of oxygen reduction on the surface of iron	I_{O2_Fe}	A/m^2	10^{-6}
Taffel slope for gold oxidation	β_{Au}	V	0.1
The Tarfel slope for iron oxidation	β_{Fe}	V	0.1
Tarfel slope of oxygen reduction on the gold surface	β_{O2_Au}	V	−0.1
Tarfel slope of reduction of oxygen on iron surface	β_{O2_Fe}	V	−0.1
Solution conductivity	δ	S/m	0.1
The thickness of liquid film	d	m	10^{-3}
Oxygen concentration at the boundary of the liquid film	c_{O2}	mol/m^3	0.2
The diffusion coefficient of oxygen	D_{O2}	m^2/s	10^{-7}

32.4 Simulation Results

The interface electrode potential, current density and oxygen concentration distribution on the electrode surface are obtained through simulation, as shown in Fig. 32.2.

In Fig. 32.2a, the potential of the iron electrode is within the range of 0.06–0.09 V, and the positive shift from equilibrium potential is more than 0.4 V. Compared with a single iron electrode, the corrosion rate of the iron-gold galvanic couple increases significantly, which is caused by the higher potential of the gold electrode interface. On the surface of the iron electrode, the electrode potential at the edge of the groove near the gold electrode is higher than that at the middle part far away from the gold electrode. At the same time, the electrode potential of the gold electrode surface near the trough is lower than that far away from the trough, indicating that the corrosion reaction is fastest at the junction of the iron electrode and the gold electrode.

For the interface current density in Fig. 32.2b, the fastest corrosion rate of iron is located at the two ends of the trough, indicating here. The main reason is that when the liquid film is thick (1 mm), the couple corrosion velocity is mainly limited by the cathode oxygen diffusion velocity. At both ends of the slot, the cathode semi-surrounds the anode, with a large area ratio between the cathode and the anode, and sufficient oxygen promotes the development of the couple corrosion. In fact, when exposed to air, corrosion occurs randomly in each slot, but the corrosion at the edge and corner is more obvious, as shown in Fig. 32.2d.

In Fig. 32.2c, the oxygen concentration on the surface of iron electrode is close to constant value, which indicates that the cathode reaction speed on the surface of iron is slow. The area between anode and cathode is larger than that between anode and cathode, and oxygen consumption is faster.

The influence of solution conductivity on electrode potential and current density was studied. The short and long axes of a single slot in the positive center were taken as objects. Draw the curve of electrode potential and current density at the edge and center of the short axis and the long axis, as shown in Fig. 32.3.

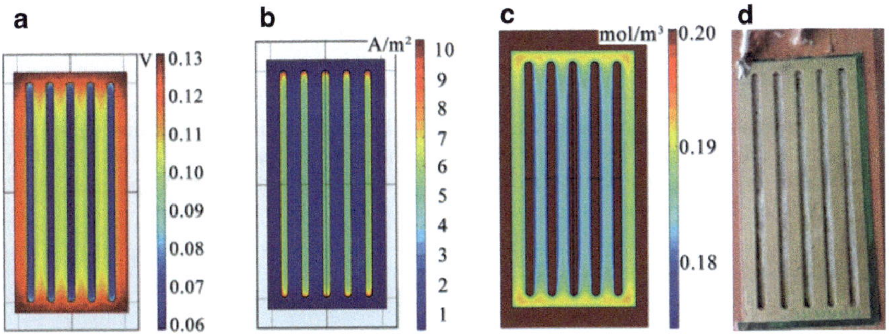

Fig. 32.2 Distribution of electrode potential (**a**), current density (**b**), oxygen concentration (**c**) and experiment results of atmospheric corrosion sensor

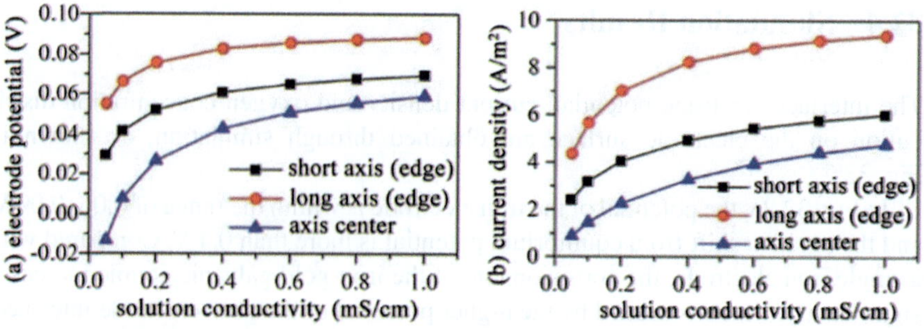

Fig. 32.3 Distribution of electrode potential (**a**), current density (**b**) and oxygen concentration (**c**) of atmospheric corrosion sensor

The electrode potential and current density at the edge of the short axis, the edge of the long axis and the center of the axis all showed a nonlinear increase with the increase of the solution conductivity, and the rate of increase gradually decreased. The relationship between solution conductivity and measured position and electrode potential and current density in galvanic couple corrosion was verified. Experiment results verified the nonlinear relationship between the corrosion current density and the solution conductivity.

32.5 Conclusion

The electrochemical numerical simulation analysis of the laminated galvanic corrosion sensor shows that the corrosion current density is nonuniform distribution, the electrode potential at the junction of anode and cathode is high, and the corrosion current is large; under the cover of thin liquid film, oxygen diffusion is the main factor limiting the corrosion rate; with the increase of solution conductivity, the electrode potential and current density continue to rise and slowly tend to be stable. Electrochemical numerical simulation technology can be used to optimize the structure and parameters of sensors for corrosion detection of smart wireless communication devices.

Acknowledgements This research was funded by National Engineering Laboratory for Ultra High Voltage Engineering Technology (Kunming/Guangzhou), grant number NEL201911, and Foundation for Young Innovative Talents in Higher Education of Guangdong, grant number 6020210025K.

References

1. Mansfeld, F., Kenkel, J.V.: Electrochemical measurements of time-of-wetness and atmospheric corrosion rates. Corrosion **33**(1), 13–16 (1977)
2. Mansfeld, F., Kenkel, J.V.: Electrochemical monitoring of atmospheric corrosion phenomena. Corrosion Science **16**(3), 111–122 (1976)
3. Shi, Y., Fu, D., Zhou, X., et al.: Data mining to online galvanic current of zinc/copper Internet atmospheric corrosion monitor. Corrosion Science **133**, 443–450 (2018)
4. Pongsaksawad, W., Viyanit, E., Sorachot, S., et al.: Corrosion assessment of carbon steel in Thailand by atmospheric corrosion monitoring (ACM). Sensors **20**(2), 23–27 (2010)
5. Mizuno, D., Suzuki, S., Fujita, S., et al.: Corrosion monitoring and materials selection for automotive eience **83**(6), 217–225 (2014)
6. Perez, N.: Elnvironments by Using Atmospheric Corrosion Monitor (ACM) sensor. Corrosion Scectrochemistry and Corrosion Science. Springer, Netherlands (2004)

Chapter 33
Analysis of Daily Public Relations in Art Colleges in the New Media Environment

Wen Jie Xu and Shui Jin

Abstract Under the new media environment, the way of information dissemination is becoming more convenient and faster, and it may only take a few seconds from the occurrence of an event to letting the world know. No organization or institution can be separated from the network of information dissemination. As a place for cultivating talents, colleges and universities also face the problem of disseminating information to the outside world and the need for daily public relations. This article takes art colleges as an entry point and explores how to conduct public relations activities through the communication characteristics of new media and the public opinion environment of art colleges. There is the particularity of the environment for cultivating talents in art disciplines. Therefore, when conducting public relations activities, it is necessary to comprehensively use public relations principles, and when necessary, in combination with crisis public relations, in order to better maintain the campus image.

33.1 Introduction

Colleges and universities are places for training talents and art colleges and universities train art talents in a targeted manner. Such special institutions often focus on student teaching and administrative management. The campus already has a very complete operating mechanism, which can be said to be well-organized. However, it ignores the continuous attention to external evaluations as much as companies and governments and does not pay attention to and participate in daily public relations activities. Therefore, in addition to attracting enough attention from the outside world during the enrolment and employment season, the outside world can only learn about the school through personal identities such as alumni, current students or faculty, without a comprehensive understanding.

Judging from the current domestic research literature, most of the research on public relations activities is based on enterprises and governments, and there is very

W. J. Xu · S. Jin (✉)
Beijing Institute of Fashion Technology, Beijing 100029, China
e-mail: jinshui@126.com

L. C. Jain et al. (eds.), *Wireless Technology, Intelligent Network Technologies, Smart Services and Applications*, Smart Innovation, Systems and Technologies 258,
https://doi.org/10.1007/978-981-16-5168-7_33

little mention of public relations activities in colleges and universities, and there is no special research on public relations activities in art colleges; secondly, research on public relations activities. It also focuses on crisis public relations, and the understanding of daily public relations only stays in the admissions promotion. The main reasons for the crisis incidents on campus are as follows: First, the composition of the members on campus is complicated, and conflicts between groups and within groups are prone to occur, and improper handling can lead to further escalation of conflicts; second, crisis awareness relatively indifferent, not deeply aware of the power of public opinion, thinking that all incidents that occur in the school have little to do with the outside world, so they did not pay attention to seizing the media's right to speak; third, the solution to the problem is relatively simple, and what the campus administrators can solve is solved by the administrators themselves through prescribed method, and the law can be judged if the problem cannot be solved. This phenomenon of emphasizing the law rather than public relations is widespread. The author believes that communication is the best way to resolve contradictions. Often, the outbreak of contradictions is that small daily problems are not communicated reasonably and accumulated to a certain threshold and burst out collectively. In addition, the school does not have a special public relations department to handle these. The problem is the part-time responsibility of the counselor or administrative teacher, so this article will discuss it.

33.2 The Characteristics of Information Dissemination in the New Media Era

33.2.1 There Are Many Communication Channels and More Comprehensive Coverage

In the past, people wanted to get news and information from newspapers, radio, television and other mass media. Nowadays, with the rapid development of the Internet and mobile devices, people communications services are no longer limited to getting information from the mass media, but also on news webpages, hot searches on Weibo, window pushes, WeChat official accounts, video websites, mobile clients, etc. Get information on similar mobile applications. According to the characteristics of media communication, these channels can report on the same incident from different aspects so that the incident can be known to people in a more comprehensive or three-dimensional manner, and it is possible to dig deeper.

33.2.2 Information Spreads Quickly but Quality Varies

Since the mass media have strict review systems before release or broadcast, editing, processing, proofreading, typesetting, or scheduling, it takes a certain amount of time for news to be communicated from the publisher to the receiver. However, the current multi-channel new media has to some extent made up for the lack of timeliness of the mass media in the past. Once an incident occurs, it will spread to all corners of the world through the Internet. Nevertheless, due to the lack of strict supervision, the news literacy of the information publisher and the influence of subjective factors, and even some interest-driven reports, there may be certain deviations or distortions or exaggerations in the reporting of the event [1]. For example, in Thailand's warning short film "Use Wisdom to See Those Unseen Things," a video of a rent-collecting wife "bullying" a vendor is on the Internet. The middle-aged women who seem to be very strong, full of air and aggressive tone are the managers of this market. In front of a pork stall, I had a dispute with the vendor, and later directly threw the vendor's weighing scale to the ground and smashed it. This scene happened to be recorded by someone who was interested and accompanied by the text: "The market proprietor bullies the vendor." Put it on the website to spread. The fact is that the merchant's scale was manipulated by the merchant himself. After the rent-collector discovered the short-suffering behavior, the scene in the video appeared only when he refused to correct it. In the recent period, China's epidemic prevention and control has achieved a staged victory, and the country has shifted from comprehensive prevention and control to a normalized prevention and control stage of resuming work and production. In contrast, politicians in the United States of America headed by Trump are driven by the party and government. And the interests of the consortium not only do not organize all available forces to focus on treatment, but also sacrifice the lives and properties of millions of people in the United States of America. At the same time, in order to divert attention from the rising unemployment rate in the United States of America and the racial discrimination and even complaints caused by the recent death of a black man by a white policeman, China has been frequently criticized. All these are the disadvantages of information communication in the new media era.

33.2.3 New Media is Highly Interactive

In the past, people only acted as receivers of information, receiving information solely from the mass media, which is often referred to as "audience." Nowadays, in the Internet age, we can participate in the spread of events in news websites, forum posts, Weibo posts, WeChat comments, etc., and role exchanges become senders or transmitters of information. In the process of real-time interaction and exchanges and discussions with others, the efficiency of information dissemination will be accelerated. In addition, everyone only uses the information they obtain to process and publish opinions, forming a certain public opinion force. This public opinion force is invisible and intangible, but in many cases, it has a reaction to the event itself, thereby promoting the development and direction of the event, and even affecting the final result of the event [2]. Let's talk about the above example: people only see the brutal behavior of market bosses treating dishonest operators through videos, but the value of people cannot be the only basis for judgment based on the pictures they see. This is not the truth of the matter regardless of whether the photographer did it intentionally or unintentionally, but people who saw this video usually think that the offending party is the charterer rather than the vendor, which leads to public opinion accusing the charterer of leaning on one side and launching a campaign on the Internet. If the market no longer patronizes the market's activities, putting it in real life is likely to cause poor management or even bankruptcy due to public opinion pressure.

33.3 Analysis of the Particularity of Art Colleges and Public Opinion Environment

33.3.1 The Competition for Art Admissions Examinations is Fierce, and the News About Negative Events on Campus is Highly Topical

As institutions of higher learning, art schools have the three major functions of cultivating talents, developing science and technology, and serving the society. But different from ordinary colleges and universities, the source of enrollment of art colleges is mainly through the art entrance examination or school examination, which has greater autonomy, and promotes teaching and education activities by cultivating artistic talents [3]. The purpose is to prosper advanced socialist culture and provide motivation to cultivate high-quality artistic aesthetics. Ordinary colleges and art colleges are both related and different. Only by grasping the characteristics of students in art colleges can we maintain the operation management and public relations activities of this special organization.

As a matter of public concern, education has always been the focus of people's livelihood issues. Students in colleges and universities are the future pillars of the motherland and the pride of the family, and many of them are artistic talents. Due to the different educational methods, most of the training of art colleges respects small and sophisticated educational concepts. However, with the improvement of people's material living standards, spiritual needs also need to be greatly satisfied. This is reflected in the study, and more students will choose to study art to enhance their aesthetic ability and vision, which makes the competition increasingly fierce. In the process of fierce competition, if someone wins the exam with an improper relationship, the natural stakeholders will hit like a flood, forming a butterfly effect. Recently, an artist surnamed Tong announced the news in a live broadcast, saying that he was not a fresh student in the college entrance examination that year, and that the college that applied for the exam that year had a regulation that only recruited fresh students and tampered with the files privately. Although he did not get admitted to the school he likes, it caused an uproar on the Internet. Regardless of the legal aspect, this has caused serious damage to the image of the Ministry of Education and the university he graduated from the perspective of public relations activities. In campus management, the words and deeds of professors and teachers have also become representatives of the campus. Once the publication is biased or does something contrary to social common sense, it will become a hot event once it breaks out [4].

33.3.2 Weak Control of Universities in Public Opinion

Compared with news newspapers, TV programs, commercial advertisements and other media, in addition to the annual art entrance examination, art schools can be used for display channels such as official websites, campus Weibo and WeChat public accounts. The attention of both the platform and the platform is very limited, and the target audience is usually examinees and parents, school students, alumni and faculty. Most people in society do not pay attention to the daily publicity of the school. For example, a press release published on the press center of the Propaganda Department of the campus official website on the theme of the 70th Anniversary of the National Day by the Beijing Institute of Fashion Technology was read only 238 times. In contrast, 5 days later, a video of the same content on CCTV News Weibo was viewed 4.43 million times. It can be seen that, in contrast to the discourse power of university propaganda and mainstream media, the ability of public opinion propaganda and public opinion monitoring is weak.

33.3.3 Art School Students Are More Individualistic

As the main body of the dissemination content, college students have the following characteristics: First, their self-awareness has gradually increased, and the management of teachers and the school is no longer a simple obedience, and they have begun to have independent thinking. This is also an era of "everyone is from the media." For dissatisfaction with some regulations or penalties, students often use the media to vent or seek help, and some even impose it on the school with the attitude of "making a big fuss, endless" pressure to achieve the desired results. Secondly, due to the immature mentality of college students, most of them are emotionally dominated in the process of viewing or handling incidents, and the comprehensive assessment of the incidents involved and the poor predictability of the consequences are often involved in more than individuals, and they are more loyal. Students will also help the bullied friend to vent their anger, leading to a mass incident. In addition, most of the art students' professional research makes the training characteristics focus on the pursuit of uniqueness and innovation, and occasionally show a certain rebellion and personality [5]. For the school, it is impossible to copy the management methods of ordinary universities to impose strict requirements on students.

33.3.4 During the Student's Study Period, the School Has the Responsibility of Supervision and Education

As a part of colleges and universities, art academies need to shoulder various responsibilities to students on and off campus in their daily operations. Especially in the domestic environment, because students are not yet fully capable of independence (most of them do not have the ability to be independent financially), but they are different from the almost completely closed elementary and middle school education, students have a lot of spare time arrangements, regardless of whether the students are in school neither outside the school is always within the visual range of teachers or campus management. In the event of emergencies or accidents, parents or the public media will hold the school responsible for the reason of being in school. However, due to the limitation of reports, the public will naturally divide the school and the students in the incident into strong and weak parties. And then judging right or wrong or giving inappropriate comments will often bring huge public opinion and accountability pressure to the school.

33.3.5 The Popularity and Reputation of Colleges and Universities Have a Large Impact on Students' Advancement and Employment

Compared with the company's brand, the quality of word-of-mouth directly affects consumer confidence and product sales, which in turn determines whether the company has a chance to survive in cruel competition. The popularity and reputation of colleges and universities will not be immediate, but it often affects student recruitment and graduates' careers, and this effect is long and lasting. At the same time, the relevant public, such as higher-level supervisory units, cooperative educational institutions, etc., will re-evaluate the school, which will bring a lot of trouble to the teaching plan and the long-term development of the school.

33.4 The Importance and Measures of Daily Public Relations Activities

33.4.1 Maintaining the Image of the Organization and Expanding Publicity Are the Basis of Public Relations Activities

When searching for news or information on the official websites of some schools, it is not difficult to find that most of the official websites of schools are useless, the information is not updated in time, or even no longer releases information. Take the Central Academy of Drama as an example, click on campus culture-campus news. The latest press release was released three years ago. Although important information such as enrollment information is still being updated, it can be seen that the school's construction of the official website, utilization and publicity are not paid enough attention. Even if the internal operation is well-organized, due to inadequate publicity or poor information, outsiders will retain their impression of the school a long time ago, or they may hear some evaluations of the school from various places. These impressions and evaluations are often accompanied by communicators' subjective colors, which do not represent the real situation of the school. Therefore, strengthening the maintenance of information channels and the promotion of image shaping is the most basic and most important in daily public relations activities.

33.4.2 The Cost of Positive Publicity in Daily Public Relations is not High

Unblocking channels, establishing a mechanism for expressing opinions, and clearing the emotions of students or faculty members can be the responsibility of the trade union or the publicity department of the student union. Promote harmony in the campus through multi-party communication between teachers and teachers, teachers and students and students and students. In addition to the promotion of activities on the official website, the general publicity film of the admissions examination website, and the official Weibo, it is necessary to pay more attention to the daily campus culture and the release of campus dynamics, so that the information inside and outside the campus is smooth, and the barriers are gradually eliminated. The evaluation of the campus by the outside world is also more objective and fair, and the school's popularity and reputation have also been improved, and this does not require too much investment and even play the role of love school and makes students feel more belonging [6].

33.4.3 Daily Public Relations and Crisis Public Relations Form Complementary Advantages

For things on campus, we must focus on prevention and ventilate in many ways [7]. When daily public relations do not work, raise the awareness of crisis public relations. We can through campus network monitoring, keyword extraction of school image and school reputation speech. Once a major campus event is mentioned, it should be taken seriously, and the matter can be investigated or appropriately intervened, and the matter should be resolved as early as possible to avoid further escalation of the problem. At the same time, we must also improve our crisis public relations capabilities. If conditions permit, we can set up a school's crisis public relations department, which is led by school leaders and composed of teachers and students. Only with one heart and one mind can problems be solved more smoothly. In terms of implementation, the school's external influence is judged, and if the influence is great, it will be dealt with publicly. If it does not constitute a great influence, it can be settled privately with the parties and the result will be made a simple statement. Compared with the past treat crises with a mentality of covering up the ugliness, this is no longer popular in this new media era. Therefore, schools must learn to use media platforms to prove their innocence, seize public opinion positions and avoid gossip and rumors blind the truth [8].

33.5 Conclusion

Public relations activities do not only exist and need when a crisis occurs but can and also need to be run all the time, day after day. Only by doing a good job of public relations and propaganda in peacetime can the crisis be resolved calmly. Public relations activities are not only the government and enterprises need to do, for colleges and universities, especially art colleges should pay more attention to the development of public relations activities.

Acknowledgements This work was supported by the Special Fund for the Construction of High-level Teacher's Group of Beijing Institute of Fashion Technology, Project No. biftxz201905

References

1. Ding L.: Ideas for dealing with public opinion crisis in colleges and universities based on the principles of public relations. Think Tank Times, 256–257 (2019)
2. Duan, H, Li, Y.W., Li, L, Huang, H.Y.: Interaction: The Outstanding Advantages of We-Media Public Relations. Public Relations World **10**, 32–39 (2019)
3. Huang X.Y.: Crisis public relations and media response to emergencies in private colleges and universities. Foreign Economic Relations and Trade **03**, 102–103 (2020)

4. Niu, J., G., Wei, T.H.: Research on emergency response to mass incidents in colleges and universities. Journal of Qingdao Vocational and Technical College **11**, 28–31 (2018)
5. Tan S.: On the characteristics of students in higher art colleges under the new situation. High Education Journal **03**, 50–51 (2017)
6. Zhao, M. L., He, C. C.: Public relations strategy for emergencies in colleges and universities in the new media era. Operation and Management **02**, 151–153 (2016)
7. Ye, W.: Re-understanding of the government's daily public relations. Inquiry **06**, 58–64 (2018)
8. Qu, J. L.: Crisis Public Relations Strategies for Colleges and Universities in the New Media Era. International Public Relations, 14 (2020)

Chapter 34
Analysis on the Installation Art

Guanyu Zhu and Shui Jin

Abstract This paper mainly discussed the innovations brought by the new media to the form and application of installation art. In this paper, new media installation artworks were listed and analyzed to figure out the material changes and expression language of installation art in the new media. New media has promoted the material development and interactive design for the installation art, which further allows installation art to integrate various interesting and innovative elements and become a unique art form. This paper reflects on how the current rapid development of computer, information technology, and interactive technology help to apply the information technology into real life and integrate installation art into human–machine interaction.

34.1 Installation Art in the New Media

Media [1] is defined as a connection between information and people. People obtain information through certain channels. Channels that carry the task of information dissemination are the media. From a production point of view, rich imaging, digital and interactive technologies, and creation have created the boom of We-media sector. People pay more and more attention to the application of visual imaging technologies and personal creativity in various fields of today's society. When it comes to the influence of new media and related technologies on art, people will think about installation art in contemporary art. Installation art is also called "ready-made art" [2]. The materials in the installation art are mostly from people's daily life and production. Artists reconstruct the materials and shapes of ready-made objects, endow them with new esthetic meanings and display them in a specific space to form a unique effect. As a physical art, installation art has had a significant impact on traditional art, as it breaks the traditional art form, and changes people's minds on art carriers. Installation art is not limited to sculpture, painting, and music. Instead, it can be expressed through any materials. In the current era, with various materials and technique choices, artists

G. Zhu · S. Jin (✉)
Beijing Institute of Fashion Technology, Beijing 10029, China
e-mail: jinshui@126.com

L. C. Jain et al. (eds.), *Wireless Technology, Intelligent Network Technologies, Smart Services and Applications*, Smart Innovation, Systems and Technologies 258,
https://doi.org/10.1007/978-981-16-5168-7_34

can present unprecedented effects in the combination of art, technology, and media. The development of technology also creates the new ways for artistic expression and communication and serves as a bridge between people and things. Therefore, artists start to ponder and image: what kind of incarnation the installation art will realize in the new media ecology based on big data, 5G mobile network and abundant materials and various thoughts [3].

34.1.1 Changes of Media

The meaning of the new media will be explored in this section. If we understand new media literally, the synonym of media is "media", which in a broad sense refers to the substance that can make connections or relationships between people, people and things, things and things, and the media is in the meaning of words More points to the storage and dissemination of information, but both have the meaning of "connection". In the Internet era, the brand-new concepts, such as internet of things (IoT), comes into being. As they connect different things and store and disseminate information, these new concepts instill "novelty" into the new media and coincides with the concept of reconstructing things and spatial connections in installation art. Based on the traditional media, new media obtain information from traditional paper media and advertising television to digital technology and products, and shift from traditional painting to "ready-made" artwork. The medium of communication and the form of expression are no longer fixed. Any academic discipline can become a channel for information storage and communication, and anything can become material or device for artistic creation. Roy Ascott, the founder of international new media art, proposed "moist media" [4]: more and more artists use nature, biology, biological processes or computer technology for artistic creation. Computers and digital technologies belong to the "silicone dry matter" in our cognition, so the biological media are the "moist media." Biological art that uses natural and biological knowledge, technology, and materials to create involves selection of the "medium" in artistic creation.

34.1.2 "Ready-Made Products" of the Times

Steampunk symbolizes the mechanical revolution in the industrial age and people's reveries about hi-tech life in future. Pop art symbolizes the unprecedented prosperity of the American and urban popular culture under the commodity economy. China's economy has been developed rapidly in recent decades, and various emerging technologies and abundant commodities have satisfied people's needs of material life. These commodities, supplies, or cutting-edge technologies for interaction and data computing all embody the features of the times. In other words, products with the elements of new media and new media software and hardware are the "ready-made" materials of installation art. At the launch of the DIP Art Fund held at the Ullens Art

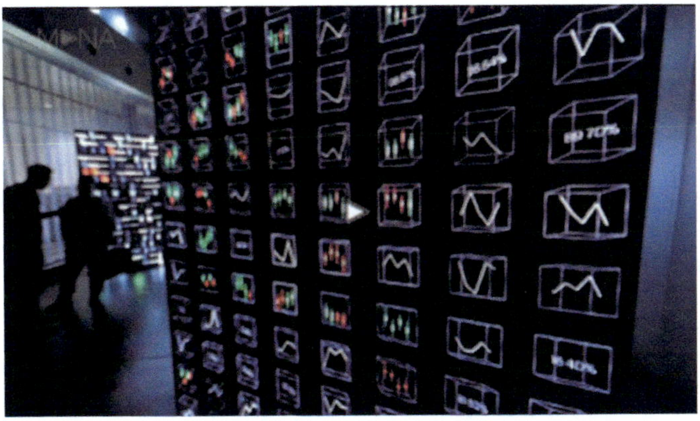

Fig. 34.1 Screenshot of video for stock exchange-code generation visual device

Fig. 34.2 Screenshot of video for stock exchange-code generation visual device

Center in Beijing, the first exhibition of "Stock Exchange-Code Generation Visual Installation" (Figs. 34.1 and 34.2) was held. The visual works presented on the three LED screens are all generated by computer codes. The flickering graphics present the ideas of the art fund in the on-site environment through direct and avant-garde art practices and create an immersive experience. Through the big screen of real-time trading, visitors can see the busy and exciting index fluctuation curve, ongoing transactions, and the red and green words in the stock market [5]. The "computer code" and "stock" elements shown on visual devices are the products of modern social development shows the visual form of modern social economic life in the form of a visual medium. Nowadays, there are countless products of science and technology and humanities in our lives. The thinking about the times have been transformed into various forms of art, like the ready-made materials of installation art. Such connections are still awaiting the discovery of artists.

34.1.3 Participation of Art in a Novel Manner

Reducing the viewer's cognition and participation costs echoes with the philosophy of installation art to a certain extent: allowing viewers to "participate" and "immerse" in the art. The current new media interaction technology is applied in many aspects in the art field to present the ideas of art more effectively. In terms of installation art, people no longer need to a solid art theory before they can appreciate works. Instead, they just need to perceive the design of the image, sound, smell, and texture in the installation design through their eyes, ears, nose, skin, and other organs. The installation art can provide special and unique experience for the viewers compared with other art forms. This form of communication with works of art reduces save people the trouble of cognitive understanding, makes the appreciation of art easier and more comfortable, and gives more space for imagination.

"Connectivity" is the core concept of installation art. Different from the art forms that require many original creation and materials, artists just need to think about the connections between existing objects, observe nature and create more connections so that viewers can obtain new experience. Playground is an interactive music art installation (Figs. 34.3 and 34.4), designed by Collectif Scale, the most experimental avant-garde art team in France. It aroused great response in the Paris Musique Club

Fig. 34.3 Screenshot of the interactive activity in the "Playground"

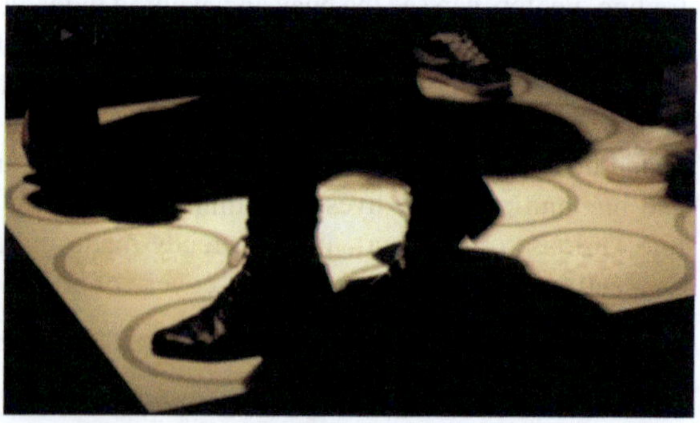

Fig. 34.4 Screenshot of the interactive activity in the "Playground"

exhibition in Paris in 2015. Playground (Figs. 34.3 and 34.4) consists of 32 snare drums, tom drums, and bass drums, and each drum is equipped with a computer-controlled drumstick. In front of the drum set array, there is an inductive control carpet. Visitors can take steps on the control carpet and ring their own beat. The more people participate, the more enthusiastic the beats. A special drum set music is hence formed [6].

"Readiness to participate" and "interesting results" are the charm of interactive installations. Even people without any artistic esthetics can "play" very happily. Art is to make life better and happier. Installations with fun and interactive effects as the core seem to have a unique mission-to simplify the complicated process and serve as the bridge between the artistic works and the participants. Driven by such a concept, interactive installations can help people explore those rendition and creative fields, such as music, painting, even animation production, and character design.

34.2 The Artistic Expression of Installation Art in the New Media

34.2.1 Concept of New Media Art

New media art covers various categories. Compared with the traditional art which transforms the media and intermediary devices, it has a certain consistency with the sensory experience brought by installation art, because ready-made products and space are often used, and there are also overlapping works styles. We generally categorize traditional art works into two-dimensional forms, such as painting and photography, and three-dimensional forms such as sculptures and installation entities, while new media art extends the concept of art to four-dimensional space, time changes, and human perception. New media art aims to transform or create a space that does not exist in a three-dimensional environment, non-existent sound tracks, or images that do not follow the passage of real time so as to bring a new experience to the viewer. In terms of installation art in contemporary art, based on the pursuit of "substance" and "essence," installation art has added new media-related technologies to the design of physical objects and spaces so that information, emotions, and brain activities can be better concretized and substantial in the works of art: It encourages people to ponder, and think about what is "virtual reality."

34.2.2 New Media Expression of Installation Art

34.2.2.1 Digital Art Language

Similar to the use of brushwork and color techniques in painting, installation art has higher requirements for display. Digital art in contemporary art aims to use digital technology in the process of creation or display and use digital imaging as a technique tool for the practice of art works. Since 1970, many terms have been used to describe this particular technique, including computer art and multimedia art. Digital art itself is placed under a more comprehensive term "new media art." Digital art can be completely computer-generated (for example, fractal and algorithm art), or it can be obtained from other sources, such as scanned photos or images drawn with vector graphics software using a mouse or drawing tablet, due to its unique expressiveness and digital calculations in the imaging effect and logical expression of installation art. A student majoring in digital media art from Jiangnan University published an art video work called "The Racecourse" on the MANA global platform for new media art (Figs. 34.5 and 34.6). In that work, the artist has transformed some photos of life and people into matrix imaging. The introduction of the work elaborates: the interaction between people and the city is happening all the time, the countless information generated by the interaction is recorded and uploaded all the time. Some people try to continuously analyze all the uploaded information based on the algorithm through pictures and even real-time monitoring, tamper the false information and convey their own views. This means that when enjoying the convenience brought about by technology, people are gradually giving up their right to think about information. Many times people feel that they are making choices and exercising free will, but it is not. There are algorithms and preferences behind shopping decisions, media news, and value transmission. "Don't let your brain become a racetrack for others' thoughts."

Fig. 34.5 "The Racecourse" artistic video works

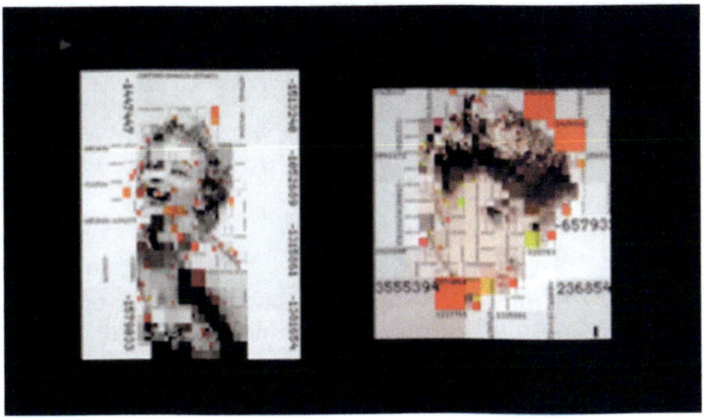

Fig. 34.6 "The Racecourse" artistic video works

34.2.2.2 Interactive Language

Interactive art can be discussed as a separate topic. In the vigorous development of new media art, many works have combined the characteristics and concepts of multiple art categories. After the concept of interaction has been developed, the artists begin to apply the various interaction means to his works. Interactive art involves the participation of the audience. Some interactive art installations allow observers or visitors to "walk in" the works to experience, and some artists require the audience to join into the artwork. Such works of art usually have mechanical brakes, linkage reactions, computers, sensors and other devices to respond to movement, heat, weather changes or other forms of input, and program them to produce sounds, pictures, dynamic effects, or somatosensory feedback for the participants. Most of the installation works of virtual internet art and electronic art are highly interactive. With the help of the concept of interaction and related digital technology, installation art can better realize the somatosensory effect and expand the period for installation art appreciation. The participation of performance activities enables artists to better express and improve "perspective of observation" in the installation works. The most popular work "CODA" (Fig. 34.7) at the Lyon Festival of Lights in France in 2019 was produced by the French avant-garde art group scale. The name of CODA is originated from ballet and music and means the ending part of a ballet or musical work. The device consists of 36 mechanical arms with two-meter-long light poles. In a matrix, the device becomes a luminous dynamic architectural structure. This installation is a futuristic transposition of ballet- people are replaced by light. This group of luminous mechanical bodies uses a unique and avant-garde dance language to create a coordinated or highly individual dance of light. Like many of their previous installations, CODA explores the existence and progressive virtualization of human. CODA, as an interactive installation work, can react according to the artist's special proposal (music scene, dance...) in real time [7].

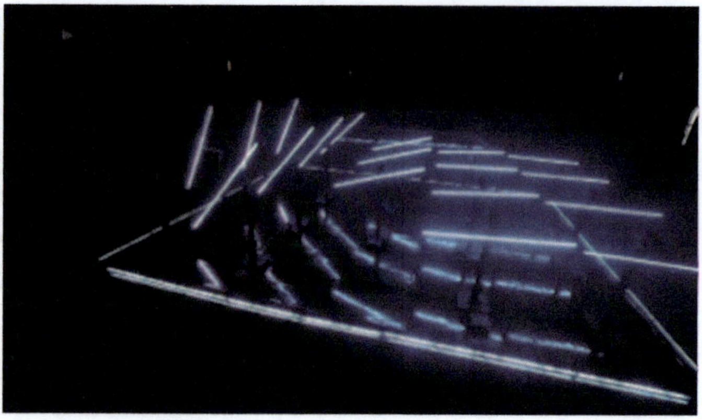

Fig. 34.7 "CODA" installation works

34.2.2.3 Mechanical Art Language

With the improvement of human production, machinery as the main tool of production has also become the expression language of art and is widely used in device design. The machinery is to help people reduce the difficulty in work or achieve tasks beyond the capabilities of humans. Daily simple and basic items, such as chopsticks, brooms, and tweezers, and complex machinery are used to create interactive and coordinated mechanical structures for production. These synchronized and complicated machinery are called a machine. From the perspective of structure and movement, mechanisms and machines are generally called machinery. As time goes by, machinery contains more profound cultural meaning and human wisdom. Nowadays, mechanical language is widely used in the field of installation art. On the one hand, the dynamic effects of the installation in the installation art are realized through linkage. On the other hand, the cultural meaning and structural beauty brought by the machine itself are also well used to present the artistic expression of contemporary installation art.

The integration of machinery and various sensing devices, and computing can also help finish many highly precise tasks. At the International Conference on Robotics and Automation (ICRA) in Montreal in 2019, the installation art work "Prosperity" produced by ELEKTRA (Fig. 34.8 and 34.9) was displayed. The main body of the work is an autonomous machine. All operations and running calculations are integrated into a microcontroller with only 256 K bytes of memory. "Prosperity" is an independent device for managing rice grains. Tweezers move from one surface to another, randomly clamp particles, and then place them in an orderly arrangement on another surface. Therefore, the rice grains are aligned and maintain their original orientation. In the past few years, Samuel St-Aubin has been devoted to his own creation and instilled another dimension into his works. He has transcended the utilitarian reality of objects, shifted the focus from focusing on surface to fundamentally destroying the relationship between objects and human. With the high precision, his

Fig. 34.8 Video screenshot of "Prosperity"

Fig. 34.9 Video screenshot of "Prosperity"

works allow us to explore the poetry of daily life and discover poetry in simple items in our daily life.

34.3 Conclusions

This paper discussed the major trajectory for the installation art in the new media. Of course, installation art is a special product as a result of the changes of media, highly saturated information algorithms, and the prevalence of artificial intelligence in the era of new media. Advances in equipment and technology have changed people's views on nature, society, and lifestyles. In the 2G era, people only knew to obtain movies, TV series and other videos through TV and hard disks; in the 4G era, the increase in network speed has made the handheld devices more versatile, and the ways of video display more diverse. Live broadcast and short video are the typical products of this era; now, with the popularization of 5G and the development of AR and VR technologies, the media tends to be a more immersive and timelier one. This has changed almost everything in people's lives, such as shopping, viewing exhibitions, interior design, and art genres. Various imaging digital technologies and the unique creativity of personal creations have created the created the boom of we-media sector. People pay more and more attention to the application of visual imaging technologies, personal value and personal creativity in various fields of today's society. No one can go against the tide of the times. Today's designers and artists also need to integrate their artistic creativity into a suitable expression language, and use different equipment, environments, technologies and other media in the new media so as to create more communication and appreciation methods for art and culture.

Acknowledgements This work was supported by the Special Fund for the Construction of High-level Teacher's Group of Beijing Institute of Fashion Technology, Project No.biftxz 201905

References

1. Baike, https://baike.baidu.com/item/%E5%AA%92%E4%BD%93. Last Accessed 17 mar 2021
2. Chen, J.: Research and Practice of Finished Objects in Installation Art Creation, pp. 115–119. Shanghai Vision (2019)
3. Yu, Q.Z.: Innovative research on the forms of installation art. Art Educ. Res. (16), 34–35 (2020)
4. Popular Science China, https://baike.baidu.com/item/%E6%96%B0%E5%AA%92%E4%BD%93/6206?fr=aladdin. Last Accessed 17 Mar 2021
5. MANA Global New Media Art Platform, https://www.manamana.net/video/detail?id=5456#!zh. Last Accessed 17 Mar 2021
6. Lu, X.: Application and development trajectory of new media art in scene display and virtual reality. Global Blue (11), 124–125 (2021)
7. Zhang, L.: Similarities and Differences of Aesthetic Characteristics between New Media Art and Traditional Art. Nanjing University of the Arts (2020)

Chapter 35
Design and Realization of General Platform for Computer Game

Taocan Zhang, Shuqin Li, Yixiao Su, Shaohua Ni, and Zhuoqun Li

Abstract This paper studies the current situation of computer game platforms and focuses on the rational use and expansion of existing game technology to build a complete general-purpose computer game platform that can load different game engines. The front-end and back-end decoupling architecture of the game platform was designed. The front-end calls the Native API at the back-end through WebSocket and uses JSON data for interaction. Realize the role of the game platform in the game rule determination and input–output interaction interface, and provide higher execution efficiency, more convenient operation mode and necessary rule evaluation for game participants. Let computer games be more fair, just and efficient. Experiments show that the system is feasible and effective, and it has practical significance for the future study of computer game.

35.1 Introduction

Machine game is one of the important research directions in the field of artificial intelligence and computer science [1]. Understanding the essence of intelligence through the process of machine game is the best experimental carrier for studying human thinking and realizing machine thinking. With the promotion of game competitions [2] in recent years, more scholars have participated in game research [3], but there are relatively few domestic research and development of general-purpose computer game platform for different chess types [4]. In the past national computer game competitions, because there was no unified competition platform, when multiple teams conducted different algorithm comparison tests, they could only independently develop programs with human–computer interaction interfaces, thus increasing the

T. Zhang · S. Li (✉) · Y. Su · S. Ni · Z. Li
School of Computer, Beijing Information and Science and Technology University, Beijing 100101, China
e-mail: lishuqin_de@126.com

S. Li · Y. Su · S. Ni · Z. Li
Sensing and Computational Intelligence Joint Lab, Beijing Information and Science and Technology University, Beijing 100101, China

L. C. Jain et al. (eds.), *Wireless Technology, Intelligent Network Technologies, Smart Services and Applications*, Smart Innovation, Systems and Technologies 258,
https://doi.org/10.1007/978-981-16-5168-7_35

workload of the battle platform interface. This prolongs the development cycle and is not conducive to rapid program development and testing. At the same time, the front-end and back-end integrated game system can only play a limited mode of man–machine game and cannot arbitrarily replace game engines that use different search algorithms [5]. Through the research results of the existing battle platforms as shown in Table 35.1, the advantages and disadvantages of the existing platforms are analyzed.

Based on the above research on the development status of machine game and platform, this paper draws on the advantages of the design and functions of the existing platform, focuses on the design of a reasonable platform architecture, and USES appropriate technology to build a perfect general computer game platform that can load different game engines, so as to ensure the fairness, fairness and efficiency of the game process.

35.2 The Architecture Design of the Computer Game Playing Platform

The platform is based on Node.js + Electron + JavaScript + Html + CSS + Python + WebSocket and other technologies to realize the whole platform architecture with front and rear terminals independent:

The front-end realized visual user interface of the battle platform based on electron and module design of general chess class, as well as the front-end chess move rule detection. Complete the visual representation of the current situation and can support the operation of the pieces on the board, at the same time, the design and implementation of the menu, timer, recovery, repentance and other common operation parts.

The back-end USES Python to implement the scheduler, which realizes the interconnection of the game program and the rule judgment of the game, completes the data saving of the situation, the analysis of the method and the judgment of the game result, realizes the rule detection of anti-cheating pieces, and it is also responsible for the output of relevant data to the front-end program. At the same time, the front-end program loads the operation to the back-end program, and the back-end program realizes the response to the front-end program.

The front-end html page calls the Native API interface of the back-end through Websocket and uses json data for interaction.

Table 35.1 Comparison of existing computer game platforms

Existing platform	Disadvantages	Advantages
Amazon (Li Zhen)	1. Supported operating systems: Windows 10, Xbox One 2. Does not support chess movement record saving 3. Does not provide API4 no timing function	1. Interface implementation effect is good 2. Support man–machine battle 3. Support reading and archiving
Invader	1. The version is backward, and the latest version was released in 2008 2. It can only be played by everyone 3. There is no API available	1. Interface implementation effect is good 2. Support man–machine battle 3. Support reading and archiving
Amazon Board Game (netease)	1. The version is backward, and the latest version was released in 2010 2. need to charge 3. support the operating system: IOS does not provide the API	1. Interface implementation effect is good
Amazons (Gregor Ulm)	1. Backdated version, the latest version, was released in 2014 2. Supported operating system: Windows 3. Complexed installation process 4. Only input coordinates to play the game 5. No API provided	1. User documentation is provided
El Juego de las Amazonas	1. The version is backward, and the latest version was released in 1995 2. Need to charge 3. Support the operating system: windows 4. Visual Basic writing 5. Does not provide API 6. No timer function	1. Provide help files 2. Support for man–machine battle 3. Support for printing and saving games
Six chess, Amazon, Sulakarta, phantom go, not go, point grid chess integrated battle platform	1. Simple page 2. No support for game between man and machine 3. No support for chess movement record 4. No support for saving chess state 5. No support for chess repenting	1. No platform dependence 2. Support network battle 3. Can display the movement record of pieces 4. Support loading engine

35.3 Design and Implementation of Computer Game Platform Front End

Based on Chromium and Node.js, the front-end framework electron [6] can use HTML, CSS and JavaScript to build applications, with good cross-platform and compatibility, suitable for scenarios with many UI changes, little volume limit and high-development efficiency. Using electron as the front-end GUI framework, this platform USES powerful web ecology to develop the interface, USES Native API to realize the platform to create room logic process, and presents the entire platform system in the form of the client. According to the idea of componentization development, the function of the platform is divided, and the front-end development is carried out with the idea of building blocks. Using object-oriented thinking, the API design of components is completed through four element attributes, methods, events and sub-window drawings of component design, which achieves a better platform effect.

The relevant logic implementation of chessboard is realized by using the prototype inheritance implementation method of JavaScript: By defining chess, a common chess class, the prototype property is used to add properties and methods to the chess object to define chess radii, boundary values, styles, the specific positions of chess pieces on the board, the display of chess pieces and the movable range of chess pieces. The whole logic process of the game is realized by defining the game logic class chess game, using the prototype property to add initial chess game and pieces to the object, drawing all pieces, drawing board and remaining pieces, switching chess sides, judging whether they are our own pieces, creating an array map of chess layout, moving pieces, judging the end of the game and other methods.

35.4 Modular Design of the Back End

35.4.1 Design and Implementation

Back-end code is written through Python language, and the calls between modules are realized through custom packages. At last, the back-end data is transmitted to the front end through the server, and the front end performs the test and outputs after the judgment is correct.

The general game playing platform integrates multi-function board operation, multiple board rules and user interaction. In order to meet the higher requirements of both sides of the game, the platform is divided into three modules: service module, room module and information module through the connection relationship of logic progression. The information module includes board, timer and player information. Besides, the board module can connect different kinds of chess.

35.4.2 Advantages of Module Division

More rigorous, professional and maneuverable.

Multi-module design to achieve broader scalability and practicality.

Integrated multiple means to promote fine-grained module division, which is conducive to separation of concerns and module reuse.

The back-end modules are divided as shown in Fig. 35.1, including the following modules.

Board module: it includes the playing rules of multiple chess types, winning conditions, review of chess faces and eating sub-rules, etc. After the player chooses the game, the corresponding checkerboard code matches the room.

Timer module: Calculate the playing time of each round and the total game time of both sides, and judge whether the stipulated time is exceeded.

Player module: Record the basic information of both sides of the game, including the current position of pieces, accessible positions of pieces and movable pieces, etc.

Room module: Refer to the above three modules. Add (delete) players, match status after starting the game, move pieces, notify players and judge the order of hands, etc., are all operated by the room module. In addition, the list of players, match status and room information are also displayed to users.

Serve module: obtain client list and room list, judge whether to join the room, and play the role of connecting front end and back end. Import the back-end code through the import room get the front-end message address, and finally synchronize the received object with the front and back ends through the JSON library.

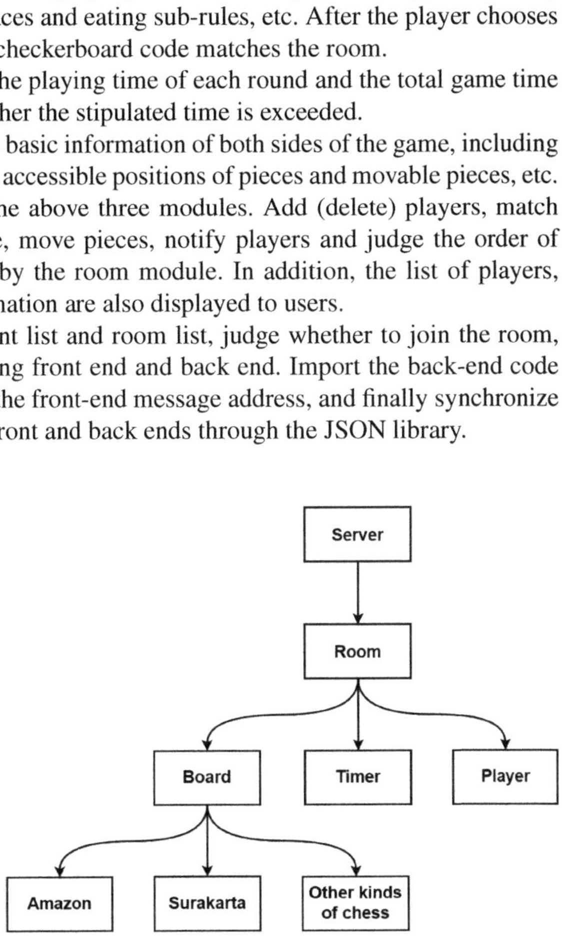

Fig. 35.1 The modules of the battle platform division

35.5 Realization of Game Platform

35.5.1 Key Technology: WebSocket

WebSocket [7] is a protocol for full-duplex communication on a single TCP connection. It makes the data exchange between the client and the server easier, allowing the server to actively push data to the client. In the WebSocket API, the browser and the server only need to complete a handshake, and a persistent connection can be created directly between the two, and two-way data transmission can be carried out.

35.5.2 Connection Between Front-End and Back-End Programs

The realization of the logic process of the front-end and back-end adjustment is shown in Fig. 35.2.

The platform uses Websocket and borrows the idea of Socket to provide a similar two-way communication mechanism between the front-end and the back-end: create a socket instance on the front-end, and provide this instance with the IP address and port number of the back-end to be connected. The backend creates another socket instance and binds the local port number to listen. When the front-end and back-end establish a connection, the two parties establish an end-to-end TCP connection, which enables two-way communication.

The url above is the address of the service opened by the native node, specifying the execution events of connection (on open), closing (on close), and message receiving (on message), accessing the HTML, and printing the ws information.

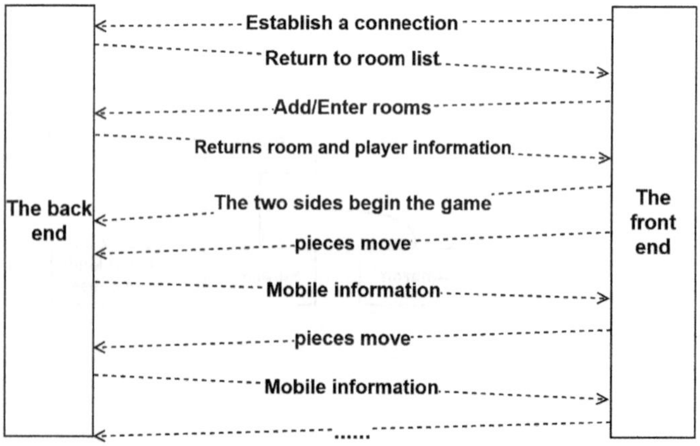

Fig. 35.2 Front-end and back-end coordination logic

Table 35.2 The native API design

Key	Value	Description
Message	Start	Game start
State	Ready	If the other party has not started, return ready, otherwise return start
Message	move	chess piece move
Location	{'from': [0, 0], 'to': [1]}	moving coordinates
kw	0	0 is a moving piece, 1 is a moving obstacle;
Message	Room_list	Room list
Room_list	[1–4]	room id array
Message	add_room	add the room
Args	[10, 10]	board size
Room_id	2	Room number
Order	1 or 2	1 is the first move, 2 is the second move
Player Id	"A"	player ID

The console prints "success connect!" to indicate that the connection is successful, and the client USES on message to process the received message.

The event parameter contains the details of this communication, and the message returned from the server will be in the event's data property.

After opening WebSocket, the server will listen in the message receive the parameter data to capture the message sent by the client, and then use "send" to send the message. The native API design is shown in Table 35.2:

35.5.3 *Technical Features of Platform Design*

35.5.3.1 The Front End and the Back End Are Separated

Separation of duties.

Interface communication between front and rear ends is realized through WebSocket;

The front end and the back end each have their own development process, build tools and test sets.

Separation of concerns, with front and rear ends becoming relatively independent and loosely coupled.

Interface specification principles.

Interface returns data to display: the front end only does rendering logic processing.

Rendering logic forbids calls across multiple interfaces.

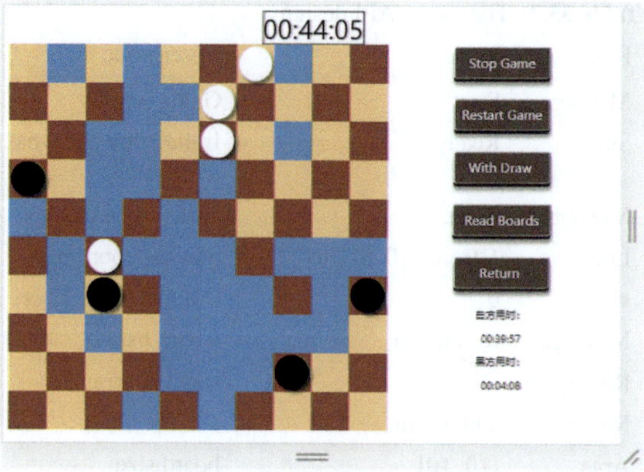

Fig. 35.3 An example of an Amazon chess match

The front end focuses on interaction and rendering logic and tries to avoid the occurrence of business logic processing.

Request and response transmission data format: JSON, JSON data as simple and lightweight as possible, to avoid the emergence of multi-level JSON.

35.5.3.2 Display the Effects of the Battle Platform

Front-end and back-end test implementation platform of the whole logic process. Figure 35.3 shows the game effect of Amazon [8] chess on the platform. The platform canvas displays the current game state. The buttons are laid out on the right side of the canvas, and the game time of the black and white sides are displayed. The total time is displayed on the top of the canvas. Realized a good game status display and platform effect. The console prints the coordinate position changes of the current chess game and other parameter information in real time.

35.6 Conclusion

Based on the separation of front-end and back-end architecture, this paper designs and implements a general-purpose computer game platform that supports extended deployment of chess types, which better reflects the state of the game and makes computer games easier to enter the public's vision. It is helpful for beginners to quickly enter the essential research of computer games and promote the popularization and promotion of computer games. Provide effective solutions for adjusting and improving computer game engine programs.

Acknowledgements This work is supported by 2020 Promote Connotation Development of Universities-Scientific Research Training Program for College Students of BISTU

(No.5102010805), by Key potential projects of Promoting Research Level program at Beijing Information Science and Technology University. (NO. 5212010937), by Normal projects of General Science and Technology research program (NO. KM201911232002), and by Construction Project of computer technology specialty (NO.5112011019)

References

1. Xu, X.H., Deng, Z.L., Wang, J., et al.: Challenging issues facing computer game research. CAAI Trans. Intell. Syst. **3**(4), 288–293 (2008)
2. University Computer Games Championship & National Computer Games Tournament Homepage, http://computergames.caai.cn/index.html. Last Accessed 21 Nov 2020
3. Silver, D., Huang, A., Maddison, C.J., et al.: Mastering the game of Go with deep neural networks and tree search. Nature **529**(7587), 484–489 (2016)
4. Wang, Y.J., Qiu, H.K., Wu, Y.Y., et al.: Research and development of computer games. CAAI Trans. Intell. Syst. **11**(6), 788–798 (2016). https://doi.org/10.11992/tis.201609006
5. Silver, D., Schrittwieser, J., Simonyan, K., et al.: Mastering the game of Go without human knowledge. Nature **550**(7676), 354 (2017)
6. Electron Homepage, https://www.electronjs.org/. Last Accessed 20 Aug 2020
7. Fette, I.: The websocket protocol. Draft-Ietf-Hybi-Thewebsocketprotocol-10.**6455**, 1–71 (2011)
8. University Computer Games Championship & National Computer Games Tournament, "Amazon competition rules" Homepage, http://computergames.caai.cn/jsgz04.html. Last Accessed 1 Sep 2020

Chapter 36
Design of Electronic Caliper for Detecting the Distance between Cable Crimping Edges

Tao Hu, Zhongfei Ye, Ming Lu, Huanfeng Ren, and Liqun Shen

Abstract The important crossing line drops, broken can lead to serious public security and network security incidents, at the same time in order to avoid the infrared, X-ray, magnetic flux leakage line detection method with high costs and low efficiency of direct measurement method, this paper designed the special equipment of pressure measuring of electronic caliper based on STM32 microcontroller unit. Generally, the cross-section of the cable after crimping is positive hexagon, and the electronic caliper provides necessary data feedback for the evaluation of the whole crimping quality by measuring and displaying the three sets of opposite margins of the regular hexagon in real time. At the same time, the database is established and intelligent evaluation is realized by means of communication with upper computer through WIFI or data storage via SD. The research in this paper is of great significance to improve the quality detection efficiency and reliability of the crimped cable.

36.1 Introduction

As high-speed railways, highways, and power grids continue to expand, the number of transmission lines is also increasing [1]. If the line is dropped or disconnected, it will lead to large public security and power grid security incidents [2]. The stability of electrical wiring bonding process is very important to ensure the quality of wiring harness. However, the method of measuring directly by workers with measuring tools has the problems of low measuring efficiency and easy work tired. Magnetic flux leakage, infrared, and other detection methods are difficult to achieve portable measurement equipment, miniaturization [3]. Based on the above reasons, it is very

T. Hu (✉) · H. Ren · L. Shen
School of Instrumentation Science and Engineering, Harbin Institute of Technology, Harbin 150001, China
e-mail: hutao@hit.edu.cn

Z. Ye · M. Lu
Electric Power Science Research Institute, State Grid Henan Electric Power Company, Zhengzhou 450052, China

L. C. Jain et al. (eds.), *Wireless Technology, Intelligent Network Technologies, Smart Services and Applications*, Smart Innovation, Systems and Technologies 258,
https://doi.org/10.1007/978-981-16-5168-7_36

necessary to design a simple automatic detection equipment that does not require too much involvement of operators in the testing process.

36.2 Design of Electronic Caliper

36.2.1 Function Design of Electronic Caliper

Measurement circuit of this article is based on the STM32 MCU multi-channel displacement measuring device, circuit parts need to be able to collect real-time pressure welding cable three groups on the margins, and PC was achieved by wireless communication and computer interface of data analysis and integration of on-chip flash and the memory for data storage, the overall structure of the system block diagram is shown in Fig. 36.1.

According to the above requirements, this paper adopts the modular development method to study the data acquisition and processing system. The system is divided into two parts: the data acquisition and processing part of the lower computer and the data receiving and analyzing part of the upper computer. The way of WIFI communication between the upper computer and the lower computer serves as the bridge of data transmission. The lower computer part includes multi-channel capacitive grid displacement sensor module [4] and microcontroller circuit module. STM32 is selected as the control chip, and modularized design is adopted for data acquisition and processing, liquid crystal display, data storage, serial communication, and wireless data transmission [5]. The main purpose is to program and realize the controller. At the same time, the microcontroller circuit module can also control the key circuit by setting different keys to realize the switch between other functions such as switch, reset, storage, etc.

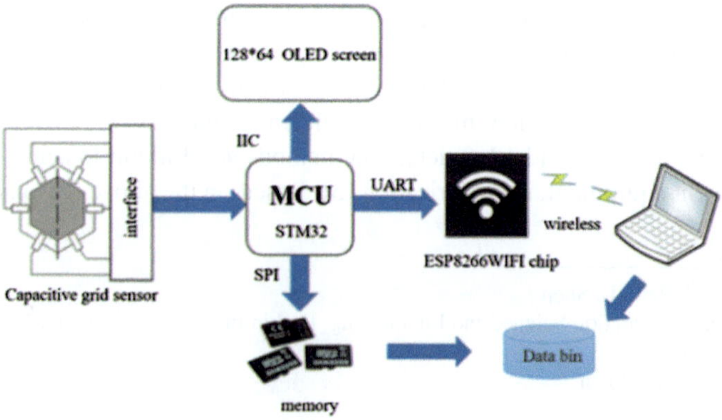

Fig. 36.1 The research scheme of the function of electronic caliper

36.2.2 *Mechanical Structure Design of Electronic Caliper*

According to the shape design of the pressed cable, the edge clamp is composed of six displacement probes, data acquisition, and display, upper and lower half-ring bracket, etc. In order to realize the edge detection of the pressed cable, the device has a shear-shaped structure that can be opened and closed, as shown in Fig. 36.2. Press the handle to open the upper and lower half rings of the caliper during measuring the edge distance detection of the crimping cable. After the cable is completely put in, release the handle, and the upper and lower half rings are precisely closed under the driving of the spring force. After reading the display data of the six probes, three sets of opposite margins of the six-prism cable are calculated.

The structure of the displacement probe is similar to the thickness gauge, and its function is similar to that of the traditional lever dial indicator. The main parts include the probe housing, movable grid fixed bracket, sliding guide, and spring. Its mechanical structure is shown in Fig. 36.3.

The probe housing should be reserved for data line holes to ensure data communication between the sensor and the circuit. In addition to protecting the internal sensor module, the probe housing also has the function of fixing the grid. The movable grid fixing bracket is fixed directly with the measuring probe, and the measuring probe is in contact with the workpiece. The sliding guide ensures that the moving grid is in a relatively parallel state when the fixed grid is working, so as to reduce the quadratic error of measurement as far as possible. The design of the spring is to ensure the probe's return force while making the reading fast and accurate. The dimensions of the probe housing and the movable grid fixed bracket are directly determined by the dimensions of the capacitive grid sensor moving grid and fixed grid selected in this paper. The design of tension spring and measuring probe must meet the corresponding mechanical principle and working life principle. At the same time, as a

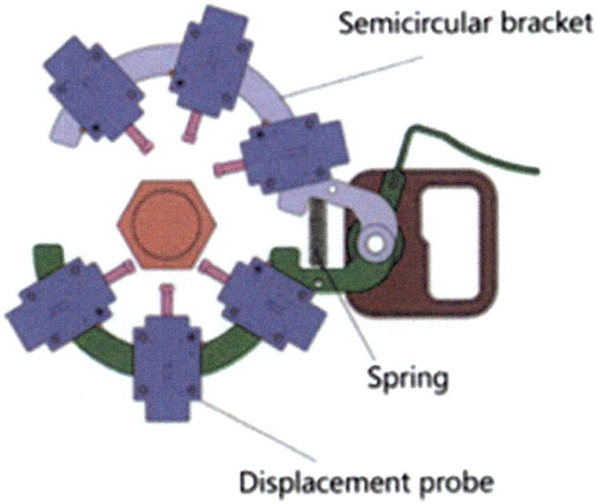

Fig. 36.2 Mechanical assembly diagram

Fig. 36.3 Probe structure diagram

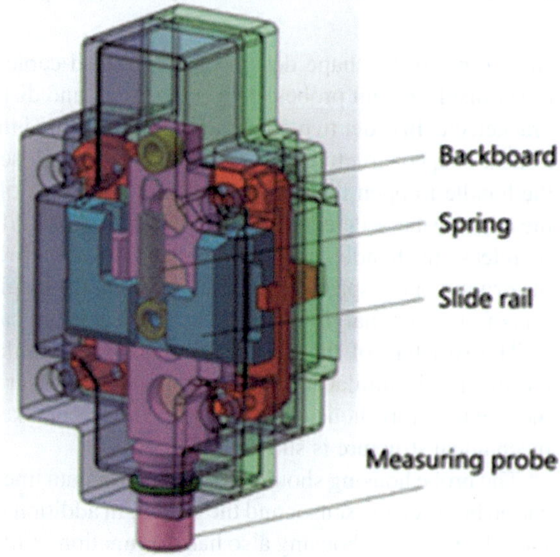

handheld device, the probe shell should be as light and compact as possible after the above parts meet the specified requirements.

The electronic caliper arm is composed of two symmetrical supporting semi-rings and springs. Compared with the probe, its design is relatively simple. Its function is to fix the relative positions of the six displacement probes and the data line extending the probe. Considering that the displacement probe can be zeroed mechanically or software, the accuracy of the caliper arm is required to be high, and the deviation of the fixed angle of the probe will affect the measurement results.

36.2.3 Circuit Design of Electronic Caliper

The microcontroller minimum system is to let the microcontroller work normally and play its functions must be a component, that is, the microcontroller can work with the least components of the system. STM32 microcontroller is widely used in the market, with stable performance and low price. In this paper, STM32F103ZET6 chip is selected as the master controller.

The display screen is monochrome OLED, as shown in Fig. 36.4. The displacement sensor is TM-003 capacitor-grid displacement sensor, as shown in Fig. 36.5.

The measuring range of this capacitive grid displacement sensor is ±999.99 mm, its resolution is 0.01 mm, and its precision is ±0.02 mm. The structure of capacitive grid ruler is divided into moving grid and fixed grid. A series of emitter electrodes of the same size and width l_0 are arranged on the moving grid. The upper end is

Fig. 36.4 Monochrome OLED

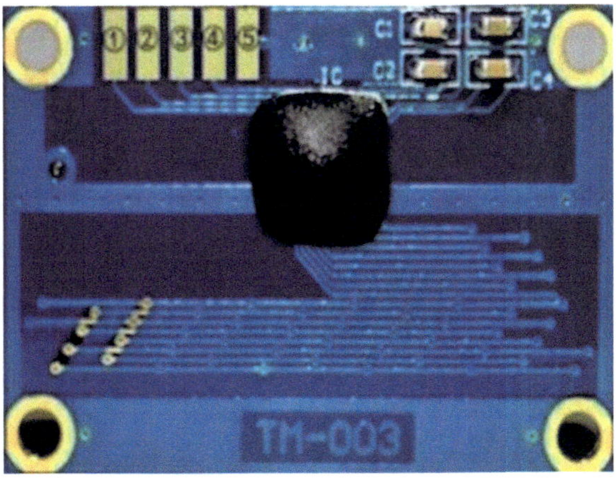

Fig. 36.5 TM-003 capacitive grid sensor

the public receiving pole. A series of reflective electrodes and shielding plates of the same size and width of $4l_0$ are arranged on the fixed grid. Different excitation voltages are applied on the emitter Plate 1–8, respectively, to generate charge on the reflecting plate through capacitive coupling and generate charge output through the public receiving pole. The charge output is usually proportional to the measured displacement. The structure of this form is simple, but the error is large if the dynamic grid fixed grid is not placed in parallel.

36.3 Measuring Principle Design of Electronic Caliper

Standard parts should be used to calibrate the caliper accurately before measurement. The design of the standard parts refers to the measurement object customization of the caliper, which is a standard six-prism, with 3 groups of fixed margins and known accuracy. When the electronic caliper paired probe is not being measured, the distance between the two probe heads, namely the initial opposable distance of the probe head, is D, and the initial position indication numbers of the two probe heads are d_{10} and $d_{20,}$ respectively, (namely the initial display data after the probe heads are electrified), as shown in Fig. 36.6. In this figure, Δd_1 and Δd_2 is the change of probe indication.

When the standard parts are calibrated, the opposite distance between the two probe heads is D_c and the readings of the two probe heads are d_{1c} and d_{2c}, respectively. At this point, when the standard part is measured, the size of the standard part is.

$$D_c = D + (d_{1c} - d_{10}) + (d_{2c} - d_{20}) \tag{36.1}$$

$$D = D_c - (d_{1c} - d_{10}) - (d_{2c} - d_{20}) \tag{36.2}$$

Among them, D_c is a set of matched margins of standard parts, which can be verified by other measuring tools and regarded as known quantity in the calculation process. The machining precision of D_c should be higher than the measuring precision of caliper probe, and the transfer of measuring datum can be realized by using D_c. d_{1c}, d_{10}, d_{2c}, d_{20} are read values for electronic caliper. As can be seen from the formula, through the calibration process, the distance D between two probe heads can be calculated when the electronic caliper is not put into the part to be tested.

The distance between the two probes is D_m. The readings of the two probes are d_{1m} and d_{2m}, respectively. Since the upper and lower probes will move up and down with the action of the operator during the measurement process, the readings of d_{1m} and d_{2m} are not fixed even for the same margin to be measured.

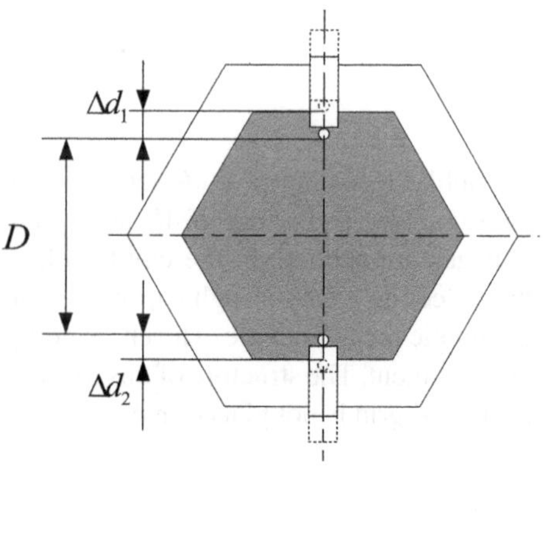

Fig. 36.6 Measuring schematic diagram

Table 36.1 Electronic caliper measurement results (mm)

Number	1	2	3	4	5	6	7	Average value
Channel 1	50.014	50.012	49.996	50.006	50.008	49.994	50.002	50.005
Channel 2	49.996	50.002	49.988	50.000	50.002	49.994	49.992	49.996
Channel 3	50.010	50.008	50.014	50.012	50.002	50.006	50.012	50.009

$$D_m = D + (d_{1m} - d_{10}) + (d_{2m} - d_{20}) \tag{36.3}$$

Substitute formula (36.2) into the above equation, and get:

$$D_m = D_c + (d_{1m} - d_{1c}) + (d_{2m} - d_{2c}) \tag{36.4}$$

It can be seen from formula (36.4) that the measurement of D_m is related to D_c, d_{1c} and d_{2c}. When D, d_{10} and d_{20} remain unchanged, it is independent of D, d_{10} and d_{20}.

36.4 Measurement Experiment of Electronic Caliper

After the above electronic caliper is manufactured, in order to verify the correctness of the above principles, with 3 grade 50 mm gauge blocks as the benchmark and the object to be tested, the variation in the length of the gauge blocks is better than 0.18 microns, which is much smaller than the detection uncertainty of the electronic caliper. Take each group of opposite probes as a channel, and perform multiple measurements on the three channels, respectively. The results are shown in the following Table 36.1.

It can be seen from the measurement results that the maximum measurement error of the above three channels is better than 0.02 mm, which meets the measurement accuracy requirements of the crimping of the transmission line.

36.5 Conclusion

In order to realize the simultaneous measurement of three groups of opposite margins of the crimped cable, a multi-channel displacement data acquisition and processing system was developed based on capacitor-grid displacement sensor and STM32 MCU in this paper. The relevant work completed is summarized as follows.

The general measuring principle of the edge of the crimped cable is designed, and the function design of the special measuring tool is realized. The design of electronic caliper structure including the final assembly and probe is realized, and the measurement precision and effect can meet the expectation. It realizes the design of electronic card gauge function circuit with STM32 as the core, integrates data collection, display, storage, and wireless communication with upper computer.

References

1. Wilde, Y., Gay, F., Piquemal, F., et al.: Measurements of single electron transistor devices combined with a CCC: progress report. In: Conference on Precision Electromagnetic Measurements Digest, pp.321–322. IEEE, Sydney (2012)
2. Zhang, L., Bao, W.B., Song, D.J., et al.: Research on cable crimping craft in mass transit vehicle. Electric Locomotiv. Mass Transit Veh. **40**(01), 61–64 (2017)
3. Rao, G.A., Kang, Y.H., W.X.J., et al.: Application of the eddy current method in the transmissionlines fault testing. J. Trans. Technol. **20**(2),14–16 (2001)
4. Wang, X.W., Qi, X., Song, Y.Q.: Capacitive transducer and its development prospect. J. Jilin Univ. (Eng. Technol. Edn.) **33**(2), 89–94 (2003)
5. Li, L., Lin, X., Zhang, X.Q., et al.: Design of remote acquisition system for debris flow data based on STM32. DEStech Trans. Comput. Sci. Eng. **16**(4), 327–335 (2019)

Chapter 37
Research on the Active Safety Monitoring Evaluation System of Key Operating Vehicles

Jing Deng, Feng Wang, Xiaobo Wu, Haiqiang Xiao, Changjun Zhao, and Changling Hua

Abstract With the substantial increase in passenger and freight volume and the increasing complexity of road conditions, transportation safety issues have become increasingly prominent. Traditional safety supervision using satellite positioning and video monitoring as the main sensing means can no longer meet actual needs. Relying on vehicle active safety intelligent prevention and control technology, the safety technical equipment of key operating vehicles in the transportation industry have been upgraded. In response to the needs of the transportation safety supervision business under the new situation, this paper conducts a research on the driving safety evaluation of key operating vehicles based on the fusion and analysis of vehicle driving and driver behavior active safety monitoring data and puts forward the evaluation index system of active safety monitoring based on the five factors of people, vehicles, roads, environment and events, which has a positive role in promoting the dynamic supervision level of road transportation.

37.1 Introduction

The rapid development of the transportation industry has injected strong impetus into economic development, but the following road traffic safety accidents also caused serious casualties and economic losses. Among them, operating vehicles account for a large proportion, especially major vicious traffic accidents, due to their large passenger and freight volume, high operational intensity, long operating time, and complex operating environment [1]. The sustained and rapid development of the national economy and the growing demand for logistics transportation have brought about the increasingly busy road traffic and the rapid growth of operating vehicles.

J. Deng · F. Wang · X. Wu (✉)
China Academy of Transportation Sciences, Beijing 100029, China

H. Xiao · C. Zhao · C. Hua
Department of Transport of Qinghai Province, Xining, Qinghai 810003, China

L. C. Jain et al. (eds.), *Wireless Technology, Intelligent Network Technologies, Smart Services and Applications*, Smart Innovation, Systems and Technologies 258,
https://doi.org/10.1007/978-981-16-5168-7_37

The traffic safety supervision mode and evaluation system based on traditional technical means are no longer applicable, so it is necessary to make breakthroughs based on advanced information technology in intelligent supervision and pre-prevention.

Since 2009, China has carried out the intelligent networking supervision of operating vehicles. Based on the integration of existing monitoring resources, the Ministry of Transport has built a national key operating vehicle networked joint control system to support cross-regional and cross-departmental joint supervision of key operating vehicles. With the comprehensive application of new information technology, and the rapid promotion of the Advanced Driving Assistance System (ADAS), active safety early warning and monitoring technology is becoming more mature. The safety supervision of operating vehicles has entered the stage of "vehicle satellite positioning + video monitoring + active safety warning," and gradually formed "pre-warning, in-process control, and post-event tracking" driving safety management closed loop.

Compared with the actual needs of road transportation traffic safety, the current road transportation traffic safety management still has some shortcomings, such as insufficient analysis and utilization of data resources, no effective safety evaluation and assessment of drivers and transportation companies. Efficient and accurate monitoring and safety evaluation of vehicles and drivers during driving can detect potential safety hazards in time, effectively prevent traffic accidents, and improve the level of traffic safety [2–4]. Safety evaluation is the basis of strengthening traffic safety supervision.

The road traffic system is a dynamic and open system. Its safety is restricted by both internal factors and external environment, it is closely related to factors such as people, vehicles, roads, environment, and management. To do a good job in vehicle safety evaluation, it is necessary to comprehensively consider the influence of multiple factors and conduct an integrated, comprehensive, and objective evaluation. In recent years, many domestic and foreign experts have carried out research on vehicle safety evaluation [5–9], analyzed the impact of drivers and other factors on traffic safety, and constructed a variety of safety evaluation index systems.

At present, the research on vehicle safety evaluation system has achieved many results, but the research on operating vehicle safety evaluation is not perfect. The transformation and upgrading of the safety technical equipment of operating vehicles in the transportation industry has triggered new regulatory requirements and has also enabled more data to be collected. Therefore, based on the data fusion and analysis of active safety monitoring of vehicle driving and driver behavior, this paper conducts an active safety monitoring and evaluation system for key operating vehicles, which is helpful to deepen the prevention-oriented supervision thinking, improve the safety supervision mechanism system, and improve the efficiency of industry supervision.

37.2 Analysis on the Nature of Traffic Accidents

The establishment of a scientific safety monitoring and evaluation system can strengthen road safety management, promote the transformation of post-event forensics to pre-prevention, and effectively avoid and reduce accidents. In order to clarify the influencing factors of traffic safety, support the selection of vehicle safety evaluation indicators, this paper analyzes the nature of traffic accidents.

(1) Suddenness. The occurrence of a traffic accident is often a momentary process, and the reaction time of the driver from visually perceiving the accident to taking emergency measures is very short, especially when the vehicle is driving at high speed.
(2) Randomness. Traffic accidents are often the result of the interaction of many factors. The change of any factor may lead to the occurrence of traffic accidents. The risk factors themselves have randomness, and the results caused by their interaction are more random.
(3) Frequency. National motor vehicle ownership continues to grow and has leaped to first place in the world. However, due to imperfect road traffic facilities, poor public awareness of traffic safety, and imperfect traffic safety management system, road traffic accidents are frequently occurring.
(4) Irreversibility. In a dynamic and open traffic system, all factors are changing randomly, so no two traffic accidents are the same from cause to result. Moreover, traffic accidents are not the expected results of human behavior and cannot be reproduced artificially, that is, they are irreversible.

37.3 Active Safety Monitoring Data Collection

Data collection is the key basic work for establishment of evaluation index system and realization of active safety and intelligent monitoring [7]. The collected data must meet the requirements of safety evaluation. Vehicle driving safety is affected by many factors such as drivers, vehicles, roads, environment, and management. Relevant data can be collected from vehicle intelligent video monitoring and alarm system, road network monitoring system, satellite positioning system, operating vehicle networked joint control system, transport administration system, and traffic management system. This section summarizes and sorts out the data required for active safety monitoring and evaluation, as shown in the following Table 37.1.

37.4 Evaluation Index System of Active Safety Monitoring

Since the road traffic system is a dynamic system composed of drivers, vehicles, roads, etc., driving safety is also affected by many factors such as environment and

Table 37.1 Active safety monitoring data

Category	Data information	Collection channel
Historical data	Driver assessment data	Operating vehicle networked joint control system
	Assessment data of transportation companies	Operating vehicle networked joint control system
	Transportation safety accident record data	Traffic management system
Static data	Vehicle basic data and vehicle management data	Operating vehicle networked joint control system, Transport administration system
	Driver basic data	Operating vehicle networked joint control system
	Basic data of transportation companies	Operating vehicle networked joint control system
	Road network information, infrastructure status data	Road network monitoring system
	Map data	Satellite positioning system
Dynamic data	Driver driving behavior data and status data	Vehicle intelligent video monitoring and alarm system
	Vehicle location data and status data	Vehicle intelligent video monitoring and alarm system, Satellite positioning system
	Road operation status data	Road network monitoring system
	Meteorological data	Operating vehicle networked joint control system

management [10]. Therefore, the active safety monitoring and evaluation of key operating vehicles is also a multi-index and multi-level evaluation problem.

37.4.1 Construction of Evaluation Index System

Active safety monitoring evaluation index system has a complex structure and numerous influencing factors, which require comprehensive consideration of the overall situation. People, vehicles, roads, environment, and events are the five major factors affecting road traffic safety, this paper starts from the above-mentioned traffic safety factors, comprehensively considers various influencing factors of transportation safety, uses the expert survey method to screen out the important and influential indicators, and gradually build a multi-level index system. The evaluation index system of active safety monitoring established in this paper is shown in Table 37.2.

Table 37.2 Active safety monitoring evaluation index system

First-level indicators	Second-level indicators	Third-level indicators	Fourth-level indicators
People	Driver status	Mental status	Fatigue driving
		Health status	Sudden illness
	Driver assessment	Historical assessment record	Record of illegal behaviors
			Records of unsafe and irregular driving behavior
			Driver ranking
		Driving habits	Driving habits
	Driver's driving behavior	Dangerous driving	Making calls, playing with mobile phones
			Hands off the steering wheel
			Not wearing a seat belt
			Not looking ahead for a long time
			Leaving the driving seat
			Smoking
		Illegal driving	Change lanes frequently or irregularly
			Overtime driving
			Overspeed driving
			Overcrowding
			Night shift (2–5 o'clock)
			Interference monitoring
			Unchecked seat belt
			Too close to the front car
Vehicle	Vehicle Type	Passenger vehicles	Large, medium, and small passenger vehicles
		Freight vehicles	Large, medium, and small Freight vehicles
		Dangerous goods transport vehicles	Class 8 dangerous goods transport vehicles
	Vehicle status	Vehicle technical status	Maintenance information
		Vehicle failure breakdown	1–4 level fault
		Vehicle loading status	No load or heavy load, Type of cargo

(continued)

Table 37.2 (continued)

First-level indicators	Second-level indicators	Third-level indicators	Fourth-level indicators
Road	Highway class	Functional grade	Expressway, first-class to fourth-class highway
	Infrastructure characteristics	Special sections	Tunnels, bridges, mountain roads, steep slopes, etc
		Other high-incident sections	Road intersections, roundabouts, etc
	Operating status	Traffic flow	Extra light, light, medium, heavy, extra heavy traffic
		Average speed	Smooth, slow, congested, etc
		Blocking events	Roadblocks (maintenance construction, traffic accidents, etc.)
Environment	Meteorological environment	Bad weather	Rain, snow, fog, wind, sand, etc
	Road environment	Road condition	Pavement water or icing
			Pavement subsidence or collapse
	Light conditions	Driving time	Day and night
		Lighting facilities	Lighting infrastructure
Event	Business management	Company-level	Level one to five
		Assessment record	Company ranking
			Overall illegal behavior records of company
			The company's overall unsafe and irregular driving behavior records
			The alarm event processing efficiency of company
	Fleet management	Vehicle management	Vehicle technical maintenance information
			Vehicle annual inspection information
		Vehicle damage	Vehicle exterior damage
			Vehicle failure

37.4.2 Active Safety Monitoring Evaluation Methods

Through multi-source data fusion analysis and application, an active safety monitoring evaluation system for key operating vehicles is built to support the road transportation industry to form a safety evaluation mechanism for drivers, fleets, and transportation companies in different dimensions.

Real-time risk evaluation of driving. Based on the data information of the active safety system and intelligent supervision platform, the driver's driving behavior and vehicle status are monitored during driving, and real-time risk evaluation is performed. When the evaluation score is lower than the preset threshold, an alarm is issued, and the same alarm will be given when there is high-risk unsafe and irregular driving behavior. The specific methods are as follows.

Integrate historical data of drivers, companies, and fleets for weighted scoring, and get the alarm threshold; Deduct points based on unsafe and irregular driving behaviors during driving to obtain real-time risk values; Factors such as the environment, road conditions, and vehicle types affect the severity of the risk. When the deduction event is triggered, check the real-time environmental factors, calculate the deduction coefficient, and determine the actual deduction value.

Trip risk evaluation. During the vehicle driving process, real-time safety monitoring and risk evaluation are performed to form a risk value curve. After the end of this trip, the risk curve is integrated to calculate the trip risk value.

Driver evaluation. The unsafe and irregular driving behaviors during the trip are counted, and the driver evaluation is updated based on the comprehensive calculation based on the historical assessment.

Company evaluation. The driver's unsafe and irregular driving behavior and the handling rate of company alarm events are included in the overall statistical data of the company, and the company evaluation results are calculated and updated.

Fleet evaluation. Based on the technical maintenance information, annual inspection information, as well as the failure and damage in operation, the vehicle management score of the trip is calculated and weighted into the overall evaluation data of the fleet, and the evaluation results of the fleet are updated.

37.5 Conclusions and Prospects

The systematic nature of road traffic safety shows the diversity of factors affecting road traffic safety and the complexity of their interaction mechanisms. Based on the analysis of traffic safety influencing factors, combined with the data collection situation, this paper builds a multi-level active safety monitoring evaluation index system. The evaluation of the driving risks of key operating vehicles, as well as multi-dimensional supervision objects such as drivers, enterprises and fleets, will help improve the dynamic supervision of road transportation, regulate driving behavior, and reduce the incidence of transportation safety accidents. In the next step, we will further refine and improve the index system and evaluation methods, conduct active safety monitoring evaluation of real vehicles, and promote the practical application of research results.

References

1. Chen, C., Zheng, H.R., Wu, Q., Gao, S., Wu, C.Z.: Evaluation of driving safety of operating vehicle drivers in special environments. Traffic Inf. Safety **31**(5), 123–127 (2013)
2. Tataanan, K., Larisa, Z.: The assessment model for economic efficiency of traffic sfety improvements. Trans. Res. Procedia **20**, 311–315 (2017)
3. Agarw, P.K., Jain, V., Bhawsar, U.: Development of a hierarchical structure to identify critical maintenance components affecting road safety. Procedia-Soc. Behav. Sci. **104**, 292–301 (2013)
4. Xiao, S.H., Li, Y.H.: Application and development analysis of automobile active safety technology. China Equip. Eng. **3**, 45–46 (2020)
5. Si, C.D., Jia, S.D., Feng, Y., Hu, X.P., Li, H.Y.: Fuzzy consistent evaluation of driver factors on road traffic safety. Highway Eng. **40**(1), 93–95 (2015)
6. Zhang, L.X., Liu, T., Pan, F.Q., Guo, T., Liu, R.C.: Analysis of the influence of driver factors on road traffic accident indicators. Chinese Saf. Sci. J. **24**(5), 79–84 (2014)
7. Sun, Y.X., Zhang, H.J., Hu, S., Pan, X.F., Dong, Y.Q., Weng, J.C.: Study on safety evaluation of highway vehicles based on dynamic monitoring data. Highway **65**(5), 223–230 (2020)
8. Cheng, J.Z., Li, Z.F., Ren, L.C., Zhai, W.Q.: Study on a fuzzy comprehensive evaluation of traffic safety based on road factors. J. Taiyuan Univ. Sci. Technol. **37**(4), 296–301 (2016)
9. Niu, S.F., Li, G.Q., Zhang, S.W.: Driving risk assessment model of operating vehicle drivers driven by satellite positioning data. China J. Highway Trans. **33**(6), 202–211 (2020)
10. Liu, S.Q., Wang, J.P., Zhang, Y.X.: Discussion on highway traffic safety evaluation index system. J. Northern Jiaotong Univ. **4**, 8–12 (1994)

Chapter 38
Research on Practical Training Curriculum Management for English Majors in Applied Undergraduate Universities Based on the "Production-Oriented Approach (POA)"

Jing Yang

Abstract Production-oriented approach (POA) is a foreign language teaching theory with Chinese characteristics created by Professor Qiufang Wen. Since its establishment, it has received eager attention from foreign language experts and teachers. This research explores the management of English training curriculum based on the POA teaching concept through teaching practice, determines specific English training projects, and constructs an English training teaching process oriented by "drive-facilitate-evaluate." A ten-week teaching experiment was conducted with 225 students from Grade Two of English major in Zhengzhou University of Industrial Technology, an applied undergraduate university in China. The teaching design, teaching plan and teaching process based on POA were demonstrated. Experiments show that the management of English training curriculum based on POA teaching concepts can help improve students' comprehensive English language application ability and creative thinking ability.

38.1 Introduction

After China's accession to the WTO, more and more foreign companies have entered China, and Chinese enterprises are gradually going global. All of these put forward new requirements for English majors. Due to the diversification of the demand for foreign language talents in society, the single "foreign language + skills" talent training model in the past can no longer meet the needs of the market economy and the development of technology. Employers increasingly need to be applied undergraduate universities to train and transport applied English talents with good English

---

J. Yang (✉)
Zhengzhou University of Industrial Technology, Zheng Zhou 451100, He Nan, China
e-mail: yangjing@zzgyxy.edu.cn

Central Philippines University, 5000 Iloilo City, Philippines

L. C. Jain et al. (eds.), *Wireless Technology, Intelligent Network Technologies, Smart Services and Applications*, Smart Innovation, Systems and Technologies 258,
https://doi.org/10.1007/978-981-16-5168-7_38

professional knowledge and skills, sound mind and body, excellent practical ability, and high cultural literacy, which requires college English teachers to change teaching ideas, to actively improve teaching methods, and to train the intelligent talents needed for society. The "Production-oriented Approach (POA)" is a learning theory and teaching method specially designed for Chinese English learners in accordance with social trend. In 2008, under the wave of reforms in English curriculum and teaching methods, professor Qiufang Wen proposed the "Output-Driven Hypothesis" [1]; after five years' practice, it was gradually expanded to the field of college English teaching in 2013; In March 2014, Professor Wen revised the original "output-driven hypothesis" into "output-driven plus input-enabled hypothesis" at the academic seminar on the development of college English teaching entitled "Situation, Goals, Ability, and Strategy" [2]; In October of the same year, at "the 7th International Symposium on English Teaching in China", the theory was officially named "Production-Oriented Approach/POA". Professor Wen proposed to construct a theoretical system of POA and distinguish the different meanings of "production" and "output": "production" includes not only the speaking and writing of "output," but also interpretation and translation [3]; the corresponding English is production, which emphasizes both the producing process and the product. The "Production-Oriented Approach" is a challenge to the separation of learning and use. It takes production as the starting point for teaching, inspires students' learning motivation and enthusiasm, and takes production as the goal to apply what they learn, combine learning with applying, and promote learning with application, highlighting the practicality of language. This theory has brought improvements in concepts and methods to practical training curriculum for English majors and has injected fresh blood into English practical teaching.

38.2 Application and Development of POA Teaching Philosophy

"Production-oriented Approach (POA)" was developed by Qiufang Wen [1–4], constructing a foreign language classroom teaching theory with Chinese characteristics. After twelve years' development, it has formed a complete theoretical system. This theory inherits the fine tradition of education in China, draws on the essence of foreign language teaching theories and practices abroad, and is based on solving the problems of "emphasizing learning and neglecting application" or "emphasizing application and neglecting learning" in foreign language teaching [3, 5–7] In 2015, Qiufang Wen constructed a POA theoretical system based on the POA teaching philosophy and the research results of her team (see Fig. 38.1).

As shown in Fig. 38.1, the teaching philosophy of POA includes "learning-centered principle," "principle of learning and applying as an organic whole," and "principle of whole person education,"; Teaching hypotheses include "output motivating," "input enabling" and "selective learning"; The teaching procedures consist

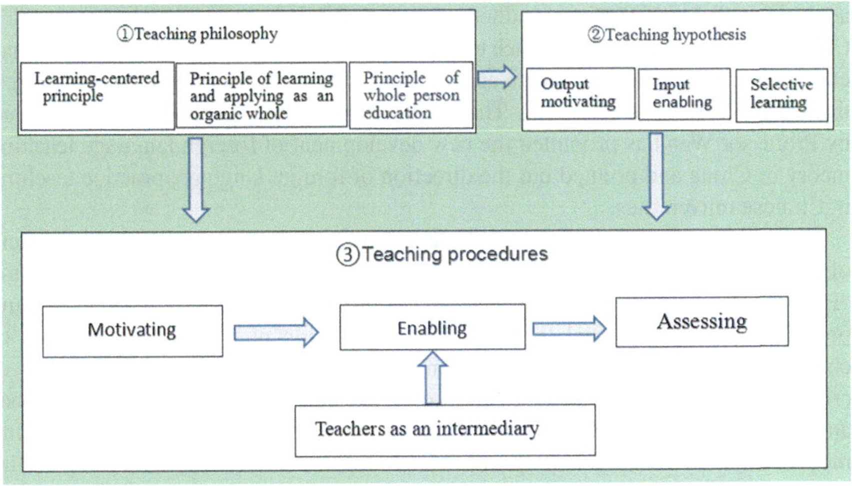

Fig. 38.1 The theoretical system of POA

of three stages: "motivating," "enabling" and "assessing". Teachers play an intermediary role in the whole teaching process [3]. In 2017, Professor Wen adapted the POA teaching theory and added "promoting learning by evaluation" in the teaching hypothesis, emphasizing the importance of evaluation and highlighting the leading role of teachers (See Fig. 38.2).

The POA teaching theory explores the four Chinese characteristics of POA from three perspectives: philosophical foundations, traditional Chinese education theories, and related Western theories: (1) Integerate curriculum theory and second language

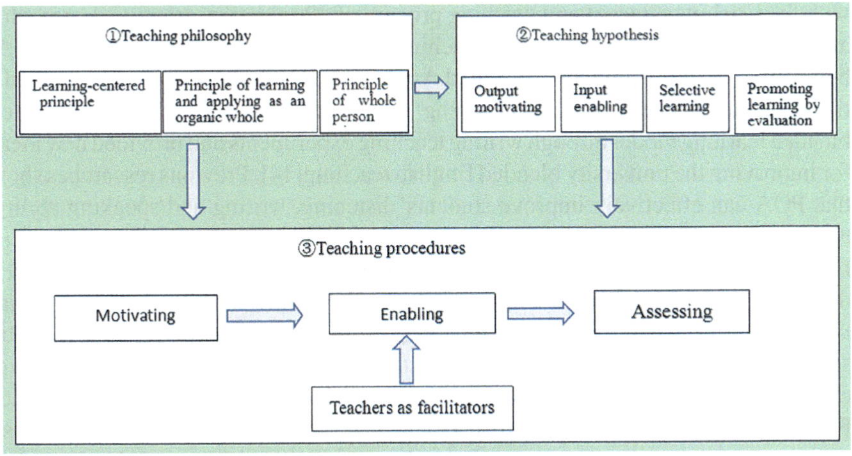

Fig. 38.2 The theoretical system of POA

acquisition theory; (2) Always adhere to "practice is the only criterion to test truth"; (3) Take a comprehensive approach to address the problem in light of China's national conditions; (4) Focus on the main aspects of teaching contradictions, and emphasize the leading role of teachers [3]. The "Production-oriented Approach" put forward by Professor Wen has promoted the new development of foreign language teaching theory in China and pointed out the direction of foreign language practice teaching in Chinese universities.

Some researchers have conducted speculative research on the theoretical system of POA, for example, Qianying Xuan conducted a teaching reflection on the "Production-oriented Approach" based on the speculative English course, and provided research material for the future exploration of the teaching method of cultivating the speculative ability of English majors [8]; Based on the theory of critical thinking model and production-oriented approach, Bojia Wang researches and analyzes the oral critical thinking class and finds that PEM critical thinking model and POA model can help improve students' overall critical thinking ability and students' argumentative essay writing level [9]. Some teachers have also explored the application of this theory in specific teaching practice, for example, Zhen Huang discussed the feasibility of "Production-oriented Approach" in English major reading teaching [10]; Ruisi Zhang discussed the applied strategies of "Production-oriented Approach" from three aspects: designing the overall thinking of teaching, developing cooperative learning mode, and designing specific tasks in reading teaching for English majors [11]; Yaru Fan tried the combined teaching of college English reading and writing under the guidance of "Production-oriented Approach," and suggested that POA should be used as a new teaching method of combining reading teaching with writing teaching to improve classroom teaching effects [12]. In addition, by combining the teaching concept of "Production-oriented Approach," some teachers and scholars have changed the traditional English teaching and constructed a new English teaching and learning model, for example, Chunhua Ren, Haiyan Li started with the teaching concept and teaching process of "Production-oriented Approach," and explored the English speaking teaching in independent college based on POA, believing that the "Production-oriented Approach" is beneficial to change the traditional teaching concept of oral teaching [13]; Yu Xu built a production-oriented blended learning model through writing teaching experiments and provided new ideas for improving the university blended English teaching[14]. Previous researches show that POA can effectively improve students' listening, writing and speaking ability, and promote students' all-round development; POA helps to fully develop the "potential" of high-quality English resources among students, so as to meet the country's demand for high-end foreign language talents; POA can enhance students' motivation and confidence in English learning [15]. From the previous researches, it can be seen that POA is mostly applied to reading teaching, writing teaching, oral teaching, etc., and few scholars apply it to practical English teaching. This study combines POA teaching process with English practical teaching so as to build an English training module teaching system suitable for student development. It is planned to conduct research from the following two questions: How do teachers build an English training course management model based on the POA teaching concept? Does the

English training curriculum management based on the POA teaching philosophy help students improve their comprehensive application ability of English language?

38.3 Management Mode of Practical Training Curriculum for English Majors

With the rapid development of our country's economy, the society's demand for applied talents is gradually increasing. Students must have both good professional standards and high professional skills. Therefore, English majors should start from the perspective of application, return to the standard for English majors, and strengthen the construction of professional connotation. In order to further strengthen practical teaching, increase the construction of practice systems at home and abroad, inside and outside the school and inside and outside the classroom, cultivate students' innovative and practical ability, it is important to create an English training curriculum management model that is compatible with professional training goals for English majors in applied-oriented universities.

38.3.1 Position Quasi-Professionals and Formulate a School-Based Talent Training Model

Taking Zhengzhou University of Industrial Technology as an example, according to the needs of national transformation and development and the orientation of applied-oriented universities, the School of Foreign Languages accurately grasps the professional positioning, and the English major has formed a modular application-oriented talent training mode system, that is "to train students to have a solid English based on the basic theory, basic knowledge, and basic skills of language, with business, translation and English education as the training direction, focusing on cultivating students' core literacy, innovative practical ability and speculative ability" and formulated the corresponding talent training program.

38.3.2 The Curriculum is in Line with the Talent Training Program

In accordance with the National Standard for Undergraduate Specialty Teaching Quality in Institutions of Higher learning (hereinafter referred to as the National Standard), the talent training program and discipline training objectives for English majors, the practical training course system shall be set up reasonably. The National Standard stipulates that the curriculum system includes five parts: general education curriculum, professional core curriculum, training orientation curriculum, practical teaching link and graduation thesis. It can be seen that practical teaching is an important part of English curriculum system. As an application-oriented university, the curriculum of English major in Zhengzhou University of Industrial Technology pays more attention to the relationship between language skill training and professional knowledge teaching, course teaching and practical teaching, and highlights the cultivation of ability and professional knowledge construction, especially the cultivation of students' communicative ability, critical thinking ability and innovative ability.

By setting up real and effective modular project training, practical English teaching can fully arouse the enthusiasm of students and stimulate their innovative thinking, so that students can actively construct new knowledge in the process of completing each "output" task, and improve their comprehensive English application ability through skills competitions such as esthetic article recitation and English speech.

In addition to improving language skills, English practical training courses combine business, tourism, translation, education and other interdisciplinary knowledge fields to meet the needs of versatile and diversified talents training, and pay more attention to elastification, flexibility, inclusiveness, and rationality.

38.3.3 Multiple Subjects Participate in Curriculum Management

Curriculum management is "the organization, leadership, supervision and inspection of curriculum compilation, implementation, and evaluation; it is jointly implemented by the education administrative department of the central government, the education administrative department of the local government, and the school functional departments and teaching institutions" [16]. Many courses of English majors have both theoretical and practical characteristics. Such courses require students to master certain language theory knowledge, and also participate in activities of relevant language practice. This requires enterprise personnel and school leaders, teachers and students to participate in the management of practical training courses and teaching design, and then participate in practical teaching according to the respective expertise of school and enterprise tutors.

Therefore, English training curriculum management can be defined as: under certain social conditions, multiple subjects participate in the process of supervising the implementation of the training curriculum.

38.4 English Professional Training Curriculum Management Under the Guidance of POA Teaching Philosophy

The practical training course for English majors pays attention to the cultivation of students' application ability and innovation ability, and forms a classroom teaching mode with students' independent learning as the center, teachers' classroom teaching as the guidance, classroom teaching activities as the carrier, ability cultivation as the goal and moral education as the orientation [17].

38.4.1 Identify Teaching Goals

As far as teaching goals are concerned, POA not only takes output as the starting point to drive students' enthusiasm for learning, but also takes output as the goal to apply what they have learned. In terms of teaching methods, it emphasizes the role of output activities in language learning and realizes the docking of input and output [6]. The POA teaching philosophy provides new course management goals and teaching ideas for the English major training courses.

First, the English professional training courses take the students' target positions and the application function of language expectations as the core, use the POA "output motivating" teaching hypothesis, and through the implementation of modular training teaching tasks, minimize the time span and increase students' information intake [18].

Second, using POA "input enabling" teaching hypothesis, organize a large number of relevant language materials and language scenarios, optimize students' knowledge structure, and strengthen the cultivation of students' practical English skills for future posts.

Third, construct a scientific and reasonable English professional training teaching system based on the three stages of the POA teaching process: "motivating," "enabling" and "assessing," which fully reflects the teaching pertinence and adaptability in compliance with social needs, job requirements and personalized learning.

38.4.2 Optimize Teaching Content

Guided by the theory of "production-oriented approach,", the English practical training course gives full play to the maximum effect of the practical training project in the limited time, so that the learning and application can be effectively connected and the maximum effect of "production" can be achieved.

In terms of teaching content arrangement, according to the requirements of the Syllabus of English Major in Colleges and Universities, it cultivates students' ability to acquire and apply knowledge, to analyze and solve problems, to put forward opinions independently and to innovate thinking.

While strengthening the training of language skills, it also pays attention to the cultivation of practical teaching and application ability, and highlights the integration of language and culture, translation, business and other fields. At the same time, the ideological and political elements of the curriculum are integrated into the classroom teaching, and the practical training content is organically combined with ideological and political education to give full play to the moral education function of the curriculum.

38.4.3 Determine the Teaching Object

This study takes Zhengzhou University of Industrial Technology as an example. The teaching objects are 225 English majors of the school. The teaching experiments are conducted in the form of training groups. Each training group consists of 20–30 students, guided by an English teacher. At the end of each training program, as a unit of the training group, students will participate in the English sitcoms, English speeches, English songs, English vocabulary, English film dubbing, and other competitions organized by the school, forming an English learning atmosphere of "promoting learning by competition," so as to improve students' "production" outcomes.

38.4.4 Enrich Teaching Methods

In the process of teaching, teachers should constantly reform teaching methods, enrich teaching approaches, actively carry out teaching research and explore teaching rules.

The practical classroom teaching adopts a combination of role-playing, script creation, film dubbing and other teaching methods, actively utilizes modern teaching methods and technologies, and establishes a school campus network and Chinese University Massive Open Online Courses (MOOC) as a platform, multimedia technology and online communication technology as the core of modern teaching means, so as to provide convenience for students to carry out independent learning and make use of POA "output-driven hypothesis" to fully mobilize students' learning enthusiasm, initiative and creativity.

38.5 Implementation Plans of Practical Training Curriculum for English Majors Based on the POA Teaching Philosophy

This research explores the management of English training curriculum based on POA through teaching practice, so as to achieve the purpose of improving students' ability of English language application and cultivating students' innovative ability. Specifically, it was implemented through the following aspects.

38.5.1 Ways of Thinking and Methods

First, on the basis of an in-depth grasp of relevant domestic and foreign research, combined with the characteristics of practical teaching for English majors, the research probes into the theoretical basis of practical English teaching through convening seminars with relevant experts and scholars, and formulates a feasible teaching plan of practical training module.

Second, the experimental teaching is conducted in a planned way. The teaching object is 225 English majors in the second year of the university. The experiment time is 10 weeks and 2 periods per week. The teaching process focuses on cultivating students' learning and research abilities, with students as the main body and teachers as the leading role, creating realistic practical situations as much as possible, so as to make students form job competition and career development skills in practice. At the same time, various forms of task-centered teaching activities are carried out. While strengthening basic training, use heuristic, discussion, discovery and research teaching methods to fully mobilize students' enthusiasm for learning and get the most out of the learning process.

Third, conduct planned and staged questionnaires, interviews and tests for students. At the same time, targeted questionnaires and interviews were conducted on teachers, and comprehensive data and materials were collected in combination with classroom observations. Finally, the data will be analyzed in both quantitative and qualitative ways. Through questionnaires and interviews, the influencing factors that are closely related to the development of students' comprehensive ability to use English. The questionnaires were distributed to 225 English majors who participated in the teaching experiment. The interviewees were 10 student representatives who had significantly improved their comprehensive ability to use English and 10 who had hardly improved after the training module. The interview centered on the results of the questionnaire survey to further understand the impact of various factors on students' comprehensive English ability.

38.5.2 Steps of English Practical Teaching

First, use the POA teaching philosophy as a guide to set the direction and overall objectives of English practical classroom teaching.

POA "principle of learning and applying as an organic whole" advocates that in classroom teaching, all language teaching activities are closely connected with application, so that there is no boundary between "learning" and "using," and "learning" is integrated with "using." It has changed the disadvantages of "teaching materials center," "text first" and "separation of learning and application" in traditional classroom teaching practice. English training module teaching is just complementary to this theory. Each module is closely related to the actual scenes in life, which greatly improves students' enthusiasm in participating in activities.

Second, determine specific English training projects and construct an English training teaching process oriented by "motivating-enabling-assessing".

In the teaching process, different from the traditional teaching method, it identifies specific English training programs according to three specific links of "motivating," that is, the teacher presents the communicative scene, the student tries to produce, and the teacher explains the teaching goal and the output task. Through the POA "input enabling" hypothesis, English training cooperative groups are established, in which students can learn selectively, solve problems together, and achieve "enabling" in the process. Finally, the POA "assessing" model is adopted to test the students' outcomes (See Fig. 38.3).

In the modular teaching of practical training, the teacher determines the appropriate output target and the corresponding output task, and designs the "output" scene around the target and task, which is used to stimulate students' motivation of learning input, so that students will always have "hunger" in the process of participating in activities and unconsciously try new communicative goals. When encountering problems, team members participate in and actively solve them together. The final completion of the output task stimulates a new round of output driven. As shown in Fig. 38.3, in the process of practical English training teaching, the three stages

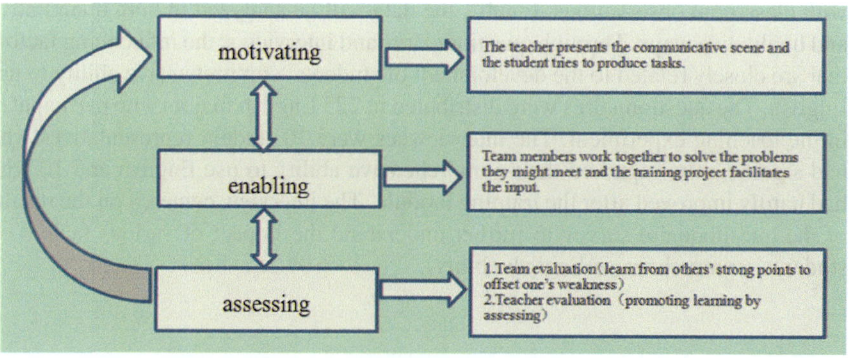

Fig. 38.3 Practical English teaching process under the guidance of POA

of "motivating-enabling-assessing" are mutually cyclic processes, which are closely linked with each other, both restricting and contributing to each other.

Third, on the basis of the first two steps, use "POA teacher-student cooperation evaluation" to innovate a new model of training classroom evaluation.

Evaluation is an intensive and in-depth stage of learning, and a good evaluation mode can encourage students to conduct targeted review in the previous stage and preview in the next stage, so as to develop good learning habits. "POA teacher-student cooperation evaluation" includes three stages: before class, in class and after class. Teachers and students participate in the whole process of evaluation [5]. That is, before class, teachers conduct detailed approval of the typical samples, teachers and students jointly evaluate the typical samples in class, and after class students modify and improve the samples according to the evaluation results.

Based on the concept and steps proposed in the "POA Teacher-student cooperative evaluation," English practical training class should first set a clear and accessible evaluation goal, let students actively participate in the evaluation, and teachers' "evaluation" and "instruction" are combined to guide students to find and solve problems, so as to achieve the goal of "promoting learning by evaluation."

38.5.3 Presentation of Training Outcomes

English training module teaching is a necessary means for cultivating application ability and operational skills, and it is an important link to transform "knowledge" into "ability." After the training, the School of Foreign Languages creates a platform for each training group to display learning results. Each group selects representatives to display the results of various training projects on the teaching platform or online platform, such as giving an English speech, singing an English song, reading a good English article, telling an English story, performing an English sitcom or perform movie dubbing, role playing, etc. Teachers and other team members evaluate the displayed works, propose problems and suggestions for improvement, and name a number of different awards, such as the best performance award, the best singer award, the best dubbing award, and the best team award.

The outcome presentation provides students with a platform to give full play to their talents and creativity, and exercise their oral English expression, which greatly improves students' enthusiasm and confidence in learning, and their independent thinking ability and group cooperation ability. The teacher supplements and summarizes the learning of the students by watching the "works" of the students, summarizes the key content of each training project, and helps the students have an overall understanding of the knowledge they have learned so as to better understand and master the knowledge they have learned. By transferring knowledge between various training projects, the core literacy of students has been cultivated, and their comprehensive English language application ability and innovative thinking have been further improved.

38.6 Conclusion

English majors in application-oriented universities must clarify the training objectives of professional talents, establish and improve the management mechanism of English teaching, and constantly innovate the management mode of practical teaching process; carry out discipline competitions and practical activities closely related to the major, enhance students' comprehensive quality and improve their core literacy; continuously improve teaching quality, and cultivate qualified innovative application-oriented talents for the society.

Based on the development and discipline orientation of English majors in application-oriented universities and according to the requirements of professional talent cultivation and the division of course modules, this study constructed a practical training course management model for English majors based on POA teaching philosophy, and formulated a specific course implementation plan. Through the model of "POA teacher-student cooperation evaluation" and the demonstration of students' learning results, the effectiveness of POA teaching philosophy on practical teaching of English majors is proved, and students' comprehensive English language application ability is improved and English practical teaching is optimized.

Acknowledgements This paper is funded by the key project of 2020 National Social Science Foundation "A Study on the Diversity and Mutual Assistance of Undergraduates' English Learning from the Perspective of Educational Ecology" (Project number: 20AYY015) and the key project of 2019 Research and Practice of Educational and Teaching Reform in Zhengzhou University of Industrial Technology "Research on English training Module Teaching based on 'production-oriented approach'" (Project number: JG-190109).

Thanks for professor Qiufang Wen's consents for using her theoretical system of POA as this paper's theoretical framework.

References

1. Wen, Q.F.: Output-driven hypothesis and english professional skills curriculum reform. Foreign Lang. World **2**, 2–9 (2008)
2. Wen, Q.F.: Output-driven-input-facilitated hypothesis: an attempt to construct the theory of college foreign language classroom teaching. China Educ. **7**(2), 3–12, 98 (2014)
3. Wen, Q.F.: Constructing a theoretical system of production-oriented approach. Foreign Lang. Teach. Res. **47**(04), 547–558, 640 (2015)
4. Wen, Q.F.: Chinese characteristics of production-oriented approach. Modern Foreign Lang. **3**, 348–358 (2017)
5. Wen, Q.F.: Teacher-student cooperation evaluation: a new evaluation form created by the production-oriented approach. Foreign Lang. Circle **05**, 37–43 (2016)
6. Zhang, W.J.: College english classroom teaching practice based on production-oriented approach. Foreign Lang. Foreign Lang. Teaching **2**, 106–114 (2016)
7. Wen, Q.H.: Research on the blended teaching reform of production-oriented" professional English: examination and evaluation. Univ. English Teaching Res. Edn. **04**, 7–13 (2018)
8. Xuan, Q.Y.: Reflection on the teaching of speculative English course based on the production-oriented approach. Contemp. Educ. Pract. Teaching Res. **23**, 155–156 (2018)

9. Wang, B.J.: Empirical research on the feedback of oral argumentation teaching on argumentative writing: based on the critical model and production-oriented theory. Foreign Lang. Teaching **40**(05), 51–56 (2016)
10. Huang, Z.: The feasibility study of the production-oriented approach in English reading teaching. J. Jiangsu Sec. Normal Univ. **8**, 115–119 (2015)
11. Zhang, R.S.: Research on the teaching practice of English majors based on the production-oriented approach. English Teacher **1**, 44–47 (2017)
12. Fan, Y.R.: An attempt to combine reading and writing in college English under the guidance of production-oriented approach. New West (Theory Edn.) **02**, 132–133 (2017)
13. Ren, C.H., Li, H.Y.: Research on college English writing teaching based on dynamic assessment model. Heilongjiang Sci. **17**, 146–147 (2017)
14. Xu, Y.: Research on production-oriented blended learning model. Cult. Educ. Data **32**, 219–221 (2019)
15. Zhang, L.L.: Research on teaching effectiveness of production-oriented approach. Modern Foreign Lang. **40**(03), 369–376, 438 (2017)
16. Gu, M.Y.: Dictionary of Education. Additional Shanghai Education Press, Shanghai (1998)
17. Huang, Y.Y.: Connotation analysis of college English teaching mode in newly-built undergraduate colleges after application-oriented transformation. Sci. Technol. Econ. Market **12**, 248–249 (2015)
18. Guo, J.H.: Modular practical training course design and English application-oriented talent training. China Higher Med. Educ. **09**, 74–76 (2009)

Chapter 39
The Influence of Ripple Depth and Ripple Taper on Bandwidth and Centre Frequency Shift Operating at Terahertz Frequency in a Coaxial Bragg Structure

Xue Yong Ding, Hong-rui Su, Lei Qiang, and Lian-sheng Wang

Abstract Based on the mode-coupling method, comparative study is carried out for the influence of different ripple depths and different ripple angles on bandwidth and centre frequency shift characteristics operating at THz high-frequency coupling mode. The results show that the bandwidth of the working mode and the competing mode increases with the increase of the slot depth, but the centre resonance frequency almost does not shift; With the increase of the slope angle, the centre frequency of the working mode almost does not shift, while the centre frequency of the competing mode will be far away from the centre frequency of the working mode, and the farther the deviation is with the increase of the slope angle. These characteristics are conducive to better suppress the band gap overlap and construct the Bragg structure with single high-order mode at THz frequency.

39.1 Introduction

The bandgap or passband formed by the frequency selective characteristics of periodic boundary conditions of the Bragg structures can be used to fabricate reflectors, filters and mode converters [1, 2]. For example, it is an important part of the Bragg resonator as a reflector, which is widely used in the cyclotron resonance maser (CARM) and free-electron laser (FEL) working at THz frequency [3–11]. Compared with cylindrical structure, the coaxial Bragg structure is more conducive to high-power microwave system and it can be designed to work in a larger size, which is convenient for machining and power capacity improvement and is conducive to mode selection. Two or more coaxial Bragg structures can form the Bragg cavity. In order to form standing wave oscillation in the cavity with its working centre frequency, the appropriate structural parameters can be selected by design, to minimize the influence of competition mode and maximize the reflectivity of the centre frequency of the working mode. It is found that the frequency band width of working mode frequency

X. Y. Ding (✉) · H. Su · L. Qiang · L. Wang
Sanya University, Sanya, Hainan 572022, People's Republic of China

L. C. Jain et al. (eds.), *Wireless Technology, Intelligent Network Technologies, Smart Services and Applications*, Smart Innovation, Systems and Technologies 258,
https://doi.org/10.1007/978-981-16-5168-7_39

response directly affects the starting vibration of the electromagnetic wave, and the wider the bandwidth is, the more favourable it is to start vibration [5].

In coaxial Bragg reflectors, not only does the initial phase difference of the inner and outer conductors have an influence on the centre frequency and bandwidth, but also the ripple depth and ripple taper of the corrugated groove also have an influence on the bandwidth and central frequency [12–16]. In this paper, the influence of ripple depth and negative ripple taper on bandwidth and centre frequency shift will be studied.

39.2 Theory Model

Figure 39.1 shows the longitudinal-sectional view of a coaxial Bragg structure with sinusoidal ripple [17, 18]. The research shows that negative taper that is shown in Fig. 39.2 has more obvious advantages than positive taper [12, 13, 15, 16]. Therefore, in this paper, we only study the negative-taper ripples structure.

The corresponding frequencies are the upper cut-off frequency f_{hc} and the lower cut-off frequency f_{lc}, respectively. The 3 dB bandwidth schematic diagram is shown in Fig. 39.3.

The centre frequency shift Δf is the absolute value of the difference between the ideal centre frequency f_0 and the actual centre frequency f_0' of the coaxial Bragg.

$$\Delta \mathrm{f} = \left| f_0 - f_0' \right|$$

Based on the coupled-mode theory, a FORTRAN programme code is performed to evaluate the Influence of ripple depth and negative ripple taper on bandwidth and centre frequency shift. For the coaxial Bragg reflector in the working mode, when the initial phase $\delta_\varphi = |\varphi_{in} - \varphi_{out}| = \pi$, the mixture spectrum of the working mode

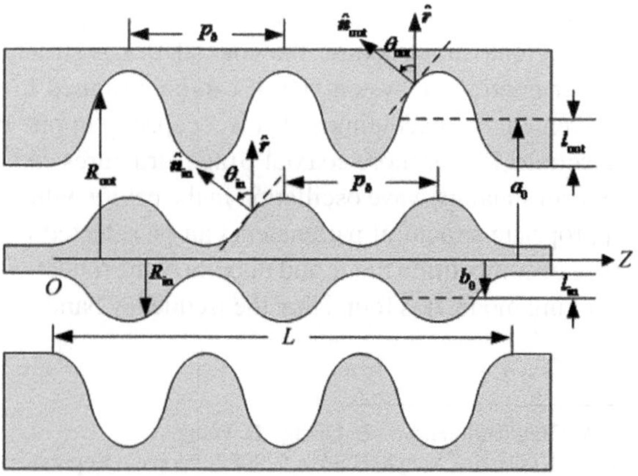

Fig. 39.1 Longitudinal-sectional view with sinusoidal ripples structure

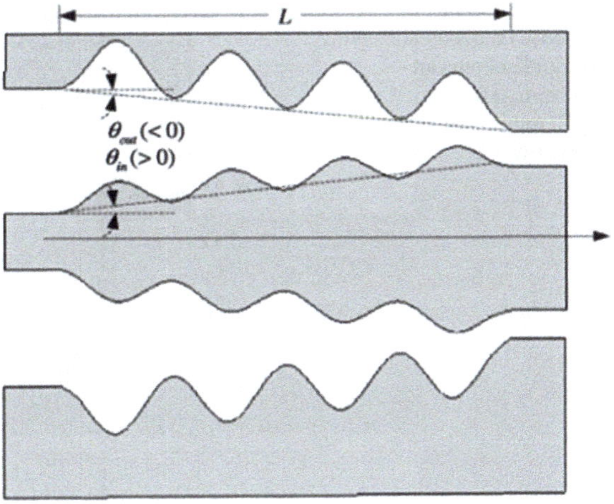

Fig. 39.2 The profile with negative-taper structure

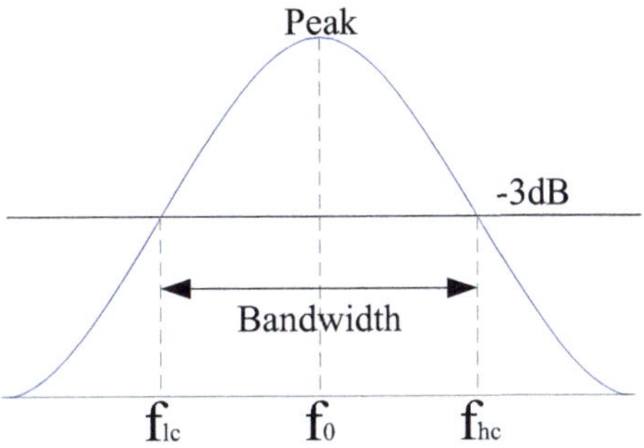

Fig. 39.3 3 dB bandwidth diagram

and the competition mode can be effectively separated, and the high reflectivity can be obtained [17, 18]. Therefore, the initial phase is set to π.

39.3 The Influence of Ripple Depth and Ripple Taper on Bandwidth and Centre Frequency Shift

Since higher-order mode operating at the THz frequency may have potential application in the CARM [11, 19, 20], as an example, we assume that the incident mode $TE_{6,1}$ with the central frequency being 0.35 THz is the desired working mode, and $a_0 = 10$ mm, $b_0 = 7.0$ mm, $l_{out} = l_{in} = 0.01$ mm, $p_b = 0.43$ mm, $L = 85.57$ mm, and the initial phase $\delta_\varphi = |\varphi_{in} - \varphi_{out}| = \pi$, respectively.

Figure 39.4 plots the reflectivity versus the frequency with different ripple depths. It is shown that in the THz frequency coupling mode, when the Hamming window is not loaded, changing the ripple depth has an impact on both the working mode

Fig. 39.4 Frequency response characteristics with the ripple depths are 0.01 mm, 0.015 mm, 0.020 mm, 0.025 mm, respectively

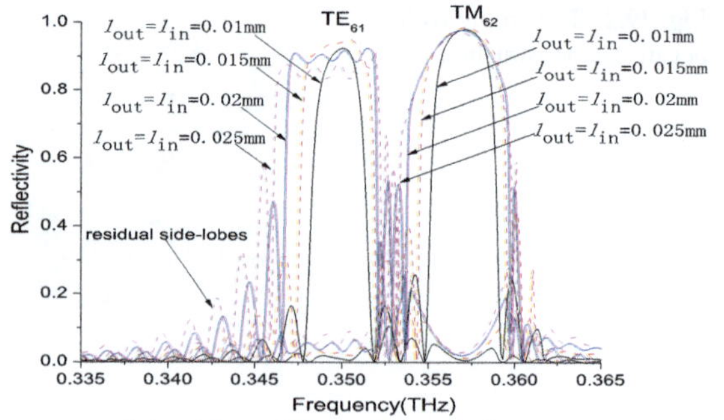

(a) without Hamming window

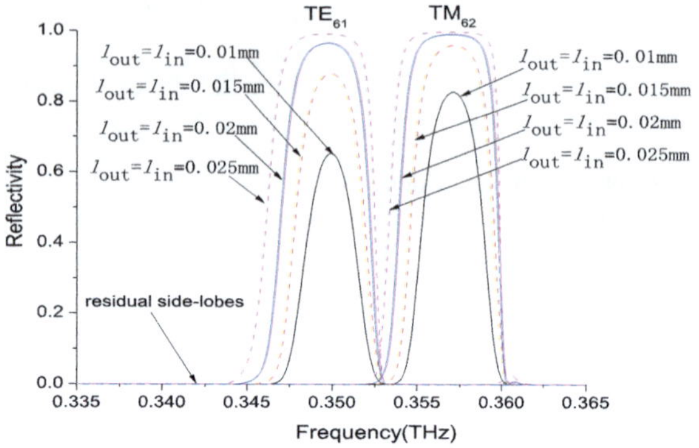

(b) with Hamming window

$TE_{6,1}$ and the competition mode $TM_{6,2}$. The bandwidth of $TE_{6,1}$ and $TM_{6,2}$ frequency response will gradually widen with the deepening of the ripple depth, and the centre frequency point will not shift. However, the reflectivity will gradually increase with the increase of ripple depth, and the band gap overlap will also worsen. No matter whether the Hamming window is loaded or not, the reflectivity of the $TM_{6,2}$ is less affected than the $TE_{6,1}$ mode. Therefore, in the THz frequency coupling mode, ripple depth and the adding window technology should be selected according to the actual needs. Figure 39.5 shows that the bandwidth change of the competing mode $TM_{6,2}$ is larger than that of the working mode $TE_{6,1}$ with the increase of ripple depth.

The performance of the coaxial Bragg structure as reflector or filter can be improved after adding negative-taper ripples [12, 13, 15, 16]. For the THz frequency coupled-mode structure, it can be seen from Figs. 39.6, 39.7 and 39.8 that the bandwidth of the working mode $TE_{6,1}$ and the competitive mode $TM_{6,2}$ will gradually widen with the increase of gradient angles. However, the bandwidth of $TM_{6,2}$ in competition mode changes greatly. The centre frequency shift of working mode $TE_{6,1}$ is small, but the centre frequency point of $TM_{6,2}$ in the competition mode will shift to larger frequency, far away from the centre frequency point of working mode $TE_{6,1}$, which is conducive to the suppression of band gap overlap, and its reflectivity is close to 1. Therefore, in order to suppress the overlap of band gap, weaken the

Fig. 39.5 Relationship between the ratio of ripple depth to ripple period and 3 dB bandwidth operating at 0.35 Thz

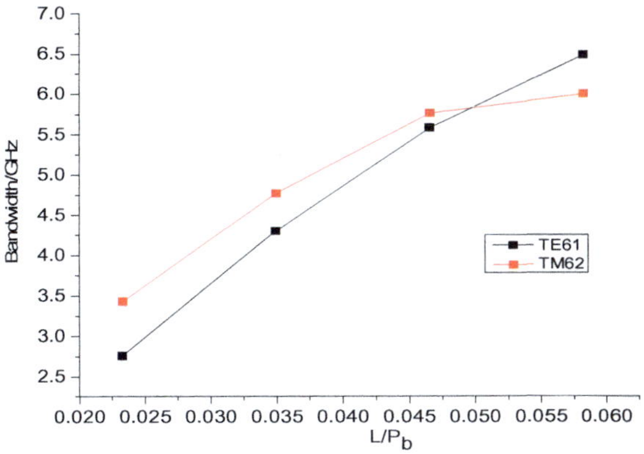

Fig. 39.6 Frequency response of the reflectivity with negative-taper ripples with the gradient angles are 0, 0.1°, 0.2°, 0.3°, 0.4°, respectively

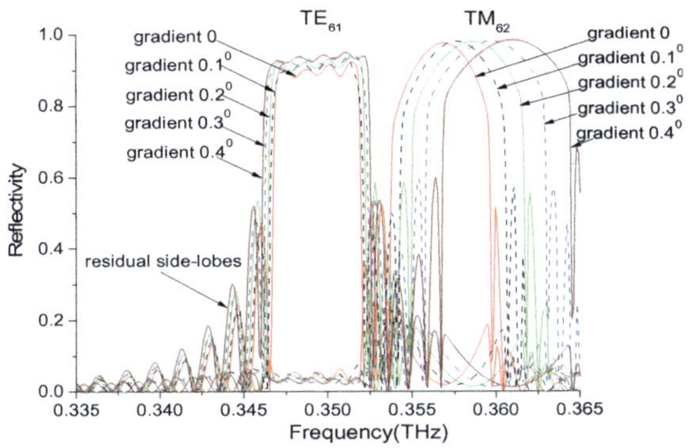

(a) without windowing

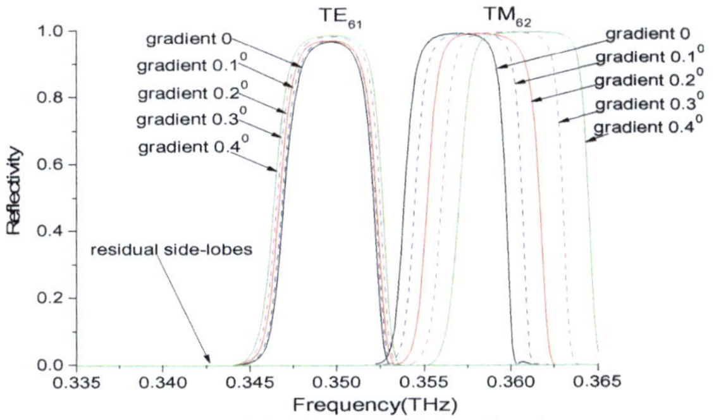

(b) with Hamming windowing

influence of competition mode and improve the selectivity of operation mode, we should choose the negative-taper ripples structure with larger gradient angle as far as possible, and set the appropriate gradient angle according to the needs.

Whether it is the change of ripple depth or the gradient angle, the windowing-technique can suppress the residual side-lobes, and they can also affect the reflectivity.

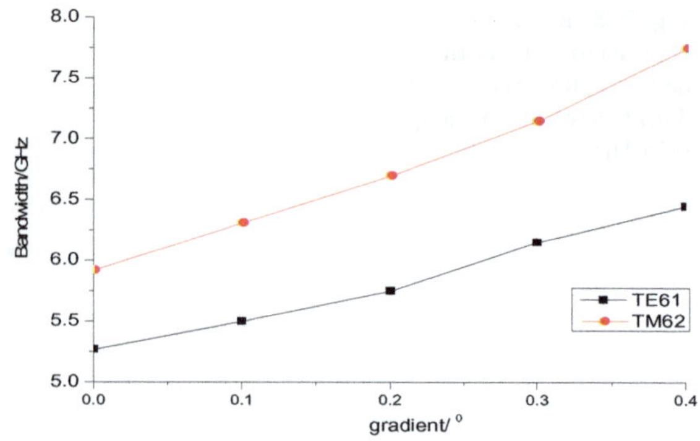

Fig. 39.7 Relationship between the different gradient angles and 3 dB bandwidth operating at 0.35 Thz in the coaxial Bragg reflector

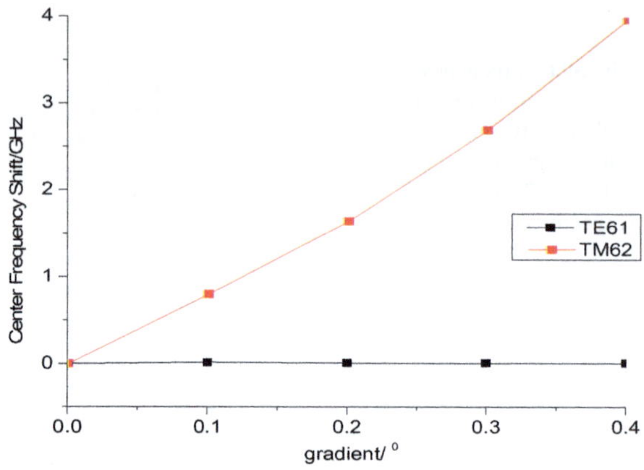

Fig. 39.8 Relationship between the different gradient angles and the centre frequency shift operating at 0.35 Thz in the coaxial Bragg reflector

39.4 Conclusions

Based on the mode-coupling theory and the FORTRAN programming language, the influence of different ripple depth and the different ripple angle on bandwidth and centre frequency shift characteristics operating at THz high-frequency coupling mode are compared and studied. By using Originlab drawing software to analyse the data, two conclusions are drawn as follows:

(1) With the increase of ripple depth, the bandwidth of working mode and competitive mode is increasing gradually, but the increasing trend of working mode bandwidth is faster than that of competitive mode, and their centre frequency almost has no shift;
(2) The bandwidth of both working mode and competition mode is gradually widened with the increase of gradient angle, however, the change of bandwidth of working mode is not obvious in competition mode. The change of gradient angle has no effect on the centre frequency shift of working mode but has great

influence on competition mode. With the increase of the gradient angle, the centre frequency shift of competition mode is larger.

Therefore, the performance of coaxial Bragg structure as reflector or filter can be widened by selecting the ripple depth and gradient angle according to the actual needs, to ensure the feasibility of the high-order single-mode operation of coaxial Bragg structure. These characteristics can optimize the mode selectivity and mode purity, to ensure the stable start-up of the operating mode. Therefore, it has great advantages and application prospects in high-power CARM or FEL oscillators operating at THz band.

Acknowledgements This work is supported by Hainan Provincial Natural Science Foundation of China (620MS063), the Provincial Science Research Foundation of Hainan (hnky2020ZD-21) and the Education Department of Hainan Province (Hnjg2019ZD-23).

References

1. Konoplev, I.V., McGrane, P., Phelps, A.D.R., Cross, A.W., Ronald, K (2005) Observation of photonic band-gap control in one-dimensional Bragg structures. Appl. Phys. Lett. **87**(12), 121104 (2005)
2. Konoplev, I.V., McGrane, P., He, W., Cross, A.W., Phelps, A.D.R., Whyte, C.G., Ronald, K., Robertson, C.W.: Experimental study of coaxial free-electron maser based on two-dimensional distributed feedback. Phys. Rev. Lett. **96**, 035002 (2006)
3. Barroso, J.J., Leite Neto, J.P.: Design of coaxial Bragg reflectors. IEEE Trans. Plasma Sci. **34**(3), 666–672 (2006)
4. Lai, Y.X., Jiang, X.M., Chen, F.Y.: Transmission bandgap of a two-dimensional coaxial Bragg structure with hybrid filling media. IEEE Access **8**, 191394–191402 (2020)
5. Konoplev, I.V., McGrane, P., Cross, A.W., Ronald, K., Phelps, A.D.R.: Wave interference and band control in multiconductor one-dimensional Bragg structures. J. Appl. Phys. **97**, 073101 (2005)
6. Glyavin, M., Zotova, I., Rozental, R., Malkin, A., Sergeev, A., Fokin, A., Rumyantsev, V, Morozov, S.: Investigation of the frequency double-multiplication effect in a sub-THz Gyrotron. J. Inf. Millimeter Terahertz Waves **41**, 1245–1251 (2020)
7. Thumm, M.: Free-electron masers versus gyrotrons: prospects for high-power sources at millimeter and submillimeter wavelengths. Nuclear Instru. Methods Phys. Res. Sect. A Accel. Spectromet. Detectors Assoc. Equip. **483**, 186–194 (2002)
8. Ginzburg, N.S., Peskov, N.Y., Sergeev, A.S., Phelps, A.D.R., Cross, A.W., Konoplev, I.V.: The use of a hybrid resonator consisting of one-dimensional and two-dimensional Bragg reflectors for generation of spatially coherent radiation in a coaxial free-electron laser. Phys. Plasmas, **9**, 2798–2802 (2002)
9. Ginzburg, N.S., Malkin, A.M., Peskov, N.Y., Sergeev, A., Zaslavsky, V.Y., Kamada, K., Soga, Y.: Tunable terahertz band planar Bragg reflectors. Appl. Phys. Lett, **95**, 043504 (2009)
10. Bovadilla, R.G.O., Molitor, D., Serrano, A.L.C., Rehder, G.P.: Optimization of RF MEMS phase shifter for microwaves applications. In: 32nd Symposium on Microelectronics Technology and Devices (SBMicro), Fortaleza, pp 1–4 (2017).
11. Geng, Z.H., Hou, X., Wang, Y.: Simulation and measurement of a W-band circular TE62 mode generator for gyrotrons. In: 2017 Eighteenth International Vacuum Electronics Conference (IVEC) (2017)

12. Ding, X.Y., Zhang, S.C., Xie, Y.C.: Characteristics of frequency response in a coaxial Bragg structure with tapered ripples. High Power Laser Part. Beams **20**(12), 2051–2054 (2008)
13. Ding, X.Y., Liu, H., Lv, Z.S.: Effect of ripple taper on coupling modes in a coaxial Bragg structure. J Infrared Milli Terahz Waves **31**, 1156–1163 (2010)
14. Ding, X.Y., Wang, L.L., Lv, Z.S.: Comparative study of coaxial Bragg structure with hybrid ripples shape. Chinese J. Radio Sci. (in Chinese) **30**(6), 1228–1234 (2015)
15. Ding, X.Y., Wang, L.L.: Comparative study of numerical simulations in coaxial Bragg Reflector with the new tapered ripples. Chinese J. Radio Sci. **26**(1), 55–61 (2012)
16. Ding, X.Y., Zhang, S.C.: Effect of ripple shape and taper on frequency response of reflectivity and transmission in a coaxial Bragg structure. J. Phys. D: Appl. Phys. **42**, 085104 (2009)
17. Lai YX, Zhang SC: Coaxial Bragg reflector with a corrugated inner rod. IEEE Microwave Wireless Comp. Lett. **17**, 328–331 (2007)
18. Lai YX, Zhang SC.: Multiwave interaction formulation of a coaxial Bragg structure and its experimental verification. Phys. Plasmas **14**, 113301 (2007)
19. Shi, D.F., Qian, B.L.,Wang, H.G.: A novel relativistic magnetron with circularly polarized TE11 coaxial waveguide mode. J. Phys. D Appl. Phys. **49**(46), SJPD00223727 1423(2016)
20. Petelin, M.I.: The gyrotron: physical genealogy. Terahertz Sci. Technol. **8**(4), 157–166 (2015)

Chapter 40
Outage Performance of Two-Stage Relay Selection in Overlay CR-NOMA Networks

Jianquan Yang, Yucheng He, Menghuan Ma, Shuting Wu, and Lin Zhou

Abstract The non-orthogonal multiple access (NOMA) technology combined with cognitive networks, can significantly improve the performance of the system. In this paper, we study the two-stages of relay selection strategy of overlay cognitive nonorthogonal performance effects of multiple access network in the model, considering the secondary network mainly provide relay to help decode forward information to obtain authorization spectrum access to cognitive relay after primary and secondary source node reception at the same time, using successive interference cancelation (SIC) strategy to decode and forward the superposition signal. The outage probability closure expression of the system under the two-stage relay selection strategy is derived, analyzes the relay numbers and distribution of factors that affect the performance of the system. The theoretical results are verified by Monte Carlo simulation.

40.1 Introduction

The combination of non-orthogonal multi-access and cognitive radio technology, known as CR-NOMA network, can improve spectrum efficiency and realize intelligent spectrum sharing [1]. Im and Lee [2] study the interrupt performance of the underly cognitive NOMA network under the condition of the imperfect successive interference cancelation.

In [3] and [4] discussed that in the overlay CR-NOMA network, the secondary source node helps the primary user to transmit information as a relay and obtains

J. Yang · Y. He (✉) · M. Ma · S. Wu · L. Zhou
Xiamen Key Laboratory of Mobile Multimedia Communications, National Huaqiao University, Xiamen 361021, Fujian, China
e-mail: yucheng.he@hqu.edu.cn

Y. He · L. Zhou
State Key Laboratory of Integrated Services Networks, Xidian University, Xi'an, Shaanxi 710071, China

L. C. Jain et al. (eds.), *Wireless Technology, Intelligent Network Technologies, Smart Services and Applications*, Smart Innovation, Systems and Technologies 258,
https://doi.org/10.1007/978-981-16-5168-7_40

authorized spectrum as a reward. Meanwhile, NOMA principle is used to superposition, decode and forward the secondary user information and the primary user information. At the receiving end, SIC technology is adopted to obtain the desired signal. In the scene of Massive Machine Type Communications (mMTC), a large number of sensors and Massive machine terminal equipment can relay to each other. The relay technology maximizes the spectrum utilization rate and extends the coverage of the network [5]. Li et al. [6] analyzes the influence of partial relay selection strategy on the performance of downlink cooperative CR-NOMA network, and verifies the influence of relay number and power allocation factor on the secondary network by deducing the closed expression of secondary network interrupt probability. Ding et al. [7] consider a downlink collaboration NOMA network. The article through two stages of relay selection strategy is derived under the interrupt probability of NOMA network compared with traditional maximum minimum relay selection strategy, the results show that compared with the traditional two-stage relay selection strategy for better diversity gain at the same time, reduce the interruption performance based on the above research, this paper presents a two-stage relay selection strategy affect the performance of cognitive NOMA overlay network interruption, the system outage probability closed expression, and the simulation results are compared to performance.

40.2 System Model

Figure 40.1 shows an overlay CR-NOMA network. (primary user network (PUN)) includes a primary transmitter PT and a primary user (PU), secondary user network (SUN) includes a secondary transmitter ST and N secondary relays $\{R_n\}_{n=1}^{N}$ and a secondary user SU. PT is far from PU, so the relay of SUN is needed to help PUN communicate. In return, SUN allows access to PUN's authorized spectrum and gets the transmission opportunity of x_2. PUN has priority decoding right in order to

Fig. 40.1 System model

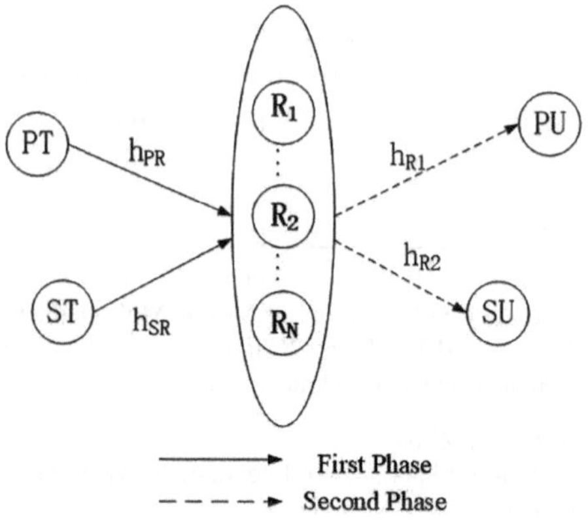

achieve communication reliability. We assume that there is no direct link between the source node and the destination due to physical obstacles or adverse channel conditions. It is assumed that all nodes in the network are equipped with a single antenna and operate in half duplex mode. All channels experience independent and identically Rayleigh fading.

In the first slot, PT, ST, respectively, send signals x_1, and to all relay nodes simultaneously. Relay always decodes signal x_1 by treating signal x_2 as noise and performs SIC to decode x_2 at receiver according to NOMA principle. Therefore, the received signal at relay is given by

$$y_{R_n} = h_{PR}\sqrt{P_T}x_1 + h_{SR}\sqrt{P_S}x_2 + n_r \tag{40.1}$$

The signal to interference plus noise ratio (SINR) to decode x_1, x_2 at is given by

$$\gamma_{R\to x_1} = \frac{\rho_T |h_{PR}|^2}{\rho_S |h_{SR}|^2 + 1} \tag{40.2}$$

$$\gamma_{R\to x_2} = \rho_S |h_{SR}|^2 \tag{40.3}$$

where P_T and P_S are the transmit power of PT, ST, respectively. Assume that the channel coefficients is denoted as $h_{PR}h_{SR}$, which are modeled as an independent circularly symmetric complex Gaussian random variable having mean zero and variance $\lambda_{PR}, \lambda_{SR}.n_r$ represents Additive White Gaussian Noise (AWGN) at the relay, $n_r \sim CN(0, \sigma_R^2)$,We define $\rho_S = P_S/\sigma_R^2$, $\rho_T = P_T/\sigma_R^2$.

In the second time slot, the selected Relay subset R_n can Successfully decode primary and secondary signal, and forward superposition signal $x_{RN} = \sqrt{\alpha_1 P_R}x_1 + \sqrt{\alpha_2 P_R}x_2$ to PU and SU. The SINR to decode signal at PU and SU is given by

$$\gamma_{PU\to x_1} = \frac{\alpha_1 \rho_R |h_{R1}|^2}{\alpha_2 \rho_R |h_{R1}|^2 + 1} \tag{40.4}$$

$$\gamma_{SU\to x_1} = \frac{\alpha_1 \rho_R |h_{R2}|^2}{\alpha_2 \rho_R |h_{R2}|^2 + 1} \tag{40.5}$$

$$\gamma_{SU\to x_2} = \alpha_2 \rho_R |h_{R2}|^2 \tag{40.6}$$

where P_R is the power of relay, assume that the channel coefficients are denoted as h_{R1}, h_{R2},which are modeled as an independent circularly symmetric complex Gaussian random variable having mean zero and variance $\lambda_{R1}, \lambda_{R2}$.

Denote $\rho_R = \frac{P_R}{\sigma^2}, \sigma_R^2 = \sigma_{PU}^2 = \sigma_{SU}^2 = \sigma^2$.

Two-stage relay selection scheme (TSRSS) achieves both goals at once, One is to ensure the realization of PU target rate, and the other is to provide the maximum rate for SU of secondary users. The first phase is to build the following relay subsets by focusing on the user PU target rate:

$$S_r = \begin{Bmatrix} n : 1 \le n \le N \\ \gamma_{R \to x_1} \ge \xi_1, \gamma_{PU \to x_1} \ge \xi_1, \gamma_{SU \to x_1} \ge \xi_1 \end{Bmatrix} \tag{40.7}$$

Denote $\xi_1 = 2^{2R_1} - 1$, R_1 is the target rate of signal x_1. The second stage is to select a relay in the constructed stepson set that can maximize the SU rate:

$$R_n^* = \arg\max_n \left\{ \min(\gamma_{R \to x_2}, \gamma_{SU \to x_2}), n \in S_r \right\} \tag{40.8}$$

40.3 Performance Analysis

In this section, we will analysis the outage probability achieved by the two-stage relay selection scheme. The Ω_1 denotes unable to build relay subset。Ω_2 denotes x_2 could not be correctly decoded by the selected optimal relay or secondary user If the relay subset is successfully built. Therefore, the outage probability can be written as:

$$P(\Omega) = P(\Omega_1) + P(\Omega_2) \tag{40.9}$$

When $\alpha_1 - \xi_1\alpha_2 > 0$, $P(\Omega_1)$ can be calculated as follow:

$$P(\Omega_1) = P(|S_r| = 0) = \prod_{n=1}^{N} [1 - (\theta_2 e^{-\theta_3})(e^{-\frac{\theta_1}{\lambda_{R1}}})(e^{-\frac{\theta_1}{\lambda_{R2}}})] \tag{40.10}$$

Or, $P(\Omega_1) = 1$, denote $\theta_1 = \frac{\xi_1}{\rho_R(\alpha_1 - \xi_1\alpha_2)}$ $\theta_2 = \frac{\lambda_{PR}\rho_T}{\xi_1\rho_S\lambda_{SR} + \lambda_{PR}\rho_T}$ $\theta_3 = \frac{\xi_1}{\lambda_{PR}\rho_T}$.
When $|S_r| > 0$, we define $x_n = \min(\gamma_{R \to x_2}, \gamma_{SU \to x_2})$ and $x_n* = \max\{x_i, \forall i \in S_r\}$.

$P(\Omega_2)$ can be calculated as follow:

$$P(\Omega_2) = \sum_{l=1}^{N} P(x_n* < \xi_2 | |S_r| = l) P(|S_r| = l) \tag{40.11}$$

where $\xi_2 = 2^{2R_2 - 1}$, R_2 Is the target rate of signal x_2.

Define $F(\Lambda_2)$ as the first probability on the right side of this equation. The probability $F(\Lambda_2)$ can be expressed as follow:

$$
\begin{aligned}
F(\Lambda_2) &= P\left\{\min(\gamma_{R\to x_2}, \gamma_{SU\to x_2}) < \xi_2 | n \in S_r, |S_r| \neq 0\right\} \\
&= \frac{\underbrace{P\left\{\min(\gamma_{R\to x_2}, \gamma_{SU\to x_2}) < \xi_2, \gamma_{R\to x_1} \geq \xi_1, \gamma_{SU\to x_1} \geq \xi_1\right\}}_{\Delta_1}}{\underbrace{P\left\{\gamma_{R\to x_1} \geq \xi_1, \gamma_{SU\to x_1} \geq \xi_1\right\}}_{\Delta_2}}
\end{aligned}
\tag{40.12}
$$

Δ_2 can be expressed as:

$$
\Delta_2 = P\left\{\gamma_{R\to x_1} \geq \xi_1, \gamma_{SU\to x_1} \geq \xi_1\right\} = \frac{1}{\lambda_{R2}}\theta_2 \mathrm{e}^{-\frac{(\theta_3\lambda_{R2}+\theta_1)}{\lambda_{R2}}} \tag{40.13}
$$

Assume that $\Delta_1 = \Delta_3 + \Delta_4$.
For $k_2 > k_1$, the probability of Δ_3 and Δ_4 can be calculated as follow:

$$
\begin{aligned}
\Delta_3 = \theta_2 e^{-\theta_3}(1 - e^{-\frac{(\lambda_{PR}\rho_T+\xi_1\rho_S\lambda_{SR})\xi_2}{\lambda_{SR}\lambda_{PR}\rho_T\rho_S}})e^{-\frac{k_2}{\lambda_{R2}\rho_R}} + \\
\theta_2 e^{-\theta_3}\left\{(e^{-\frac{k_1}{\lambda_{R2}\rho_R}} - e^{-\frac{k_2}{\lambda_{R2}\rho_R}}) + \theta_2\varsigma_1(\mathrm{e}^{-\frac{(\alpha_2\rho_R\lambda_{R2}+\lambda_{SR}\theta_2\rho_S)k_2}{\lambda_{SR}\theta_2\rho_S\lambda_{R2}\rho_R}} - \mathrm{e}^{-\frac{(\alpha_2\rho_R\lambda_{R2}+\lambda_{SR}\theta_2\rho_S)k_1}{\lambda_{SR}\theta_2\rho_S\lambda_{R2}\rho_R}})\right\}
\end{aligned}
\tag{40.14}
$$

$$
\Delta_4 = \theta_2^2\,\varsigma_1 e^{-\theta_3}\left\{e^{-\frac{(\lambda_{SR}\theta_2\rho_S+\alpha_2\rho_R\lambda_{R2})k_1}{\lambda_{R2}\lambda_{SR}\theta_2\rho_S\rho_R}} - e^{-\frac{(\lambda_{SR}\theta_2\rho_S+\alpha_2\rho_R\lambda_{R2})k_2}{\lambda_{R2}\lambda_{SR}\theta_2\rho_S\alpha_2\rho_R}}\right\} \tag{40.15}
$$

where $\varsigma_1 = \frac{\lambda_{SR}\rho_S}{(\alpha_2\rho_R\lambda_{R2}+\lambda_{SR}\theta_2\rho_S)}$ and $k_1 = \frac{\xi_1}{(\alpha_1-\alpha_2\xi_1)}$, $k_2 = \frac{\xi_2}{\alpha_2}$.
Therefore, the overall outage probability can be expressed as:

$$
P(O) = \sum_{l=0}^{N}\binom{N}{l}(F(\Lambda_2))^l[1-\Theta]^{N-l}\Theta^l \tag{40.16}
$$

40.4 Numerical Results

In this section, the outage performance of CR-NOMA network under Rayleigh channel is verified by simulation. System parameters are set to $\lambda_{SR} = \lambda_{R2} = 2$, $\lambda_{PR} = \lambda_{R1} = 1$, $R_1 = 0.8$ bit/s, $R_2 = 1.5$ bit/s. $P_r = 20$ dBm, $P_S = 20$ dBm.

Figure 40.1 shows that two-stage relay selection strategy can significantly reduce the outage probability of CR-NOMA network with the increase in the number of relays. The results are verified by Monte Carlo simulation. In Fig. 40.2, it describes the effect of the power allocation factor on the interrupt performance of the system, and we obtain the optimal power allocation factor assigned to the main signal (Fig. 40.3).

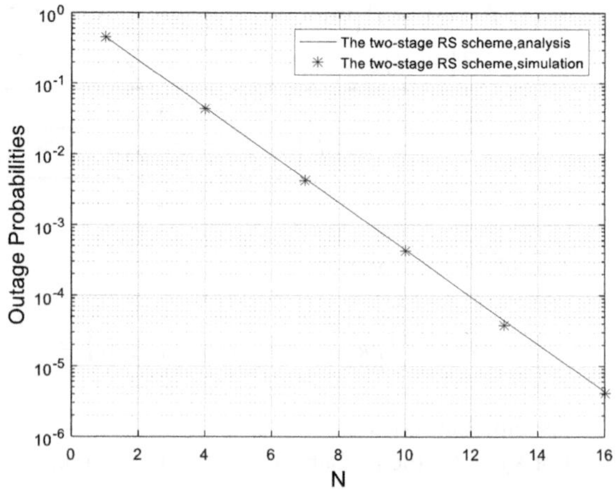

Fig. 40.2 The relationship between system outage probability and the number of relays

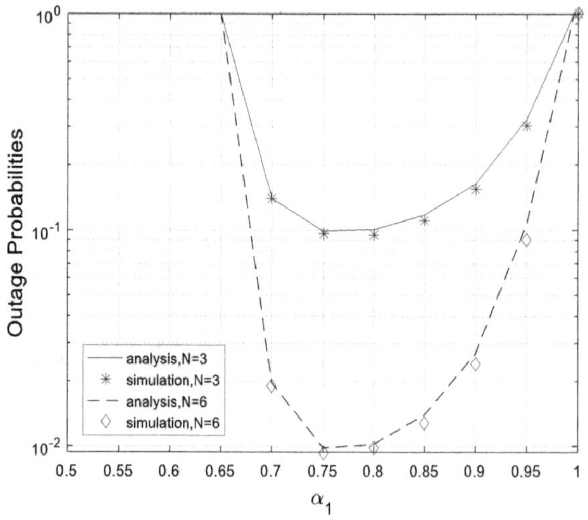

Fig. 40.3 Relationship between system outage probability and power distribution factor

40.5 Conclusion

This paper considered the effect of two-stage relay selection strategy on the performance of overlay CR-NOMA network. The closed expression of system outage probability is derived and verified by simulation. The results show that the system performance can be improved by increasing the number of relays and selecting the appropriate power distribution factor.

References

1. Arzykulov, S., Tsiftsis, T.A., Maham, B., Abdallah, M.: On the outage of underlay CR-NOMA networks with detect-and-forward relaying. IEEE Trans. Cognit. Commun. Netw. **5**(3), 795–804 (2019)
2. Im, G., Lee, J.H.: Outage probability for cooperative NOMA systems with imperfect SIC in cognitive radio networks. IEEE Commun. Lett. **23**(4), 692–695 (2019)
3. Luo, L., Li, Q., Cheng, J.: Performance analysis of overlay cognitive NOMA systems with imperfect successive interference cancellation. IEEE Trans. Commun. **68**(8), 4709–4722 (2020)
4. Le, Q.N., Nguyen, N., Yadav, A., Dobre, O.A.: Outage performance of full-duplex overlay CR-NOMA networks with SWIPT. In: IEEE Press. 2019 IEEE Global Communications Conference (GLOBECOM), pp. 1–6. Waikoloa, USA (2019)
5. Tian, L., Yan, C., Li, W., Yuan, Z., Cao, W., Yuan, Y.: On uplink non-orthogonal multiple access for 5g: opportunities and challenges. China Commun. **14**(12), 142–152 (2017)
6. Li, C., Guo, D., Guo, K., Qin, Y., Xu, R.: Outage performance of partial relay selection in underlay CR-NOMA networks. In: IEEE Press. 2019 28th Wireless and Optical Communications Conference (WOCC), pp. 1–5, Beijing, China (2019)
7. Ding, Z., Dai, H., Poor, H. V.: Relay selection for cooperative NOMA. IEEE Wireless Commun. Lett. **5**(4), 416–419 (2016)

Chapter 41
Outage Probability for CR-NOMA Systems with Imperfect SIC Under Two-stage Relay Selection Scheme

Liangmei Zhang, Yucheng He, Linsen Yi, and Lin Zhou

Abstract In this paper, we investigate the outage probabilities of secondary destination nodes in cognitive radio non-orthogonal multiple access (CR-NOMA) systems with imperfect successive interference cancellation (SIC) under two-stage relay selection scheme (TSRS). CR-NOMA technology, which integrated with cognitive radio and NOMA, can significantly raise the level of spectrum utilization and improve the system performance. On the other hand, in order to enhance the secondary network performance, lots of relay nodes are implemented in the secondary network to cooperate and relay the message of users to the secondary users. Considering imperfect SIC at receiver, we derive accurate outage probabilities expressions of secondary destination nodes in the case of all channel coefficients in this system are subjected to Rayleigh distribution. We discuss the effects of imperfect SIC factor, interference temperature constraint, and the maximum transmit power on the system outage performance thoroughly. The relevant simulation results verify the derived expressions of the outage probabilities are in great consistent with Monte Carlo simulation results.

41.1 Introduction

With the development of wireless network application scenarios (e.g., Internet of things (IoT) and massive machine-type communication (mMTC)), the next generation wireless communication technology has higher requirements for spectrum efficiency and users' quality of service (QoS). Non-orthogonal multiple access (NOMA)

L. Zhang · Y. He (✉) · L. Yi · L. Zhou
Xiamen Key Laboratory of Mobile Multimedia Communications, National Huaqiao University, Xiamen 361021, Fujian, China
e-mail: yucheng.he@hqu.edu.cn

Y. He · L. Yi · L. Zhou
State Key Laboratory of Integrated Services Networks, Xidian University, Xi'an 710071, Shanxi, China

L. C. Jain et al. (eds.), *Wireless Technology, Intelligent Network Technologies, Smart Services and Applications*, Smart Innovation, Systems and Technologies 258,
https://doi.org/10.1007/978-981-16-5168-7_41

is regarded as a significant candidate for the next generation wireless communication [1]. Different from the traditional orthogonal multiple access (OMA) technology, NOMA can accomplish multiplexing on different resource blocks (e.g., time/frequency), which greatly improves the spectrum efficiency. According to NOMA protocol, the source node sends the superimposed signal with different power level, and the receiver takes successive interference cancellation (SIC) to decode the information of users [2]. Cooperative NOMA can effectively improve the security performance of system and enhance the transmission reliability [3]. The influence of diverse relay selection schemes on system performance in cooperative NOMA is studied in [4]. Simulation results indicate the performance of two-stage relay selection scheme (TSRS) is better than the traditional maximum minimum relay selection scheme in cooperative NOMA system.

Cognitive radio (CR) as another effective approach to improve the spectrum efficiency has been extensively concerned. In [5], the outage probabilities expressions of secondary users in downlink cognitive radio non-orthogonal multiple access (CR-NOMA) relay network with imperfect SIC are discussed.

In this paper, we study an underlay CR-NOMA network with imperfect SIC under TSRS. We obtain the closed forms of outage probabilities. We discuss the effects of the maximum transmit power, imperfect SIC factor, and interference temperature constraint on CR-NOMA network at the same time.

41.2 System Model

We study an underlay cooperative CR-NOMA network as displayed in Fig. 41.1. This network comprises a primary receiver (PR), a secondary transmitter (S), two secondary users (SU_1 and SU_2), and a number of relay nodes ($R_n, n = 1, 2, \ldots, N$). Assuming the interference caused by primary transmitter (PT) to the secondary network can be ignored.

Considering the severe shadow fading, node S to node SU_i has no direct link, and the relay node R_n decodes and forwards (DF) the intended messages to the users

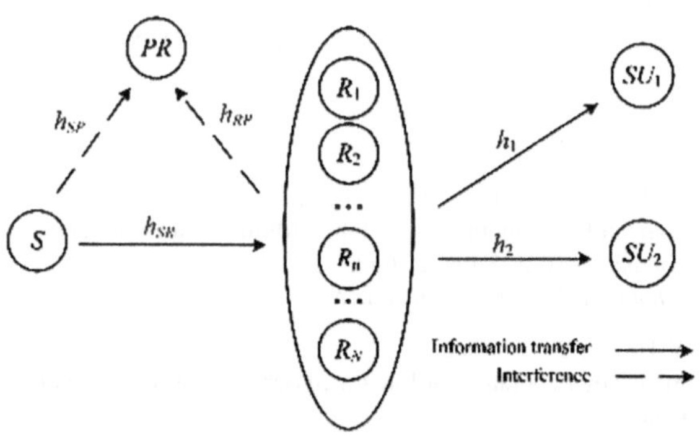

Fig. 41.1 System model

cooperatively. Assuming all nodes are equipped with a single antenna and work in half-duplex transmission mode, all the channels undergo quasi-static independent but not identically distributed (i.i.n.d) Rayleigh fading. Therefore, the channel coefficients of the links $S \rightarrow PR, R_n \rightarrow PR, S \rightarrow R_n, R_n \rightarrow SU_1, R_n \rightarrow SU_2$ be expressed as $h_{jk} \sim CN(0, \lambda_{jk}), (jk \in \{SP, R_nP, SR_n, R_nU_1, R_nU_2\})$.

In the first time slot, S broadcasts superposed signal $x_s = \sqrt{a_1 P_S} x_1 + \sqrt{a_2 P_S} x_2$ to the relay nodes, where x_1 and x_2 are the intended messages to SU_1 and SU_2, respectively, i.e., $E\{x_i^2\} = 1, i \in \{1, 2\}$; P_S is the transmit power at the S; a_i is the power allocation factor for x_i and $a_1 + a_2 = 1$; considering SU_1 is further away from the relay nodes than SU_2, according to the NOMA protocol, we order $a_1 > a_2$. Considering the interference caused by S to PR, the transmission power of S is constrained as follows:

$$P_s = \min\left\{P_S^m, \frac{I}{|h_{SP}|^2}\right\} \tag{41.1}$$

where P_S^M is the maximum value of transmission power and I is the value of interference temperature constraint.

Hence, the relevant received signal expression at relay nodes can be given by:

$$y_{R_n} = h_{SR_n}\left(\sqrt{a_1 P_S} x_1 + \sqrt{a_2 P_S} x_2\right) + \omega_{R_n} \tag{41.2}$$

Applying NOMA, SIC is carried out at R_n, according to $a_1 > a_2$; therefore, the selected relay first decodes the message of SU_1, then subtracts this section from received signal to detect the message of SU_2. Therefore, the signal to interference plus noise ratio (SINR) to detect x_1 at R_n is shown as:

$$\gamma_{R_n}^{\{x_1\}} = \frac{a_1 \rho_S |h_{SR_n}|^2}{a_2 \rho_S |h_{SR_n}|^2 + 1} \tag{41.3}$$

Considering imperfect SIC, the SINR to detect x_2 is shown as:

$$\gamma_{R_n}^{\{x_2\}} = \frac{a_2 \rho_S |h_{SR_n}|^2}{a_1 \rho_S |g_{SR_n}|^2 + 1} \tag{41.4}$$

where $\rho_S = \frac{P_S}{\sigma^2}$ is the transmit signal to noise ratio (SNR) at S, $g_{SR_n} \sim CN(0, \eta\lambda_{SR_n})$,$0 \le \eta < 1$, and η is the factor of residual interference caused by imperfect SIC.

From Eqs. (41.3) and (41.4), we can derive that the message x_1 and x_2 will be successfully detected by relay in following conditions: $\frac{1}{2}\log\left(1 + \gamma_{R_n}^{\{x_i\}}\right) \ge R_i^*$,where R_i^* denotes the targeted rate for SU_i.Considering the interference caused by R_n to PR, the transmission power of R_n is limited as follows:

$$P_{R_n} = \min\left\{P_{R_n}^M, \frac{I}{|h_{RP}|^2}\right\} \tag{41.5}$$

where $P_{R_n}^M$ is the maximum value of transmission power.

In the second time slot, the relay recodes and forwards superposed signal to the users.

Hence, the relevant received signal at users is given by:

$$y_{R_n U_i} = h_{R_n U_i}\left(\sqrt{a_1 P_{R_n}} x_1 + \sqrt{a_2 P_{R_n}} x_2\right) + \omega_{R_n U_i} \tag{41.6}$$

where $\omega_{R_n U_i}$ denotes the AWGN at SU_i with $\omega_{R_n U_i} \sim CN\left(0, \sigma^2\right)$.

Therefore, the SINR to decode x_1 at SU_1 is given by:

$$\gamma_{\mathrm{SU}_1}^{\{x_1\}} = \frac{a_1 \rho_{R_n} \left|h_{R_n U_1}\right|^2}{a_2 \rho_{R_n} \left|h_{R_n U_1}\right|^2 + 1} \tag{41.7}$$

The SINR to decode x_1 at SU_2 is given by:

$$\gamma_{SU_2}^{\{x_1\}} = \frac{a_1 \rho_{R_n} \left|h_{R_n U_2}\right|^2}{a_2 \rho_{R_n} \left|h_{R_n U_2}\right|^2 + 1} \tag{41.8}$$

Considering imperfect SIC, the SINR to detect x_2 is given by:

$$\gamma_{SU_2}^{\{x_2\}} = \frac{a_2 \rho_{R_n} \left|h_{R_n U_2}\right|^2}{a_1 \rho_{R_n} \left|g_{R_n U_2}\right|^2 + 1} \tag{41.9}$$

where $g_{R_n U_2} \sim CN\left(0, \eta \lambda_{RU_2}\right)$.

In this paper, we discuss outage performance with imperfect SIC under TSRS scheme. This scheme is described as follows: Firstly, we select the relay subset that can satisfy the requirements of SU_1 achievable rate and ensure both users can decode x_1 successfully in the second time slot simultaneously, i.e., $S_r = \{R_n : E_n, 1 \leq n \leq N\}$, where E_n is the joint Boolean expression that satisfies the above conditions for a dedicated relay, that is: $E_n = \left\{\gamma_{R_n}^{\{x_1\}} \geq 2^{2R_1^*} - 1, \gamma_{R_n}^{\{x_2\}} \geq 2^{2R_2^*} - 1, \gamma_{SU_1}^{\{x_1\}} \geq 2^{2R_1^*} - 1, \gamma_{SU_2}^{\{x_1\}} \geq 2^{2R_1^*} - 1\right\}$. Secondly, the relay which can maximize the achievable rate of SU_2 is selected from this subset for forwarding the users' message, which can be expressed as: $n_{TSRS}^* = \arg\max_n\left\{\min\left\{\frac{1}{2}\log\left(1 + \gamma_{R_n}^{\{x_2\}}\right), \frac{1}{2}\log\left(1 + \gamma_{SU_2}^{\{x_2\}}\right)\right\}, R_n \in S_r\right\}$.

41.3 Outage Probability Analysis

We derive the outage probabilities of users in this subsection. For SU_1, the outage will arise when there is no relay can decode the message of user correctly, i.e., $|S_r| = 0$. For SU_2, the outage will take place in two circumstances. The first circumstance is that the relay cannot detect either x_1 or x_2 correctly. The second circumstance is that SU_2 cannot decode its owner message x_2 even if the relay decodes both x_1 and x_2 successfully.

We assume relay nodes are aggregated closely so that the links from any relay to the same users' node are independently and identically distributed. Therefore, the outage probability of SU_i is received by:

$$P_{SU_1}^{out} = P(|S_r| = 0) = (1 - P_1)^N \tag{41.10}$$

$$\begin{aligned} P_{SU_2}^{out} &= P\{|S_r| = 0\} + P\{\gamma_{n^*} < \varepsilon_2, |S_r| > 0\} \\ &= (1 - P_1)^N + \sum_{l=1}^{N} \binom{N}{l} (1 - P_1)^{N-l} (P_1 - P_2)^l \end{aligned} \tag{41.11}$$

where $P_1 = P\{E_n\}$, and the expressions of P_1 and P_2 are given by:

$$\begin{aligned} P_1 &= P\{E_n\} = P\left\{\gamma_{R_n}^{\{x_1\}} \geq \varepsilon_1, \gamma_{R_n}^{\{x_2\}} \geq \varepsilon_2, \gamma_{SU_1}^{\{x_1\}} \geq \varepsilon_1, \gamma_{SU_2}^{\{x_1\}} \geq \varepsilon_1\right\} \\ &= \left(1 - e^{-\frac{\rho_I}{\rho_S^M \lambda_{SP}}}\right)\left(1 - e^{-\frac{\rho_I}{\rho_{R_n}^M \lambda_{RP}}}\right) e^{-\frac{k_1}{\rho_{R_n}^M \lambda_{RU_1}} - \frac{k_1}{\rho_{R_n}^M \lambda_{RU_2}}} \\ &\times \left(e^{-\frac{k_4}{\rho_S^M \lambda_{SR}}} - k_5 e^{-\frac{k_4}{\rho_S^M k_5 \lambda_{SR}} + \frac{k_2}{\rho_S^M k_3 \xi \lambda_{SR}}}\right) + \varphi_1 \times \left(1 - e^{-\frac{\rho_I}{\rho_S^M \lambda_{SP}}}\right) \\ &\times \left(e^{-\frac{k_4}{\rho_S^M \lambda_{SR}}} - k_5 e^{-\frac{k_4}{\rho_S^M k_5 \lambda_{SR}} + \frac{k_2}{\rho_S^M k_3 \xi \lambda_{SR}}}\right) \\ &\times e^{-\frac{k_1}{\rho_{R_n}^M \lambda_{RU_1}} - \frac{k_1}{\rho_{R_n}^M \lambda_{RU_2}} - \frac{\rho_I}{\rho_{R_n}^M \lambda_{RP}}} + \left(1 - e^{-\frac{\rho_I}{\rho_{R_n}^M \lambda_{RP}}}\right) e^{-\frac{k_1}{\rho_{R_n}^M \lambda_{RU_1}} - \frac{k_1}{\rho_{R_n}^M \lambda_{RU_2}}} \\ &\times \left(\varphi_2 e^{-\frac{k_4}{\rho_S^M \lambda_{SR}} - \frac{\rho_I}{\rho_S^M \lambda_{SP}}} - k_5 \times \varphi_3 e^{\frac{k_2}{\rho_S^M k_3 \xi \lambda_{SR}} - \frac{k_4}{\rho_S^M k_5 \lambda_{SR}} - \frac{\rho_I}{\rho_S^M \lambda_{SP}}}\right) \\ &+ \varphi_1 e^{-\frac{k_1}{\rho_{R_n}^M \lambda_{RU_1}} - \frac{k_1}{\rho_{R_n}^M \lambda_{RU_2}} - \frac{\rho_I}{\rho_{R_n}^M \lambda_{RP}}} \\ &\times \left(\varphi_2 e^{-\frac{k_4}{\rho_S^M \lambda_{SR}} - \frac{\rho_I}{\rho_S^M \lambda_{SP}}} - k_5 \times \varphi_3 e^{\frac{k_2}{\rho_S^M k_3 \xi \lambda_{SR}} - \frac{k_4}{\rho_S^M k_5 \lambda_{SR}} - \frac{\rho_I}{\rho_S^M \lambda_{SP}}}\right) \end{aligned}$$

$$\begin{aligned} P_2 &= P\left\{E_n, \gamma_{SU_2}^{\{x_2\}} \geq \varepsilon_2\right\} \\ &= P\left\{\gamma_{R_n}^{\{x_1\}} \geq \varepsilon_1, \gamma_{R_n}^{\{x_2\}} \geq \varepsilon_2, \gamma_{SU_1}^{\{x_1\}} \geq \varepsilon_1, \gamma_{SU_2}^{\{x_1\}} \geq \varepsilon_1, \gamma_{SU_2}^{\{x_2\}} \geq \varepsilon_2\right\} \end{aligned}$$

$$
\begin{aligned}
&= \left(1 - e^{-\frac{\rho_I}{\rho_S^M \lambda_{SP}}}\right)\left(1 - e^{-\frac{\rho_I}{\rho_{R_n}^M \lambda_{RP}}}\right) e^{-\frac{k_1}{\rho_{R_n}^M \lambda_{RU_1}}} \\
&\times \left(e^{-\frac{k_4}{\rho_S^M \lambda_{SR}}} - k_5 e^{-\frac{k_4}{\rho_S^M k_5 \lambda_{SR}} + \frac{k_2}{\rho_S^M k_3 \xi \lambda_{SR}}}\right) \times \left(e^{-\frac{k_4}{\rho_{R_n}^M \lambda_{RU_2}}} - k_5 e^{-\frac{k_4}{\rho_{R_n}^M k_5 \lambda_{RU_2}} + \frac{k_2}{\rho_{R_n}^M k_3 \xi \lambda_{RU_2}}}\right) \\
&+ \left(1 - e^{-\frac{\rho_I}{\rho_S^M \lambda_{SP}}}\right) \times \left(e^{-\frac{k_4}{\rho_S^M \lambda_{SR}}} - k_5 e^{-\frac{k_4}{\rho_S^M k_5 \lambda_{SR}} + \frac{k_2}{\rho_S^M k_3 \xi \lambda_{SR}}}\right) \\
&\times \left(\varphi_4 e^{-\frac{k_4}{\rho_{R_n}^M \lambda_{RU_2}} - \frac{\rho_I}{\rho_{R_n}^M \lambda_{RP}} - \frac{k_1}{\rho_{R_n}^M \lambda_{RU_1}}} - k_5 \times \varphi_5 e^{-\frac{k_4}{\rho_{R_n}^M k_5 \lambda_{RU_2}} - \frac{\rho_I}{\rho_{R_n}^M \lambda_{RP}} - \frac{k_1}{\rho_{R_n}^M \lambda_{RU_1}} + \frac{k_2}{\rho_R^M k_3 \xi \lambda_{RU_2}}}\right) \\
&+ \left(1 - e^{-\frac{\rho_I}{\rho_{R_n}^M \lambda_{RP}}}\right) e^{-\frac{k_1}{\rho_{R_n}^M \lambda_{RU_1}}} \times \left(e^{-\frac{k_4}{\rho_{R_n}^M \lambda_{RU_2}}} - k_5 e^{-\frac{k_4}{\rho_{R_n}^M k_5 \lambda_{RU_2}} + \frac{k_2}{\rho_{R_n}^M k_3 \xi \lambda_{RU_2}}}\right) \\
&\times \left(\varphi_2 e^{-\frac{k_4}{\rho_S^M \lambda_{SR}} - \frac{\rho_I}{\rho_S^M \lambda_{SP}}} - k_5 \times \varphi_3 e^{-\frac{k_4}{\rho_S^M k_5 \lambda_{SR}} - \frac{\rho_I}{\rho_S^M \lambda_{SP}} + \frac{k_2}{\rho_S^M k_3 \xi \lambda_{SR}}}\right) \\
&+ \left(\varphi_2 e^{-\frac{k_4}{\rho_S^M \lambda_{SR}} - \frac{\rho_I}{\rho_S^M \lambda_{SP}}} - k_5 \times \varphi_3 e^{-\frac{k_4}{\rho_S^M k_5 \lambda_{SR}} - \frac{\rho_I}{\rho_S^M \lambda_{SP}} + \frac{k_2}{\rho_S^M k_3 \xi \lambda_{SR}}}\right) \\
&\times \left(\varphi_4 e^{-\frac{k_4}{\rho_{R_n}^M \lambda_{RU_2}} - \frac{\rho_I}{\rho_{R_n}^M \lambda_{RP}} - \frac{k_1}{\rho_{R_n}^M \lambda_{RU_1}}} - k_5 \times \varphi_5 e^{-\frac{k_4}{\rho_{R_n}^M k_5 \lambda_{RU_2}} - \frac{\rho_I}{\rho_{R_n}^M \lambda_{RP}} - \frac{k_1}{\rho_{R_n}^M \lambda_{RU_1}} + \frac{k_2}{\rho_R^M k_3 \xi \lambda_{RU_2}}}\right)
\end{aligned}
$$

and $\varepsilon_i = 2^{2R_i^*} - 1$, $\rho_I = \frac{I}{\sigma^2}$, $k_1 = \frac{\varepsilon_1}{(a_1 - a_2\varepsilon_1)}$, $k_2 = \frac{\varepsilon_2}{a_2}$, $k_3 = \frac{a_1\varepsilon_2}{a_2}$, $k_4 = \max\{k_1, k_2\}$, $k_5 = \frac{\xi k_3}{1+\xi k_3}$,

$$
\begin{aligned}
\varphi_1 &= \frac{\rho_I \lambda_{RU_1} \lambda_{RU_2}}{\left(k_1 \lambda_{RP}\left(\lambda_{RU_1} + \lambda_{RU_2}\right) + \rho_I \lambda_{RU_1} \lambda_{RU_2}\right)}, \ \varphi_2 = \frac{\rho_I \lambda_{SR}}{k_4 \lambda_{SP} + \rho_I \lambda_{SR}}, \\
\varphi_3 &= \frac{\rho_I k_3 \xi k_5 \lambda_{SR}}{(k_4 \lambda_{SP} \xi k_3 + \rho_I k_3 \xi k_5 \lambda_{SR} - k_2 k_5 \lambda_{SP})} \\
\varphi_4 &= \frac{\rho_I \lambda_{RU_1} \lambda_{RU_2}}{\rho_I \lambda_{RU_1} \lambda_{RU_2} + k_4 \lambda_{RP} \lambda_{RU_1} + k_1 \lambda_{RP} \lambda_{RU_2}}, \\
\varphi_5 &= \frac{\xi k_3 k_5 \rho_I \lambda_{RU_1} \lambda_{RU_2}}{\left(\xi k_3 k_5 \rho_I \lambda_{RU_1} \lambda_{RU_2} + \xi k_1 k_3 k_5 \lambda_{RP} \lambda_{RU_2} + \xi k_3 k_4 \lambda_{RP} \lambda_{RU_1} - k_2 k_5 \lambda_{RP} \lambda_{RU_1}\right)}
\end{aligned}
$$

41.4 Numerical Results

In this subsection, we demonstrate the relevant Monte Carlo simulation results to verify the correctness of our analytical expressions of the users' outage probabilities, and the number of Monte Carlo simulations is 10^6. Suppose, the default setting of system parameters is as follows: $N = 5$, $\lambda_{SP} = \lambda_{RP} = 0.5$, $\lambda_{SR} = \lambda_{RU_2} = 1$, $\lambda_{RU_1} = 2$, $a_1 = 0.8$, $I = 25$ dBm, $R_1^* = 0.8\,\text{bit/s/Hz}$, $R_2^* = 2$ bit/s/Hz, $P_S^M = 30$ dBm, $P_{R_n}^M = 30$ dBm, and $\sigma^2 = 0$ dBm.

In Fig. 41.2, we study the effect of η on outage probabilities by plotting the outage probability versus I in dB. We study three cases of the level of residual interference, i.e.,$\eta = 0.01$, $\eta = 0.05$, and $\eta = 0.1$. It is displayed that the users' outage probability diminishes as I ascends, and the outage probability diminishes as η diminishes. Figure 41.3 shows the effect of the maximum available transmit power on outage probabilities. It is stated that the outage probabilities of the users diminish as ρ_S^M ascends, and the outage probabilities of users diminish as η diminishes.

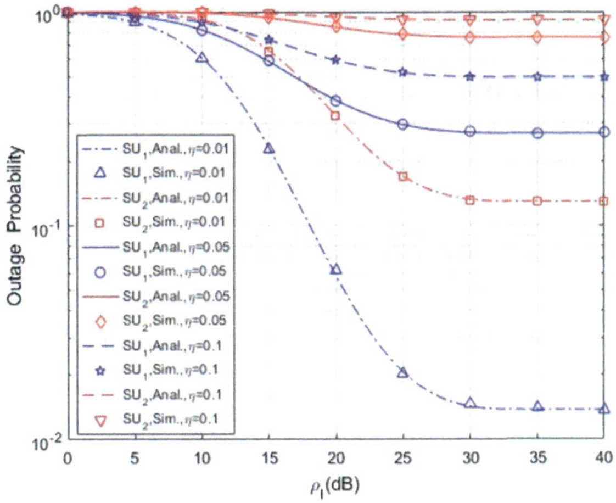

Fig. 41.2 Interference constraint versus outage probability

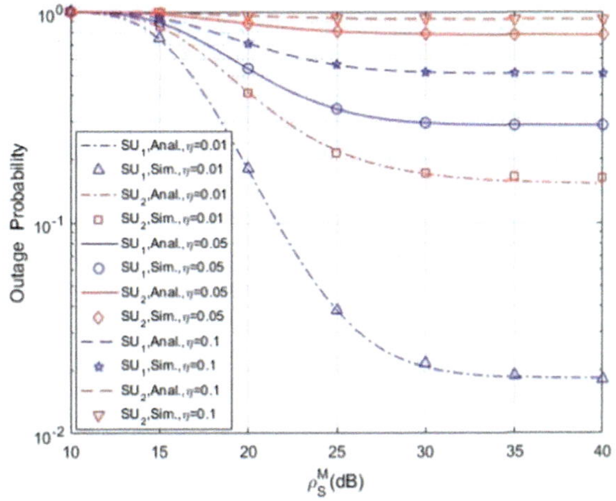

Fig. 41.3 Maximum transmit power versus outage probability

41.5 Conclusions

This article discusses the outage performance of a CR-NOMA system with imperfect SIC under TSRS. Considering the interference caused by secondary source and the relay nodes to the primary receiver, we obtain the expressions of outage probabilities inerrably. It is demonstrated that our analytical expressions are in accordance with the simulation results well. And for the further research direction, we can take the contents of energy collection at relay node into consideration.

References

1. Ding, Z., Lei, X., Karagiannidis, G.K.: A survey on non-orthogonal multiple access for 5g networks: research challenges and future trends. IEEE J. Sel. Areas Commun. **35**(10), 2181–2195 (2017)
2. Ding, Z., Dai, H., Poor, H.V.: Relay Selection for cooperative NOMA. IEEE Wireless Commun. Lett. **5**(4), 416–419 (2016)
3. Liu, Y., Ding, Z., Elkashlan, M.: Non-orthogonal multiple access in large-scale underlay cognitive radio networks. IEEE Trans. Veh. Technol. **65**(12), 10152–10157 (2016)
4. Lv, L., Chen, J., Ni, Q.: Cooperative non-orthogonal multiple access in cognitive radio. IEEE Commun. Lett. **20**(10), 2059–2062 (2016)
5. Im, G., Lee, J.H.: Outage probability for cooperative NOMA systems with imperfect SIC in cognitive radio networks. IEEE Commun. Lett. **23**(4), 692–695 (2019)

Author Index

L. C. Jain et al. (eds.), *Wireless Technology, Intelligent Network Technologies, Smart Services and Applications*, Smart Innovation, Systems and Technologies 258,
https://doi.org/10.1007/978-981-16-5168-7

9789811651670